Acoustic Waves and Acoustic Devices

Acoustic Waves
and Acoustic Devices

Edited by **Sonny Lin**

𝒞LANRYE
INTERNATIONAL

New Jersey

Published by Clanrye International,
55 Van Reypen Street,
Jersey City, NJ 07306, USA
www.clanryeinternational.com

Acoustic Waves and Acoustic Devices
Edited by Sonny Lin

International Standard Book Number: 978-1-63240-010-9 (Hardback)

Contents

Permissions

List of Contributors

Preface

Acoustics is an advanced field which enjoys a never ending youth and novel innovations in this field are introduced by either the search for a better comprehension, or by technological innovations. Micro-fabrication methodologies introduced a complete novel class of microdevices, which exploit acoustic waves for several different tasks, specifically for sensing purposes and information processing. Performance enhancements are achievable by optimized modelling tools, which are able to deal with more complicated configurations, and through more enhanced strategies of fabrications and of integration in technological systems, for example wireless communications. This book primarily deals with modelling and fabrication methods for microdevices, including unusual configurations and phenomena. Theoretical studies and modelling strategies are discussed, for phenomena ranging from the revelation of cracks to the acoustics of the oceans.

This book is the end result of constructive efforts and intensive research done by experts in this field. The aim of this book is to enlighten the readers with recent information in this area of research. The information provided in this profound book would serve as a valuable reference to students and researchers in this field.

At the end, I would like to thank all the authors for devoting their precious time and providing their valuable contribution to this book. I would also like to express my gratitude to my fellow colleagues who encouraged me throughout the process.

Editor

Design and Fabrication of Microdevices

Modeling and Design of BAW Resonators and Filters for Integration in a UMTS Transmitter

Matthieu Chatras , Stéphane Bila, Sylvain Giraud, Lise Catherinot,
Ji Fan, Dominique Cros, Michel Aubourg, Axel Flament,
Antoine Frappé, Bruno Stefanelli, Andreas Kaiser, Andreia Cathelin,
Jean Baptiste David, Alexandre Reinhardt, Laurent Leyssenne and Eric Kerhervé

Additional information is available at the end of the chapter

1. Introduction

Bulk-Acoustic Wave (BAW) resonators and filters are highly integrated devices, which represent an effective alternative for narrow-band components (up to 5% fractional bandwidth) up to few GHz [1].

This chapter presents the integration of a BAW filter and of a BAW duplexer in a UMTS transmitter. The first section details one dimensional and three-dimensional techniques for the modeling and the design of BAW resonators. The second section proposes a synthesis approach for dimensioning BAW filters and the third section illustrates the approach with the characterization of several fabricated prototypes. Finally, The UMTS transmitter incorporating a BAW filter and a BAW duplexer is described with a particular emphasis on the performances of these devices.

2. Model and design of bulk acoustic wave resonators

2.1. Modeling of a BAW resonator in 1 dimension

The proposed method compares the impedance of a piezoelectric resonator obtained by both an electrical equivalent model and a piezoelectric model. By this way, it is possible to obtain the values of the electrical model as functions of all geometrical and material characteristics. The two models and their relation are described in the following sections.

2.1.1. *Resonator impedance based on electrical (BVD) model*

A piezoelectric resonator can be modeled by the lossless BVD (Butterworth Van Dick) model as shown in Figure 1(a)) [2], [3]. C_o is the geometric capacitance of the structure and the L_m C_m series circuit (called the motional arm) represents the mechanical resonance (motional behavior). According to the circuit in Figure 1 (a), two resonances are obtained, and the equivalent impedance Z_{BVD} of this circuit can be easily derived:

$$f_s = \frac{1}{2\pi\sqrt{L_m C_m}} \tag{1}$$

$$f_p = f_s\sqrt{\frac{C_m + C_o}{C_o}} \tag{2}$$

$$Z_{BVD} = \frac{j(\omega L_m + 1/\omega C_m)}{1 - \omega^2 C_o L_m + C_o/C_m} \tag{3}$$

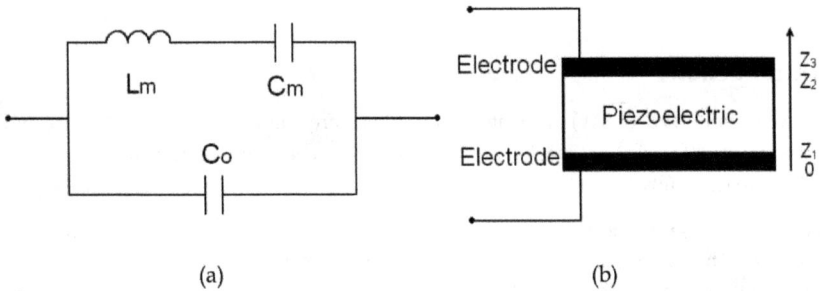

Figure 1. (a) Equivalent circuit of the lossless BVD model. (b) One-dimensional structure piezoelectric resonator.

2.1.2. *Resonator impedance based on piezoelectric equations*

The description of a piezoelectric resonator, made of a single piezoelectric layer and two thin electrodes as depicted in Figure 1 (b), is considered. Using piezoelectric fundamental equations [4], one can write:

$$T(z) = cS(z) - eE(z) \tag{4}$$

$$D = eS(z) + \varepsilon E(z) \tag{5}$$

$$S(z) = \frac{\partial u(z)}{\partial z} \tag{6}$$

$$-\rho^2 u(z) = \frac{\partial T(z)}{\partial z}$$ (7)

In the above equations, T is the mechanical stress tensor, S is the strain tensor, E is the electric field, D is the electrical displacement vector (C/m2), ϱ is the density, u is the mechanical displacement vector, c is the elastic stiffness tensor calculated at constant electric field, e is the piezoelectric tensor (C/m2), and ε is the relative permittivity.

Using the boundaries conditions for T (z) and u(z), we have:

$$T_1(0) = 0 \; ; \quad T_3(z_3) = 0$$ (8)

$$T_1(z_1) = T_2(z_1) \; ; \quad u_1(z_1) = u_2(z_1)$$ (9)

$$T_2(z_2) = T_3(z_2) \; ; \quad u_2(z_2) = u_3(z_2)$$ (10)

The potential difference u on the piezoelectric layer can be obtained by integrating Ez (the electric field) on the thickness of the considered layer:

$$U = -\int_{z2}^{z1} Ez(z)dz$$ (11)

By definition, the current I is the temporal derivation of the charge Q at the surface of the electrodes, which in sinusoidal mode is equivalent to:

$$I = jwQ$$ (12)

The continuity of the normal component of the electrical displacement vector D at the interface piezoelectric-metal makes it possible to express the charge Q as a function of D and of the surface of the metal electrodes:

$$Q = Dz \; S$$ (13)

The expression of the current becomes then:

$$I = jwS \, Dz$$ (14)

Consequently, the impedance of the piezoelectric layer can be obtained in function of the thickness of the used materials and the dimensions of the resonator, as written in equation (15):

$$Z_{eq} = \frac{U}{I} = \frac{[(z_2 - z_1) - \alpha e_2(r_{21} - r_{11}) - \beta e_2(r_{22} - r_{12})] / \varepsilon_2}{j\omega A}$$ (15)

Where e_2 is the piezoelectric tensor of the piezoelectric layer, A is the surface (area) of the electrodes, ε_2 is the permittivity of the piezoelectric layer, and r_{11}, r_{12}, r_{21}, r_{22}, α, and β are expressions in function of known constants.

The impedance can be evaluated knowing the material properties and dimensions. This method can also be used for more complex structures, such as SCF (Stacked Crystal Filters), or CRF (Coupled Resonator Filters) resonators or filters, accounting for all the layers.

2.1.3. Equivalence of one dimensional models

Using a least squares method, the two expressions in equations (3) and (15) can be equated. By this way, the values of C_o, C_m and L_m for different resonator areas A and for the thicknesses of each layer can be obtained.

For example, the expression of L_m as a function of the surface A and the thickness t of the top electrode is presented. For fixed values, $[A_1\ A_2\ A_3 \cdots\cdots A_i]$ and $[t_1\ t_2\ t_3 \cdots\cdots t_i]$, we obtain corresponding values of L_m. For each same thickness t_i, we can consider that L_m is only varying according to the surface A. As shown in equation (16), we can find a polynomial function in the variable A fitting the values of L_m. The polynomial coefficients (W_i, X_i, Y_i and Z_i) are now independent of the surface A and depend only on the thickness t. In a second time, using the same method, the coefficients can be expressed as a function of variable t.

$$L_m = W_i A^n + X_i A^{n-1} + \cdots\cdots + Y_i A + Z_i$$
$$= \left(\sum_{j=1}^{N} a_j t^{j-1} \right) A^n + \left(\sum_{j=1}^{N} b_j t^{j-1} \right) A^{n-1} + \cdots\cdots + \left(\sum_{j=1}^{N} c_j t^{j-1} \right) A + \left(\sum_{j=1}^{N} d_j t^{j-1} \right) \tag{16}$$

This identification method can also be applied to the other variables, such as the thickness of the loading layer. The three elements of the BVD model (C_o, C_m and L_m) can be determined as functions of the piezoelectric resonator dimensions (thickness of each layer and surface of resonator). According to the expressions of C_o, L_m, and C_m, surfaces and thicknesses can be optimized with an electrical software for designing a BAW resonator or a BAW filter.

2.2. Model in 3 dimensions

Even though Film Bulk Acoustic Resonator simulation with 1D model enables to quickly compute complex frequency response as filters, it becomes too restrictive when spurious modes in lateral dimensions have to be predicted. The 3D Finite Elements Method (FEM) enables to investigate the effect of the electrode shape on the spurious modes that are present in the electrical impedance. In order to reduce or to suppress these modes, solutions have to be investigated.

2.2.1. Finite element model

Piezoelectricity is a phenomenon which couples electrical and mechanical domains. It can be modeled into coupled equations:

$$T_{ij} = c_{ijkl} * S_{kl} - e_{ijk} * E_k \tag{17}$$

$$D_k = e_{kl} * E_l + e_{ijk} * S_{ij} \qquad (18)$$

With: T: mechanical stress (Pa), E: electric field (V/m), S: mechanical strain, D: electric displacement(C/m²), c: stiffness tensor (Pa), ε : permittivity tensor (F/m) and e: piezoelectric tensor (C/m²)

Unfortunately, one-dimensional approximation becomes too restrictive when we need to predict spurious modes that may appear with lateral direction mode coupling or with resonators electrical or mechanical cross coupling. Thus a 3D simulation tool is needed to compute the resonator in three dimensions.

In the following sections, examples of 3D FEM computation will be proposed and described highlighting the advantages of 3D simulations.

2.2.2. Mechanical displacement modes

We have analyzed the suspended resonator structure presented in Figure 2. The structure is clamped on lateral sides (no mechanical displacement in space directions), the bottom electrode is grounded and a potential constraint is applied to the top electrode.

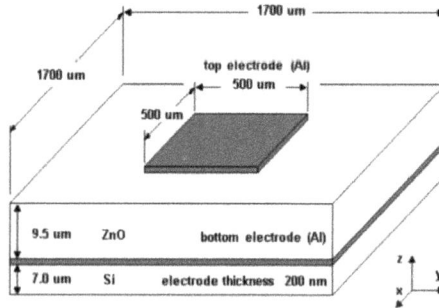

Figure 2. 3D suspended resonator structure.

In order to identify spurious modes, the mechanical displacement for four modes has been calculated, as displayed in Figure 3.

The first mode is the so-called thickness extensional mode. It corresponds to the maximum energy coupling and to minimum mechanical losses. This mode is the one taken into account in 1D analytical models.

The second mode is a cavity mode. In FEM, in order to keep the problem size finite, the physical domain needs to be truncated. This truncation introduces artificial boundaries where artificial boundary conditions are considered. Domain truncation causes reflection of the waves on clamped lateral sides. Then a standing wave may appear at certain frequencies. Those spurious modes are modeling errors and will be suppressed by mesh apodization.

Standing wave modes are harmonic thickness extensional modes due to acoustic wave reflection on top electrode edges. They appear as spurious modes in the electrical response. In order to use this electrical response for oscillator or filter applications, we need to suppress these modes. Different solutions can be considered.

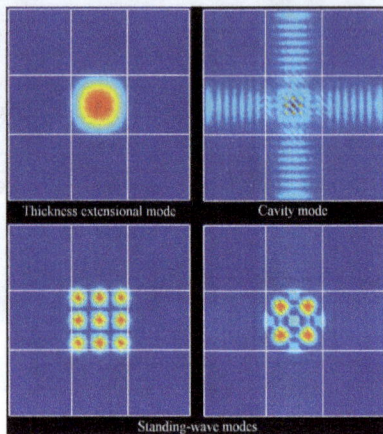

Figure 3. Top view of the suspended resonator mode shapes at resonant frequencies (The only top electroded region is the square in the middle).

2.2.3. Suppression of standing wave modes

2.2.3.1. Apodization

The first solution consists of cut a triangular part of the square top electrode and to paste it on another edge as shown in Figure 4. Then a quadrilateral electrode is obtained with no parallel sides [5].

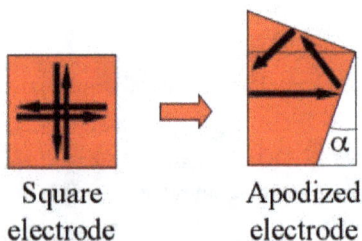

Figure 4. Apodization applied to a square electrode

Standing waves cannot appear since multiple reflections are not constructive. One can notice from Figure 5 that standing wave modes are not or weakly coupled as the apodization angle increases. A 3D simulation software enables the analysis of an entire apodized resonator and is very useful in this case for determining an optimum apodization angle.

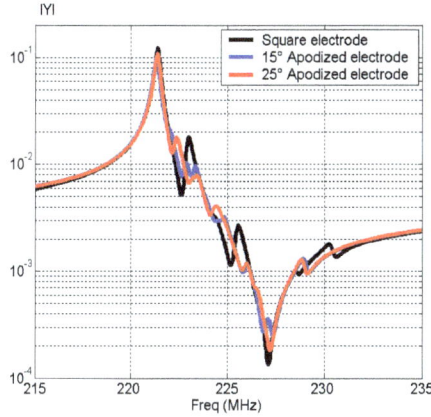

Figure 5. Apodized top electrode electrical admittance for two apodization angle values

2.2.3.2. Edge loading

Another solution proposed in [6] is considered. A narrow region is deposited at the edge of the suspended resonator top electrode as described in Figure 6. This thickened region constitutes a frame that matches the acoustic impedance and suppresses reflection on top electrode edges. Within certain optimum range for the edge region width, the resonator operates in a mode where the mechanical displacement is constant on the top electrode surface (Figure 7).

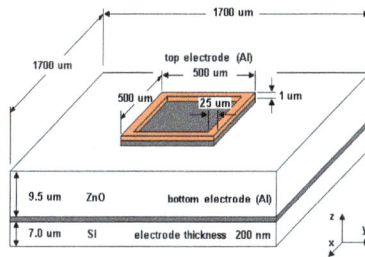

Figure 6. Framed top electrode suspended resonator 3D structure

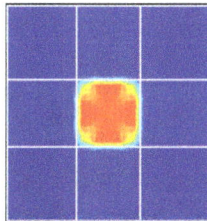

Figure 7. Framed top electrode suspended resonator thickness extensional mode

It can be observed, in Figure 8, that the electrical response is standing wave modes free. Standing waves on the top electrode are not coupled. Moreover, 3D resonant frequency tends toward 1D result.

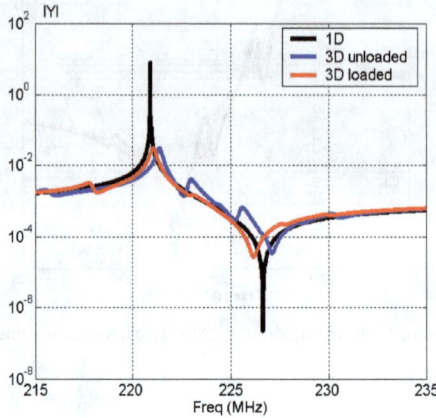

Figure 8. Framed top electrode suspended resonator admittance

2.2.3.3. Cavity modes suppression by resonator apodization

In order to identify spurious modes due to domain truncation, we have changed the distance from the top electrode edge to the structure edge (A=500 μm or 400 μm in Figure 9). We have found that resonant conditions and resonant frequencies for cavity modes change with cavity dimensions.

The mechanical displacement for those modes is displayed in Figure 9.

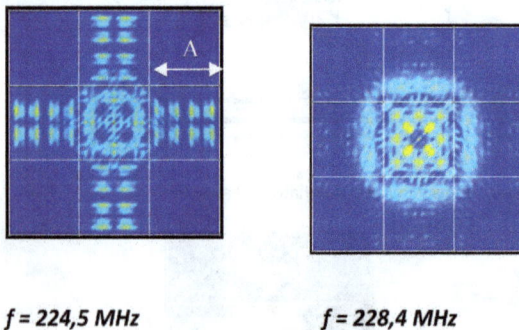

f = 224,5 MHz   *f* = 228,4 MHz

Figure 9. Mechanical displacement for two cavity modes (Top view)

While the mode which appears at $f = 224,5\ MHz$ is obviously a cavity mode, we can have doubts about the one at $f = 228,4\ MHz$. In order to remove all doubts, we have applied the apodization technique to the resonator edges (Figure 10).

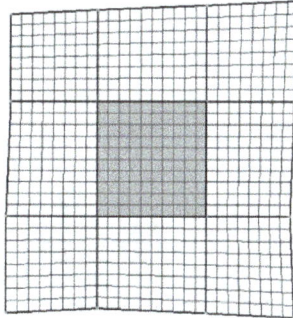

Figure 10. Apodized cavity (Top view) and mesh

We obtained an asymmetrical structure with no parallel sides. The irregular shape avoids phase reflections on the structure edges with constructive interferences, reducing the occurrence of standing waves.

One can observe in Figures 11, 12 and 13 that this technique enables to obtain an electric response free of spurious cavity modes.

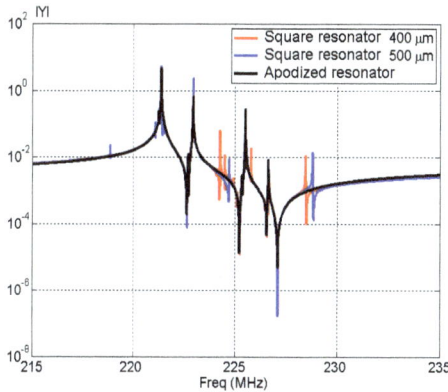

Figure 11. Suspended resonator electrical admittance

A full 3D FEM tool is very useful to analyze and to predict the behavior of complex structures. One can take into account non-homogenous structure and non-linear materials. Obviously it is the only method to compute 3D geometries. The shape of the mechanical displacement can be obtained to identify real or non-physical modes. Solutions have been proposed with apodized shape or with edge loading to remove parasitic modes.

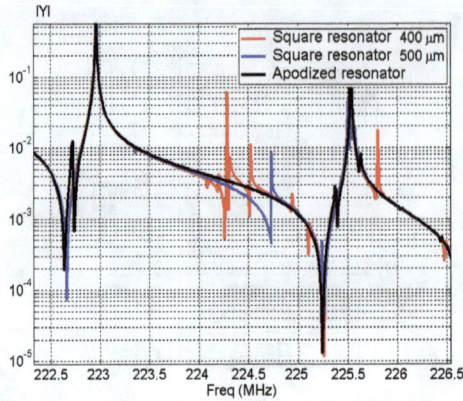

Figure 12. Suspended resonator electrical admittance around 224,5 MHz

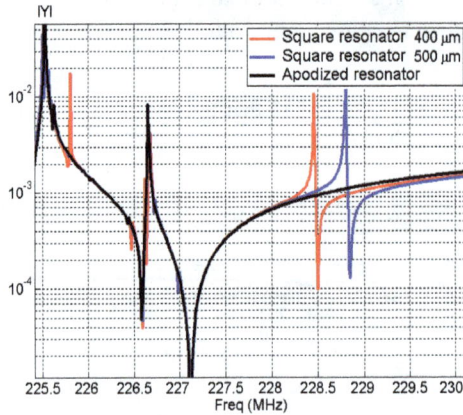

Figure 13. Suspended resonator electrical admittance around 228,4 MHz

3. Design of bulk acoustic wave filters

3.1. Implementation

The MBVD (Modified Butterworth Van Dyke) model, presented in Figure 14, can been used to compute the behavior of a BAW resonator [7]. Compared to the BVD model, the MBVD model incorporates resistances, which take into account losses in the piezoelectric material and in the electrodes.

This equivalent circuit resonates for two particular frequencies:

$$f_s = \frac{1}{2\pi\sqrt{Lm.Cm}} \tag{19}$$

$$f_p = f_s \sqrt{1 + \frac{Cm}{Co}} \tag{20}$$

fs and fp are known as the series and parallel resonant frequencies and correspond respectively to a minimum and to a maximum of the electrical impedance. fs and fp are related to the electromechanical coupling coefficient k_t^2 by:

$$k_t^2 = \frac{\pi^2}{4} \frac{f_p - f_s}{f_p} \tag{21}$$

Moreover, one can define Q_s and Q_p the quality factors of series and parallel resonances:

$$Q_s = \frac{2\pi f_s \, Lm}{Rm} \tag{22}$$

$$Q_p = \frac{1}{2\pi f_p \, Co \, Ro} \tag{23}$$

For dimensioning a BAW filter, each resonator is represented by its MBVD model and the synthesis of the BAW filter is carried out by optimizing each resonator with respect to a specified target.

BAW filters are usually implemented arranging series and shunt resonators. Basic resonator arrangements, namely ladder and lattice configurations, are defined as shown in Figures 15 and 16.

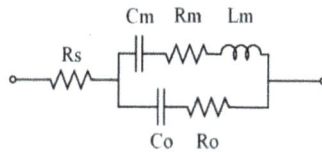

Figure 14. MBVD model of a BAW resonator

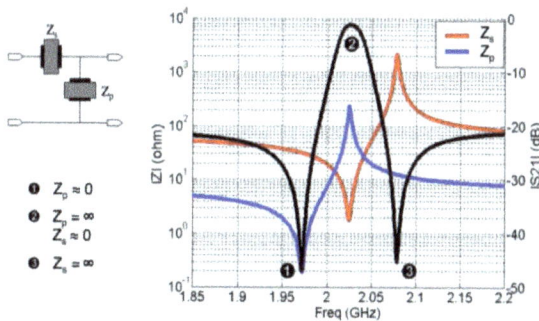

Figure 15. Electrical impedance of shunt and series resonators for providing a band-pass filter with a ladder network

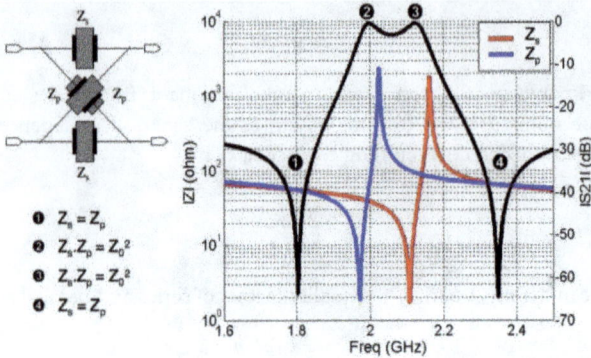

Figure 16. Electrical impedance of shunt and series resonators for providing a band-pass filter with a lattice network

The ladder configuration presents a high rejection close to the filter passband but a poor out of band rejection. On the other hand, the lattice filter exhibits higher out of band rejection but a poor rejection close to the filter passband. Combining these two configurations, one can obtain a mixed ladder-lattice filter with very good properties, as shown in Figure 17. S_{21} parameter is the forward transmission coefficient of the filter.

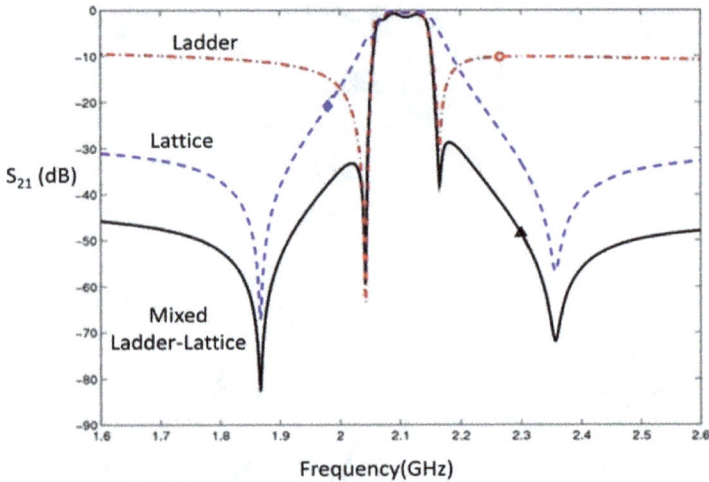

Figure 17. Behavior of the mixed ladder lattice filter.

3.2. Synthesis

Regarding the synthesis, each resonator is characterized by several fixed technological parameters obtained by electromechanical modeling or process characterization:

- k_t^2, the electromechanical coupling coefficient
- Q_s, the quality factor at series resonant frequency
- Q_p, the quality factor at parallel resonant frequency
- ε_r, the piezoelectric material permittivity
- $R_{\square}$, the square resistance of electrodes

The remaining variable parameters are:

- A, the surface of the top electrode ($A = L.W$ where L and W are the length and the width of the top electrode respectively, the aspect ratio L/W remaining generally the same for all electrodes)
- d and l: the thickness of the piezoelectric layer and the thickness of the loading layer respectively, which control the series resonant frequencies (f_s) of series and parallel resonators.

The lumped elements of the MBVD model are related to these parameters through the following expressions:

$$C_0 = e_0 e_r \frac{A}{d} \tag{24}$$

$$C_m = C_o \left[\left(\frac{f_s}{f_p} \right)^2 - 1 \right] \tag{25}$$

With

$$f_p = \frac{f_s}{1 - \frac{4}{\pi^2} k_t^2} \tag{26}$$

$$L_m = \frac{1}{C_m \left(2\pi f_s\right)^2} \tag{27}$$

$$R_m = \frac{2\pi L_m f_s}{Q_s} \tag{28}$$

$$R_o = \frac{1}{Q_p C_0 2\pi f_p} \tag{29}$$

$$R_s = R_{\square} \cdot \frac{L}{\omega} \tag{30}$$

One can note that Lm, Cm and Co can also be approximated directly by polynomial expressions of the layer thicknesses and resonator area as explained in section 2.1.3.

The thicknesses of the piezoelectric and loading layers and the area of each resonator are optimized with respect to the specifications. This optimization can be driven by the minimization of a cost function defined by the filtering pattern [8].

3.3. Electromagnetic co-simulation

The previous synthesis relies on MBVD models of BAW resonators, which do not take into account metallic losses or couplings due to interconnections and access ports. Since metallic lines used for connecting resonators have irregular geometries depending on the arrangement of resonators, models for such elements cannot be implemented in a synthesis tool. Nevertheless a simulation is possible a posteriori with the layout of the filter in order to estimate additional losses or to check eventual couplings due to metallic lines.

The layout of the filter to be realized can be drawn with an electromagnetic (EM) software, e.g. Momentum included in Agilent ADS [9]. Using such a EM software, all electrostatic and electromagnetic phenomena are characterized considering the geometry and the physical characteristics of stacked layers.

The electrostatic part of MBVD resonators (Rs, Ro and Co) is directly taken into account in the distributed model (related particularly to the area of resonators). However, the motional part (Rm, Cm and Lm) is modeled by lumped elements connected through internal ports as shown in the example given in Figure 18.

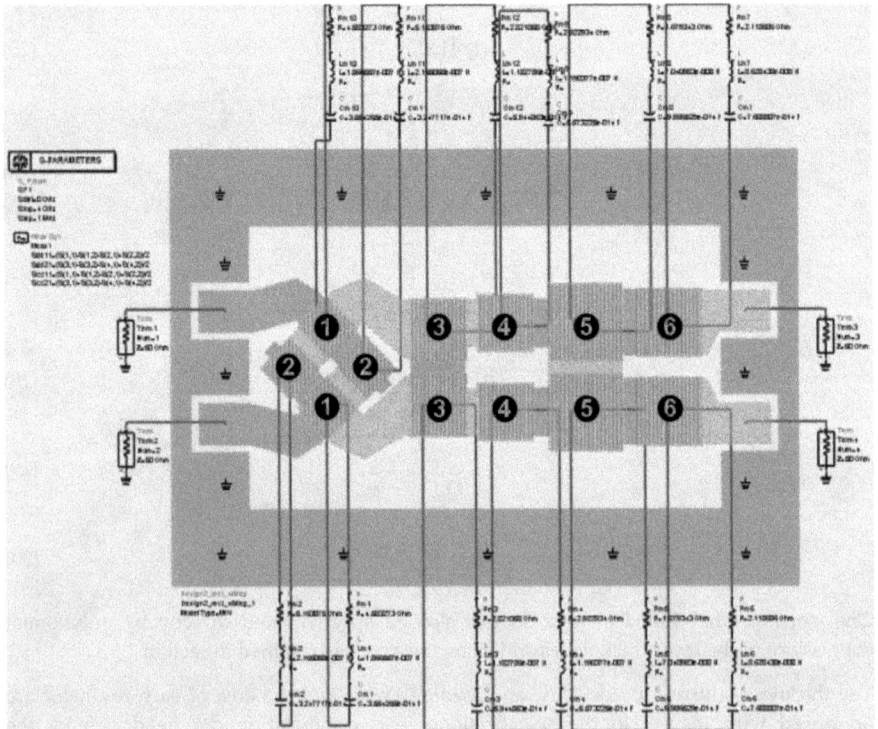

Figure 18. UMTS filter co-simulation

Figure 18 shows the co-simulation of a mixed ladder lattice filter in Agilent ADS/Momentum environment, including the layout and the motional parts directly derived from the previous synthesis.

4. Fabrication and characterization of BAW resonators and filters

4.1. Single resonator

Several BAW devices have been designed using the method proposed previously. Molybden is chosen as electrode material and the sputtering method is used. Therefore, AlN films have been deposited in (002) direction with the c-axis perpendicular to the substrate surface [10-13]. The optimized resonators and filters with the proposed method have been fabricated by CEA-Leti [14]. Figures 19 and 20 show a Solidly Mounted Resonator (SMR) resonator and a comparison between the simulated result and the measurement data respectively. The apodization of the top electrode was used to avoid parasite modes. One can observe a good agreement between simulations and measurements. S_{21} and S_{11} parameters are respectively the forward transmission and the reflection coefficients of the filter.

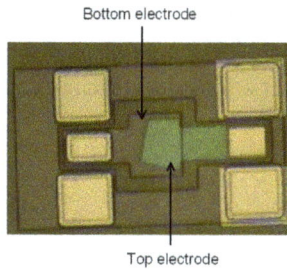

Figure 19. Structure of SMR resonator

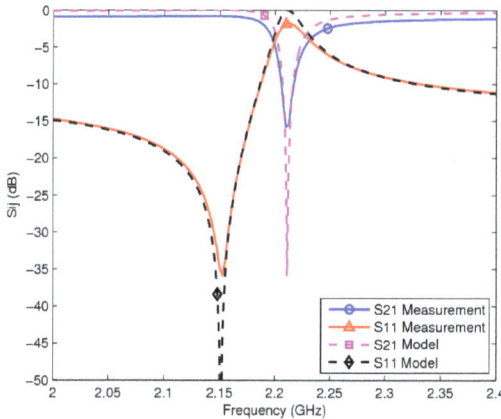

Figure 20. Comparison between the simulated result and the measurement data for SMR resonator

The resonant frequency is 2.212 GHz and anti-resonant frequency is 2.152 GHz, which leads to a coupling coefficient k_t^2 equal to 6.69%. The quality factor of this resonator is around 200.

4.2. 2-pole and 3-pole ladder filters

2-pole and 3-pole filters shown in Figure 21 were fabricated by CEA-Leti. Measurements and simulations for each filter are presented respectively in Figures 22 and 23. One can observe a good agreement between the measured responses and the simulated ones. The measured response of the 2-pole filter presents a passband of 55 MHz centered at 2.13 GHz (fractional bandwidth of 2.58 %) and the insertion losses are about 1.5 dB. Similarly, the 3-pole filter shows a 54 MHz passband at 2.13 GHz (FBW: 2.53 %) and insertion losses of approximately 2 dB.

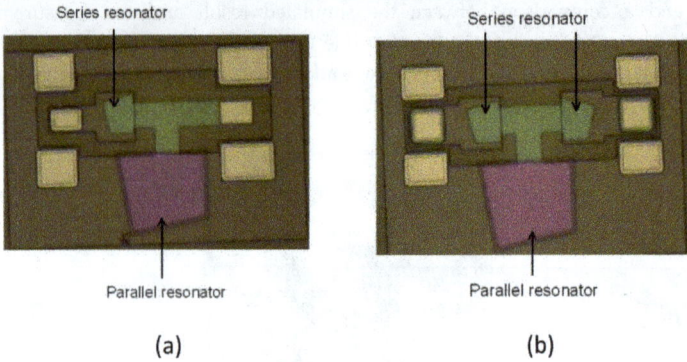

(a)   (b)

Figure 21. (a) Fabricated 2-pole filter. (b) Fabricated 3-pole filter (top view).

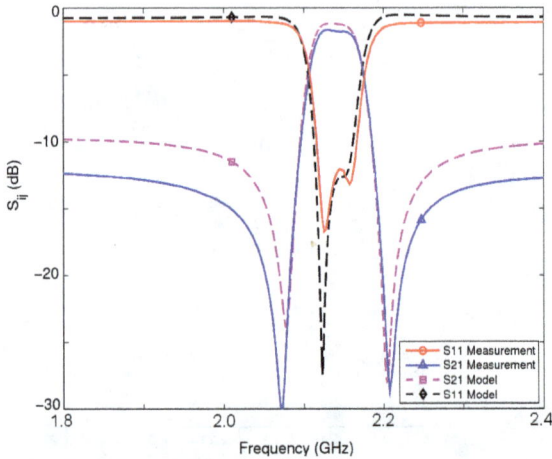

Figure 22. Measured and computed responses (model with losses) of the 2-pole filter

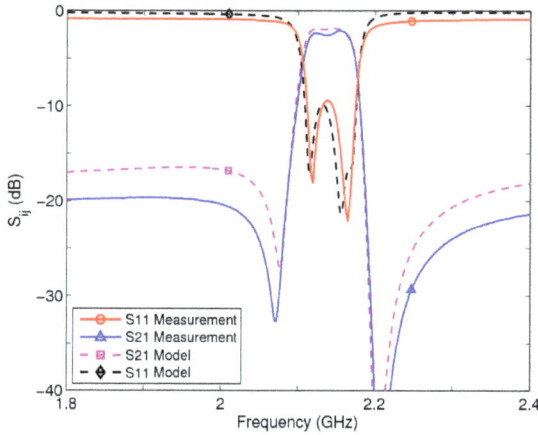

Figure 23. Measured and computed responses (model with losses) of the 3-pole filter

4.3. Differential ladder and lattice filters

Bandpass filters have been synthesised with differential structures (100Ω input/output impedance). These filters have been fabricated by CEA-LETI and UPM (Universidad Politécnica de Madrid) [15]. Each resonator is deposited on a Bragg mirror (SiN/SiOC) and is built with an Aluminium Nitride (AlN) piezoelectric layer and two Iridium (Ir) electrodes. With this technology, top electrode thickness is defined in order to act like a loading layer [16, 17] and to reach desired resonant frequency. Ir electrodes are utilized in order to enhance electromechanical coupling [18]. Iridium is a metal that presents a high-density [19] which leads to a high acoustic impedance, a low electric resistivity, and a specific crystal structure that promotes the growth of AlN films of excellent piezoelectric activity [20].

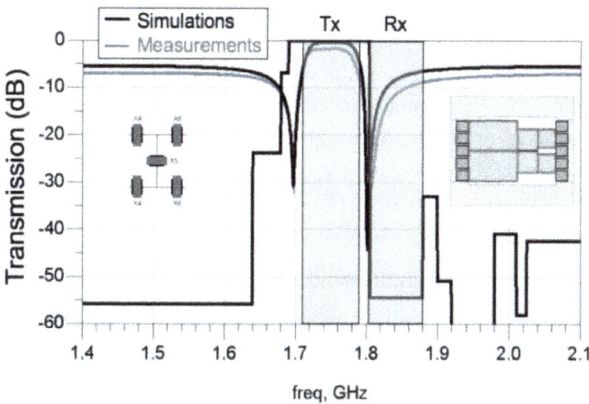

Figure 24. 1.5-stage ladder filter compared to UMTS standard

One can notice that there is a good agreement between simulated and measured responses in Figures 24 and 25. As electrical losses of Iridium were not characterized before simulation, they were not taken into account. Measured insertion losses are about 3dB for ladder structure and 6dB for lattice structure. A good out of band rejection is obtained, only limited by measurement noise.

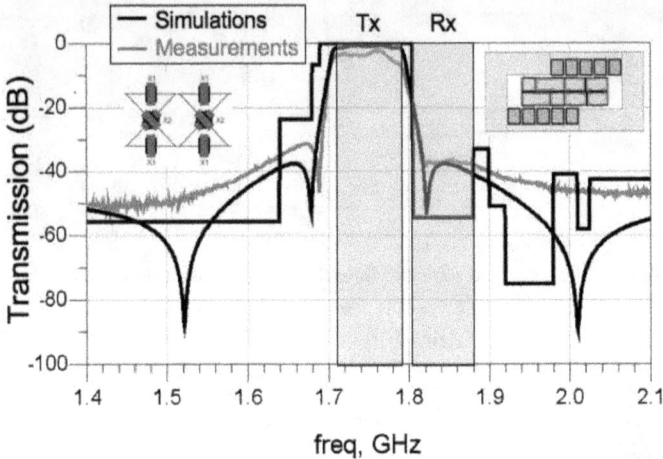

Figure 25. 2-stage lattice filter compared to UMTS standard

5. Integration of BAW components in a transmitter module

This section deals with the development of a BAW-SMR technology for application to mobile multi-standard communication terminals. UMTS and DCS1800 were chosen as targeted standards to validate a bi-standard BAW-based RF architecture, presenting both reconfigurability of active parts and integration of BAW technology.

This section presents the first demonstration of a complete UMTS transmitter including a CMOS 90nm digital RF signal generator connected to a BAW filter, a Power Amplifier (PA) stage and a BAW duplexer. Although the active part was originally designed for multi-mode purposes, UMTS standard was chosen to illustrate the proof of concept. The purpose is to emphasize the use of BAW-based structure for wireless mobile communications.

In the first section the proposed multi-standard transmission chain will be detailed, and a focus on the transmit BAW filter will be discussed. The design and measurement results of the stand-alone filter will then be presented before talking about the advantages of this filter with the RF signal generator. The BAW duplexer will be described and the measurement results of this filter will be detailed. Finally, the full transmitter test bench will be highlighted and the measurement results commented on.

5.1. RF signal generator

5.1.1. Third-order delta-sigma modulator

The architecture presented in Figure 26 is clearly aimed at future software defined terminals by pushing the band or standard specific components as close as possible to the antenna. The targeted standard is UMTS, one of the main mobile communication standards in Europe using 1920 – 1980 MHz frequency band for TX and 2110-2170 MHz frequency band for RX. The architecture could easily be extended to additional standards such as DCS or PCS at the cost of extra BAW filters for the appropriate frequency bands.

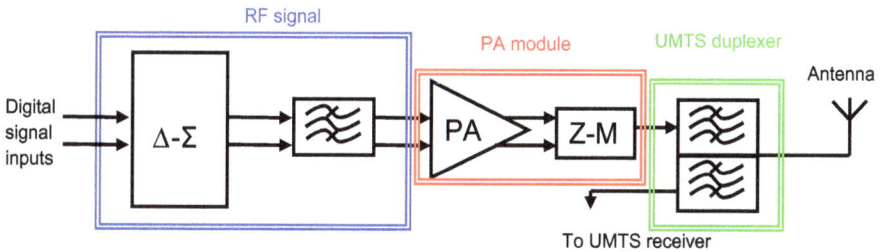

Figure 26. Modules definitions for first single standard UMTS transmitter

Two oversampled low-pass ΔΣ stages representing I and Q channels, work synchronously to generate a high sample rate 1-bit output stream [21] that feeds a digital image-reject mixer [22]. This latter stage produces a high frequency sampled two-level RF signal which exhibits very good in-band performances and a quantization noise shaping due to ΔΣ modulation [21]. A first stage of filtering precedes the power amplifier whose matching network converts a nominal single ended load impedance of 50 Ω to the optimal impedance for the active stage. A duplexer is inserted in order to isolate the receiver from the transmitter while enabling them to share a common antenna.

The UMTS transmit filter is designed on a ladder-lattice topology with 50 Ω differential input - 100 Ω differential output ports to benefit from advantages of both structures. This filter aims at meeting the standard emission mask by lowering the out-of-band quantization noise. Figure 27 shows the output spectrum of the digital RF signal generator and indeed justifies that the transmitter requires a large amount of out-of-band filtering at the output of the ΔΣ modulators. The most stringent specification for UMTS is the required rejection (50 dB) in the RX band (2110 MHz - 2170 MHz). It is very difficult to reach such a high rejection at a very close bandwidth from the center frequency band. A typical lattice BAW based filter exhibits a typical attenuation of roughly 40-45dB. Consequently, the insertion of a duplexer in the UMTS transmission chain is required to achieve this high attenuation in the RX band as well as its isolation between the TX and RX paths. The design and measurement results of these BAW-based structures are presented in the following sections.

Output of the digital upconverter

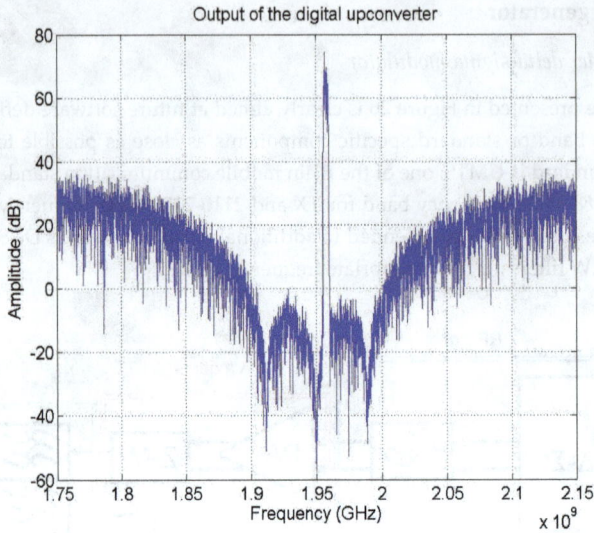

Figure 27. Spectrum of a $\Delta\Sigma$ modulated signal, up-converted around the RF carrier.

5.1.2. BAW filter

5.1.2.1. Filter synthesis

The synthesis of a UMTS filter has been performed [23], where a passband is needed for the 1920 – 1980 MHz range. Important constraints of the UMTS standard [24] are a high rejection and a high selectivity. To achieve these performances, a differential mixed ladder-lattice filter topology [25] with 100Ω differential impedance has been proposed. As explained in section 3, a mixed ladder-lattice filter as shown in Figure 28, provides sharp band edges and a good rejection in the stopband.

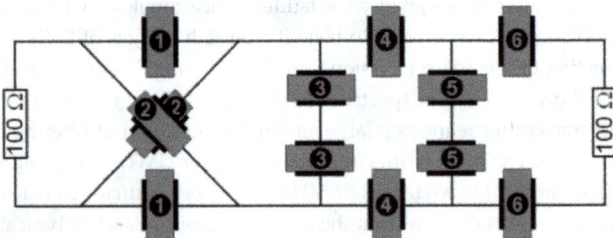

Figure 28. UMTS filter topology

After synthesis, optimal technological data for unloaded and loaded resonators are found as listed in Table 1. Related MBVD elements of each resonator are listed in Table 2 and the scattering parameters are given in Figure 29.

	fs (MHz)	fp (MHz)	K²eff (%)	Qs	Qp
Unloaded	1954	2004	6.2	500	300
Loaded	1898	1946	6.1	500	300

Table 1. Unloaded and loaded resonators given technological data

		X1	X2	X3	X4	X5	X6
Elements values	Rm (Ω)	4.58	5.16	2.82	2.92	1.68	2.12
	Lm (nH)	186.6	216.5	118.3	119	70.4	86.3
	Cm (fF)	35	32	59	56	100	77
	R0 (Ω)	0.39	0.43	0.23	0.25	0.14	0.18
	Co (pF)	0.68	0.63	1.16	1.07	1.94	1.47
	Rs (Ω)	0.19	0.17	0.19	0.20	0.23	0.22

Table 2. MBVD elements values of optimised UMTS filter

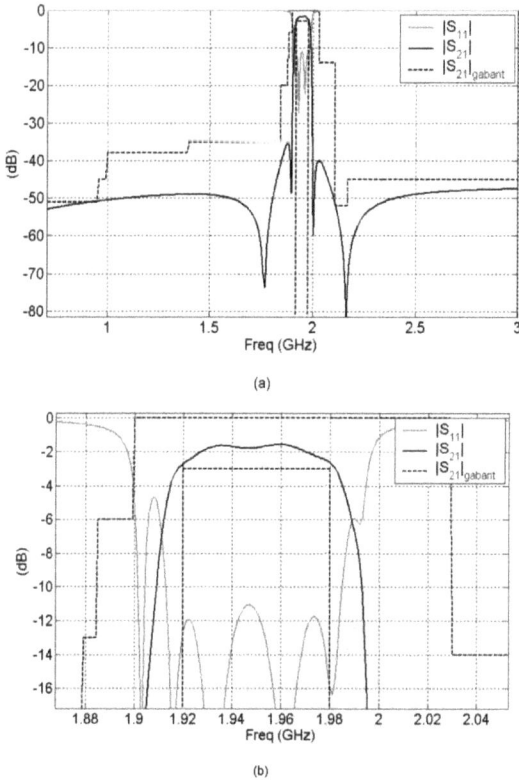

(a)

(b)

Figure 29. Scattering parameters of the UMTS filter: (a) wide band performance, (b) in band performance, compared to UMTS standard

One can observe, in Figure 30, that scattering parameters completely fulfill the specifications. Moreover, the layout of the filter has been co-simulated with a 2.5D electromagnetic software in order to estimate additional losses and to characterize eventual couplings due to metallic lines. As shown in Figure 30, the co-simulation is still in line with the specifications.

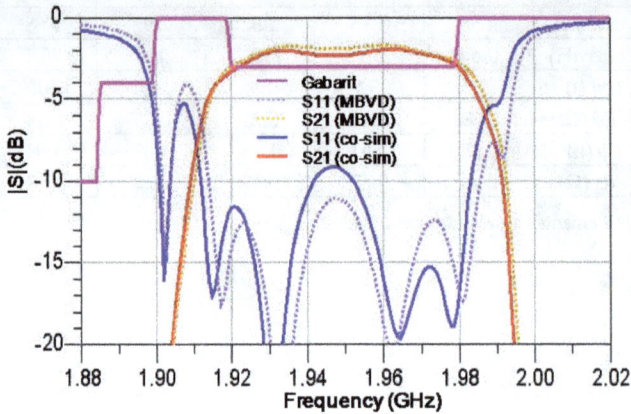

Figure 30. EM co-simulation of the UMTS filter, compared to UMTS standard

5.1.2.2. Fabrication and measurements

The filter has been fabricated by CEA-Leti. Each resonator is deposited on a Bragg mirror (SiN/SiOC), using an Aluminium Nitride (AlN) piezoelectric layer, two Molybdenum (Mo) electrodes and a Silicon Oxide (SiO2) loading layer. A photo of the filter is presented in Figure 31. The measured unpackaged quality factor at resonance frequency is close to 800 for the series resonators and close to 300 for the parallel ones. The measured response of the filter is presented in Figure 32. Insertion losses are around 3 dB and the required rejection and selectivity are fulfilled. The bandwidth is slightly reduced due to a lower resonant frequency for series resonators during fabrication

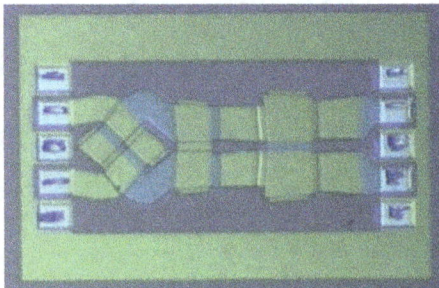

Figure 31. Fabricated filter with differential accesses.

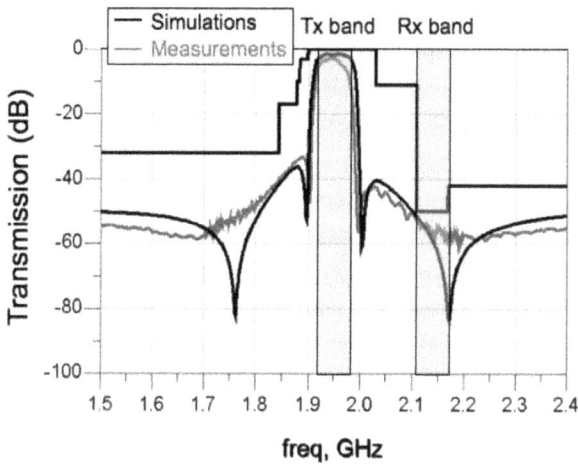

Figure 32. Measurement of the UMTS filter, compared to UMTS standard

5.2. Complete UMTS transmitter

5.2.1. RF Signal generator

The all digital RF signal generator architecture is presented in Figure 33. A first stage oversamples the baseband I and Q signals to reach the modulator frequency. The principle of $\Delta\Sigma$ modulation and digital mixing has been explained in section 5.1 and one can find further explanations in [21]. The generator output signal is a very high speed 1-bit signal with bandpass shaped quantization noise centered on the standard transmit band. The differential output buffer is able to drive a differential 50Ω load with a 1V power supply, at very low output impedance of 0.6Ω. As the BAW filter needs a 50Ω differential input impedance match, 25Ω resistors need to be inserted between the signal generator and the BAW filter. This very simple passive and unfortunately unmatched matching network degrades the filter transfer function, introducing ripple in the pass band and lowering out-of-band attenuation. Both the RF signal generator and the BAW filter are molded with a resist on a module which is then soldered on a larger PCB. The output of this subset circuit is 100 Ω differential [26].

The baseband input signals are generated by an external Arbitrary Waveform Generator (Tektronix AWG 420), in which a Matlab sequence has been programmed to generate a WCDMA modulated signal on I and Q channels. The external clock reference comes from a synthesizer delivering a clock signal at 2.6GHz with an output level of –5dBm. This clock is the nominal frequency clock of the modulators and of the digital image-reject mixer and fully determines the center frequency of the RF signal. Using a 2.6GHz, the $\Delta\Sigma$ generator outputs an RF modulated signal at 1.95GHz using the first image-band.

The measurement results (Figure 34) consist of a spectrum analysis of output signals to evaluate the benefits of using BAW filtering in a $\Delta\Sigma$ modulation approach. It clearly shows that the out-of band quantization noise nearby the central bandwidth has been reduced below the noise level of the spectrum analyzer in this setup, due to the near-band high rejection of the ladder BAW filter configuration. Emission specifications in DCS are satisfied whereas more than 20dB of attenuation is still needed in UMTS RX band. Far-band filtering (thanks to lattice configuration) is also efficient, resulting in more than 35dB attenuation at low frequencies. In the first image band, the measured ACLR are 43 and 42dB, respectively at 5MHz and 10MHz offsets. The EVM is 3.7% and the measured channel power is -27dBm. This low value is due to the use of the first image and the loss in the series resistors.

The strongest constraint in UMTS transmission architecture is the very high level of rejection needed in the UMTS reception band which is very close to the transmission band. The measurement results show that a ladder-lattice BAW filter is not sufficient to fulfill this specification. A BAW-based duplexer, providing further isolation between both signal paths is consequently mandatory and will be described in the following section.

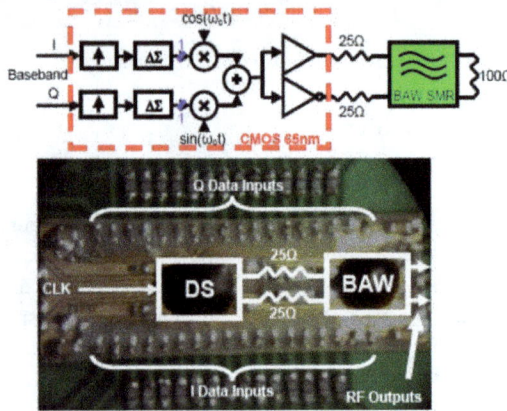

Figure 33. Signal generator and BAW filter architecture (left), photograph of the DS modulator and BAW assembly

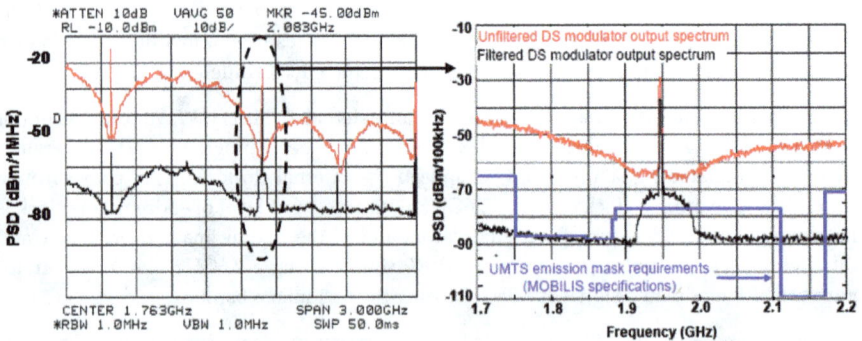

Figure 34. Delta-sigma digital generator output spectrum before (red) and after (black) BAW filtering

5.2.2. BAW duplexer

The BAW duplexer has been designed using a mix of the exposed BAW filter methodology and co-simulation with electromagnetic tools [27], [28], [29]. The topology of this duplexer is shown in Figure 35.

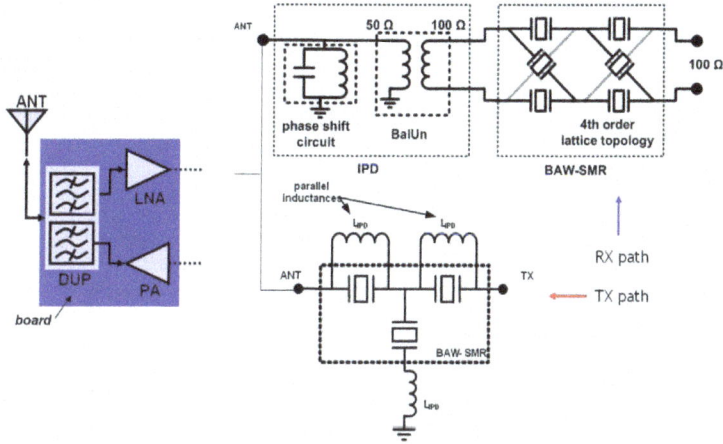

Figure 35. BAW duplexer topology

The BAW duplexer was made with two BAW filters and a glass substrate containing high quality passive elements (IPD from ST Microelectronics [30]). Each BAW filter occupies 1mm². The TX and RX filters were flip-chipped on the 3.9x3.9mm² substrate, as shown in Figure 36.

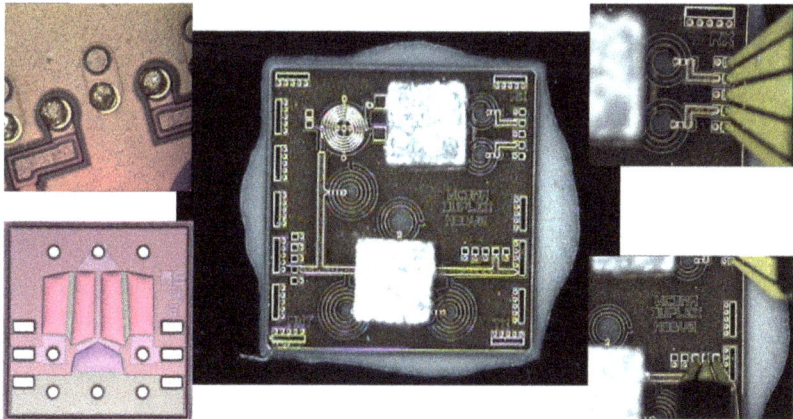

Figure 36. Left: Photograph of bumps and micropackaged TX filter. Center and right: On probe measurement of the BAW duplexer

The on-probe measurement of the duplexer is performed in two steps, for TX and RX paths respectively. This is due to calibration restrictions: two ports are 50 Ω single ended and the third one is a 100 Ω differential port, and on probe calibration of the vectorial analyzer is not guaranteed in these conditions. Figure 37 shows the comparison between probe measurement and backward simulation (taking into account of shifted components compared to expected elements in initial simulation). It shows globally a good fitting. The

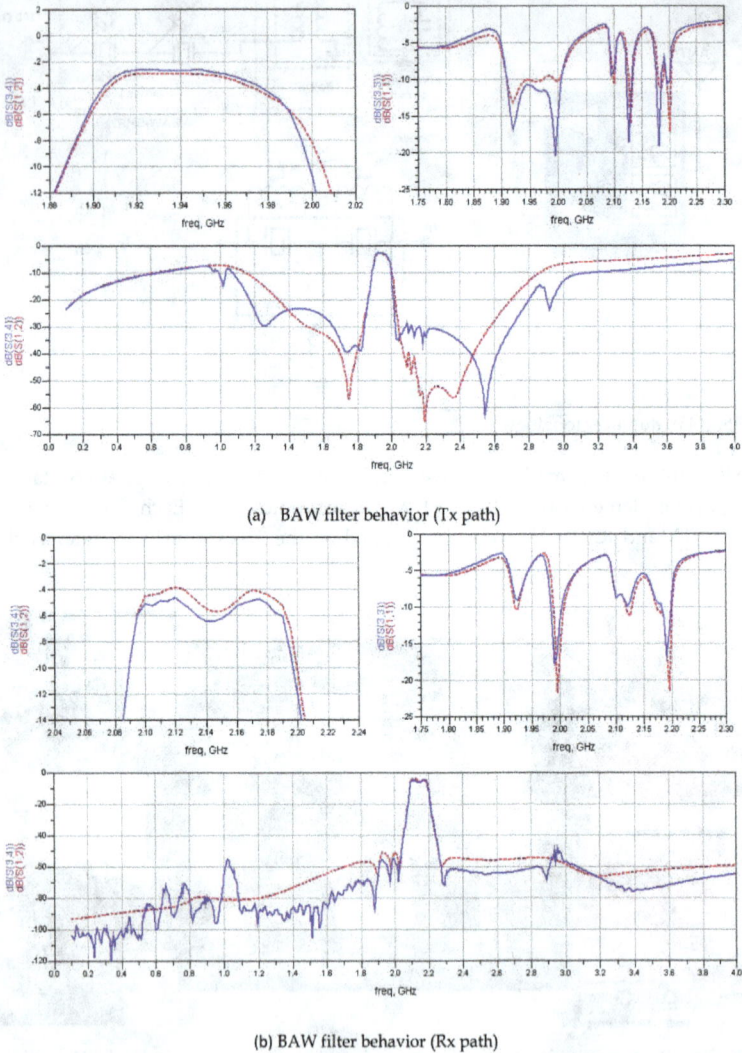

(a)   BAW filter behavior (Tx path)

(b) BAW filter behavior (Rx path)

Figure 37. On probe measurement (blue) and simulation (BAW) – (a) TX path, (b): RX Path

only strong difference is the TX isolation at the RX frequency, which is due to grounding effects; it disappears when the ground is bounded. It should be noticed that because of the one-path measurement conditions, the other path is open, leading to a strong mismatch in the adjacent band. Moreover, the unexpected low coupling factor of the resonators and the especially balun variations explain the relatively low performances of this duplexer.

When mounted on board as shown in Figure 38, the BAW duplexer exhibits approximately the same performances as on probe, with a better out-of-band rejection. The RX rejection in the TX path is better than 40dB, the TX rejection in the RX path is better than 45dB, the TX insertion loss evolves from 2.6dB to 4.6dB in the upper border of the band, and the RX insertion loss falls from 4.5 to 6dB mainly due to the balun mismatch.

Figure 38. On board mounting of the BAW duplexer.

5.2.3. Complete transmitter test bench and measurement results.

Figures 39 and 40 present the measurement setup. Two test benches have been implemented: one with the RF signal generator feeding the transmit BAW filter and the PA module, and the second one is made of the latter test bench which has been completed by the BAW duplexer to get the complete transmitter.

At the differential output of the RF signal generator board, the BiCMOS7RF differential input – single ended output PA is connected. Isolators are placed on each signal path between the two boards. The PA exhibits a peak power gain of 13dB at 1.7GHz and almost 500MHz of -3dB bandwidth with an output compression point of 27.5dBm showing that this stage will not contribute to non-linearity with output power below 20dBm. The power gain at 1.95GHz is 10.3dB. Figure 41 presents the measured output spectrum for the signal at the output of the PA module.

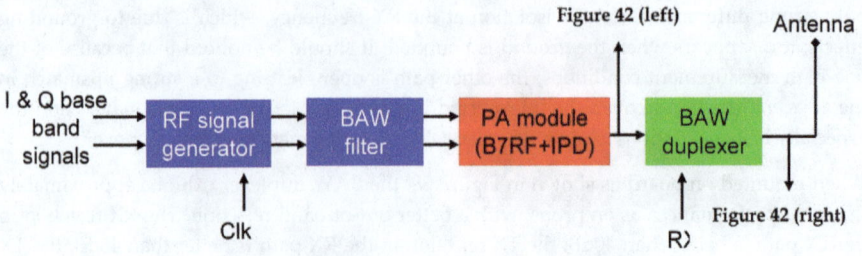

Figure 39. Test fixture for the full transmitter

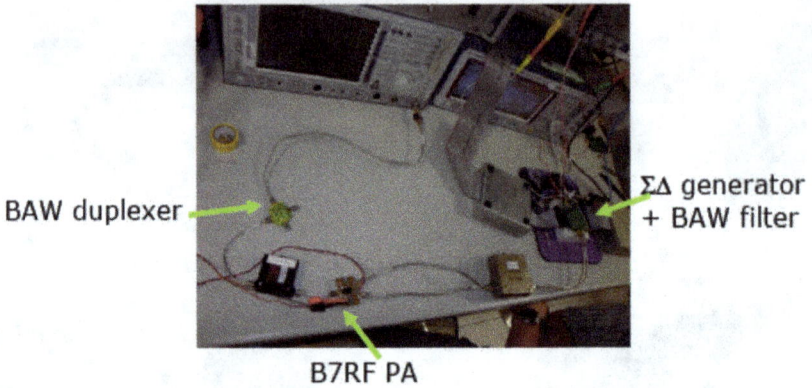

Figure 40. Measurement setup for filtered ΔΣ+ B7RF PA + BAW duplexer (full transmitter)

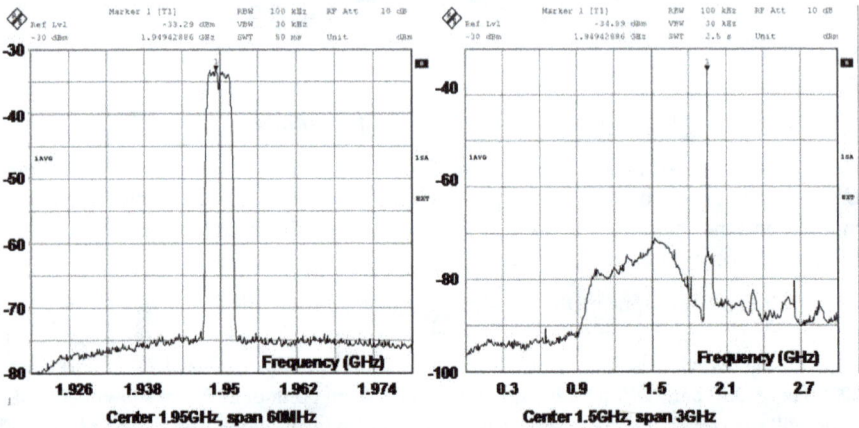

Figure 41. Output spectrum of the power gain stage, zoomed on the bandwidth (left) and full span (right)

One can observe that as the signal level at the input of the PA is low, there is no "visible" distortion on the signal (when looking in-band), and the ACLR level is preserved. When investigating a wider frequency window, we see the Sigma-Delta like lower side skirts re-growing in the band due to the wide-band amplification of the B7RF PA. Finally, a 42dB ACLR at 5 and 10MHz from the carrier for a -17.3dBm integrated power is obtained. EVM has been measured to be 4%.

At the single-ended output of the B7RF power amplifier, an isolator and the BAW duplexer are connected on the Tx input. The output chain measurements are done on the antenna pin of the duplexer demo-board, while the differential output toward the Rx path is shunted to 50 Ω loads. In figure 42, the filtering role of the duplexer Tx path on the overall transmitted signal is clearly shown. The ACLR constraints are generally preserved through the complete transmission chain. Moreover, the low-side spectrum re-growth observed in figure 42 is swept over by the BAW duplexer Tx filtering path. Finally, a 41dB and 41.7dB ACLR respectively at 5 and 10MHz from the carrier for a -20.8dBm integrated power is obtained. EVM has been measured to be 5%.

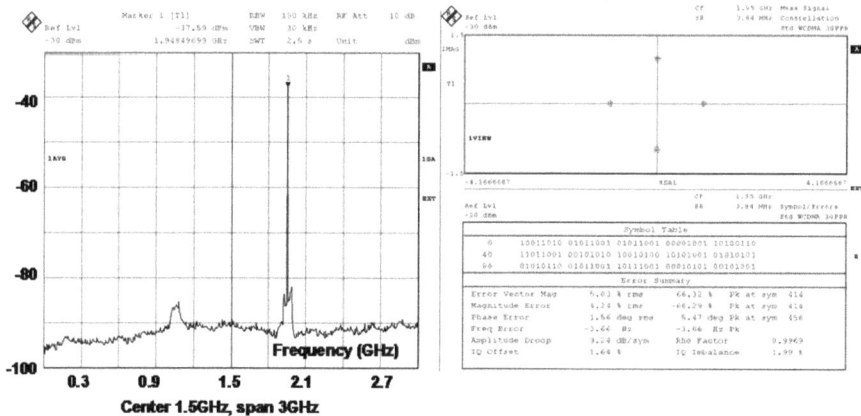

Figure 42. Measured wide-band output spectrum and EVM feature for the full transmitter (filtered ΔΣ + B7RF PA + BAW duplexer)

The complete transmitter is in line with the ACLR and EVM specifications. The transmitter was not able to fulfill the transmit power specifications due to some discrepancies on the distributed gain over the chain. Nevertheless, spurious emission requirements are almost fulfilled, thanks to the BAW high out-of-band rejection. The goal of the presented results was to demonstrate a novel type of Software Defined Radio transmitter architecture for W-CDMA (and DCS) standard(s) with BAW based filter and W-CDMA duplexer. Extra gain should be inserted into the whole transmission chain at the price of an increase in EVM and non-linearity degradation (lowering the ACLR).

6. Conclusion

In this chapter several methods for BAW devices simulation have been presented and investigated. Regarding the efficiency and the rapidity of the needed computation, designers could choose the 1D, the co-simulation piezoelectric equations and 2.5 D simulations or the full 3D Finite Element Method. The co-simulation equations and 2.5 D seems to be the best compromise between simulation time and quality of results. The detailed methods have been validated on several types of filters (Ladder, Lattice and mixed Ladder-Lattice). Optimization method and tools for filters have also been proposed to compute high quality filters (low insertion losses, high rejection, high pole number…).

BAW based filters and duplexers have then been presented and integrated in a complete UMTS transmitter highlighting the advantages of BAW devices.

Other piezoelectric materials are investigated and appear as promising solutions for realizing wide-band filters or sensors [31], [32].

Author details

Matthieu Chatras, Stéphane Bila, Sylvain Giraud, Lise Catherinot,
Ji Fan, Dominique Cros and Michel Aubourg
XLIM, UMR CNRS 7262, University of Limoges, Limoges, France

Axel Flament , Antoine Frappé , Bruno Stefanelli and Andreas Kaiser
IEMN, UMR CNRS 8520, Villeneuve d'Ascq, France

Andreia Cathelin
STMicroelectronics, TR&D, Crolles, France

Jean Baptiste David and Alexandre Reinhardt
CEA-LETI, Grenoble, France

Laurent Leyssenne and Eric Kerhervé
IMS, UMR CNRS 5818, Université de Bordeaux, Talence, France

7. References

[1] K.M.Lakin, G.R. Kline, K.T. Mccarron, "High Acoustic Resonators and Filters", 1993 IEEE Microwave Symp. Digest, 3, pp 1517-1520

[2] K. Y. Hashimoto, "RF Bulk Acoustic Wave Filters for Communications", Artech House, 2009.

[3] A. Shirakawa, JM. Pham, P. JarrY, E. Kerherve, "FBAR Filters Synthesis Method and Reconfiguration Trends", Chapter 3 of Microwave Filters and Amplifiers book, pp.19-47, Research Signpost, 2005.

[4] Lakin, K. M., Kline, G. R., and Mccarron, K. T., "Development of Miniature Filters for Wireless Application," *IEEE Transaction on Microwave Theory and Techniques*, Vol.43, 2933-2939, 1995

[5] K. M. Lakin and K. G. Lakin, « Numerical Analysis of Thin Film BAW Resonators », *Proceedings of Ultrasonics Symposium*, Vol. 1, pp. 74-79, October 2003.

[6] R. Thalhammer, R. Aigner, « Energy loss mechanisms in SMR–type BAW devices », *Microwave Symposium Digest*, 2005 IEEE MTT-S International 12-17 June 2005

[7] J. F. Rosenbaum "Bulk Acoustic Wave Theory and Devices", Boston: Artech House, 1988

[8] L. Catherinot, S. Giraud, M. Chatras, S. Bila, D. Cros, T. Baron, S. Ballandras, P. Monfraix, L. Estagerie "A general procedure for the desing of BAW filters", International Journal of RF and Microwave Computer-Aided Engineering, September 2011, Vol. 21, Issue 5, pp 458-465.

[9] http://www.home.agilent.com

[10] Lanz, R.; Muralt, P. "Solidly mounted BAW filters for 8 GHz based on AlN thin films", IEEE Symp. on ultrasonics, 2003, Vol. 1, pp. 178-181.

[11] C. Cibert, C. Champeau, M. Chatras, D. Cros and A. Catherinot « Pulsed laser deposition of aluminum nitride thin films for FBAR application » Applied Surface Science 253 (2007) pp. 8151-8154

[12] H.P.Loebl, M.Klee, C.Metzmacher, W.Brand, R.Milson And P.Lok "Piezoelectric Thin AlN Film for Bulk Acoustic Wave (BAW) Resonators" Materials Chemistry and Physics, Vol. 79, pp. 143-146, 2003

[13] Ylilammi, J.Ella, M.Partanen and J.Kaitila "Thin Film Bulk Acoustic Wave Filter" IEEE Transactions on Ultrasonics, Ferroelectrics, and Frequency Control, vol. 49, No. 4, pp. 535-539, April 2002

[14] http://www.leti.fr/en

[15] http://www.upm.es/internacional

[16] E. Iborra, M. Clement, J. Olivares, S. Gonzalez-Castilla, J. Sangrador, N. Rimmer, A. Rastogi, B. Ivira, and A. Reinhardt, "BAW resonators based on AlN with Ir electrodes for digital wireless transmissions." 2008 IEEE Ultrason. Symp. Proc., (2008) pp. 2189-2192.

[17] B. Ivira, P. Benech, R. Fillit, F. Ndagijimana, P. Ancey, and G. Parat, "Self-Heating Study of Bulk Acoustic Wave Resonators Under High RF Power" IEEE Trans. Ultrason. Ferr. Freq. Control, 55, (2008). pp. 139-147.

[18] A. Reinhardt, F. de Crécy, M. Aïd, S. Giraud, S. Bila, and E. Iborra, "Design of Computer Experiments: A powerful tool for the numerical design of BAW filters" 2008 IEEE Ultrason. Symp. Proc., (2008) pp. 2185-2188.

[19] A. Devos, E. Iborra, J. Olivares, M. Clement, A. Rastogi, and N. Rimmer, "Picosecond Ultrasonics as a Helpful Technique for Introducing a New Electrode Material in BAW Technology: The Iridium Case", 2007 IEEE Ultrason. Symp. Proc., (2007) pp. 1433-1436.

[20] J. Olivares, M. Clement, E. Iborra, S. González-Castilla, N. Rimmer, and A. Rastogi, "Assessment of Aluminum Nitride Films Sputtered on Iridium Electrodes", Ultrasonics Symposium 2007, pp 1401-1404.

[21] A. Frappé, A. Flament, B. Stefanelli, A. Kaiser, A. Cathelin, "An all-digital RF signal generator using high-speed $\Delta\Sigma$ modulators", IEEE Journal of Solid-State Circuits, art. No. 15, Vol.44, Oct. 2009

[22] Vankka, J. Sommarek, J. Ketola, I. Teikari, M. Kosunen and K. Halonen, "A Digital Quadrature Modulator with on-chip D/A Converter," IEEE Journal of Solid-State Circuits, Vol. 38, No. 10, pp. 1635-1642, Oct. 2003.

[23] S. Giraud, S. Bila, M. Chatras, D. Cros, M. Aubourg, "Bulk acoustic wave filter synthesis and optimisation for UMTS application", EuMW 2009, Rome

[24] 3GPP UE Radio Transmission and Reception (FDD) TS 25.101. Available at http:\\www.3gpp.org\

[25] A. Shirakawa, P. Jarry, J.-M. Pham, E. Kerhervé, F. Dumont, J.-B. David, A. Cathelin, "Ladder-Lattice Bulk Acoustic Wave Filters: Concepts, Design, and Implementation", International Journal of RF and Microwave Computer-Aided Engineering, 5 June 2008, pp.476

[26] A. Flament, S. Giraud, S. Bila, M. Chatras, A. Frappe, B. Stefanelli, A. Kaiser, A. Cathelin, "Complete BAW filtered CMOS 90nm digital RF signal generator", Joint IEEE North-East Workshop on Circuits and Systems and TAISA Conference, 2009

[27] E. Kerhervé, J.B. David, A. Shirakawa, M. El Hassan, K. Baraka, P. Vincent, A. Cathelin, "SMR-BAW duplexer for W-CDMA application", Journal of Analog Integrated Circuits and Signal Processing,2010.

[28] A. Shirakawa, P. Jarry, J.M. Pham, E. Kerherve, F. Dumont, J.B. David, A. Cathelin, "Ladder-Lattice BAW Filters: Concepts, Design and Implementation" , International Journal of RF & Microwave Computer Aided Engineering (RFMiCAE), 2008

[29] P.Bradly, R.Ruby, J.Larsoniii, Y.Oshmyansky and D. Figueredo "A Film Bulk Acoustic Resonator (FBAR) Duplexer for USPCS Handset Applications" IEEE International Microwave Symposium Digest 2001, Vol.1, P367-370

[30] Calvez, C.; Person, C.; Coupez, J.; Gallée, F.; Gianesello, F.; Gloria, D.; Belot, D.; Ezzeddine, H. "Packaged hybrid Si-IPD anrenna for 60 GHz applications", EuMW 2010, pp683-686.

[31] M. Chatras, L. Catherinot, S. Bila, D. Cros, S. Ballandras, T. Baron, P. Monfraix, L. Estagerie "Large Band Pass BAW Filter for Space Applications"IEEE, IFCS, San Francisco, May 2011

[32] T. Baron, J. Masson, D. Gachon, J.P. Romand, S. ALzuaga, L. Catherinot, M. Chatras, S. Ballandras. "A Pressure Sensor based on HBAR micromachined structure" IEEE IFCS, New Port(USA), june 2010.

Chapter 2

High-Overtone Bulk Acoustic Resonator

T. Baron, E. Lebrasseur, F. Bassignot, G. Martin, V. Pétrini and S. Ballandras

Additional information is available at the end of the chapter

1. Introduction

Piezoelectricity has been used for the development of numerous time&frequency passive devices [1]. Among all these, radio-frequency devices based on surface acoustic waves (SAW) or bulk acoustic waves (BAW) have received a very large interest for bandpass filter and frequency source applications. Billions of these components are spread each year around the world due to their specific functionalities and the maturity of their related technologies [2]. The demand for highly coupled high quality acoustic wave devices has generated a strong innovative activity, yielding the investigation of new device structures. A lot of work has been achieved exploiting thin piezoelectric films for the excitation and detection of BAW to develop low loss RF filters [3]. However, problems still exist for selecting the layer orientation to favor specific mode polarization and select propagation characteristics (velocity, coupling, temperature sensitivity, *etc.*). Moreover, for given applications, deposited films reveal incapable to reach the characteristics of monolithic substrates [4].

For practical implementation, BAW is applied for standard low frequency (5 to 10MHz) shear wave resonators on Quartz for instance. SAW, Film Bulk Acoustic Resonator (FBAR) and High overtone Bulk Acoustic Resonator (HBAR) devices are applied for standard radio-frequency ranges and more particularly in S band (2 to 4GHz). HBAR have been particularly developed along different approaches to take advantage of their extremely high quality factor and very compact structure. Until now, many investigations have been carried out using piezoelectric thin films (Aluminum Nitride – AlN, Zinc Oxide – ZnO) atop thick wafers of silicon or sapphire [5] but recent developments showed the interest of thinned single-crystal-based structure in that purpose [6]. Although marginal, their application has been mainly focused on filters and frequency stabilization (oscillator) purposes [7], but the demonstration of their effective implementation for sensor applications has been achieved recently [8]. These devices maximize the Q factor that can be obtained at room temperature using elastic waves, yielding quality factor times Frequency products ($Q.f$) close or slightly above 10^{14}, *i.e.* effective Q factors of about 10,000 at 1GHz in theory (practically, Q factors in excess of 50,000 between 1.5 and 2GHz were experimentally achieved [9])

HBAR-based sensors exploit two principal features yielding notable differences with other sensing solutions. The first one is related to the anisotropy of piezoelectric crystals on which these devices are built, which allows one for selecting crystal cut angles to optimize their physical characteristics. It is subsequently possible to choose cut angles to favor or minimize the parametric sensitivities of the considered wave propagation. The second remarkable feature of these devices concerns the use of piezoelectric excitation/detection of acousto-electric waves which allows for wireless interrogation in radio-frequency ranges such as ISM bands centered at 434MHz, 868MHz, 915MHz or even 2.45GHz.

This chapter presents HBAR principles and related applications. Specific acoustic and electrical behaviors of HBAR are discussed and the different ways devoted to the manufacture of these devices also are presented. Applications of HBAR such as oscillator stabilization, intrinsically temperature-compensated devices and sensors are finally reported. Further developments required to promote the industrial exploitation of HBAR are discussed to conclude this article.

2. HBAR principles

HBARs are constituted by a thin piezoelectric transducer above a high-quality acoustic substrate, as shown in figure 1. The piezoelectric transducer generates acoustic waves in the whole material stack along its effective electromechanical strength. Stationary waves are established between top and bottom free surfaces according to normal stress-free boundary conditions. As no electrical boundary condition arises at this surface, all the possible harmonics of the fundamental mode can exist. However, only the even modes of the transducer are excited as the only ones meeting the electrical boundary conditions applied to the transducer.

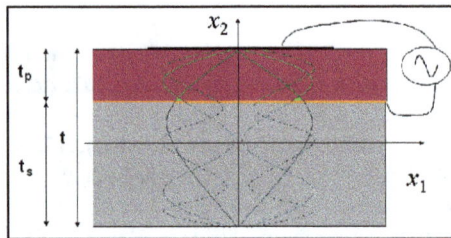

Figure 1. Schematic of HBAR

The electrical response of a HBAR can thus be interpreted as the modulation of the transducer resonance by the whole-stack bulk modes, presenting a dense spectrum of discrete modes localized around the resonance frequencies of the only piezoelectric transducer, as shown in figure 2. Since the substrate thickness is much larger than that of the piezoelectric film, most energy is stored in the substrate, and thus, the quality (Q) factor is dominated by the acoustic property of the substrate. When the thickness of the substrate decreases, the device tends to behave as a FBAR (corresponding to t_s=0 in fig.1). Depending on both the material and the cut orientations of piezoelectric transducer, pure longitudinal or pure shear waves or combinations of these basic polarizations can be excited.

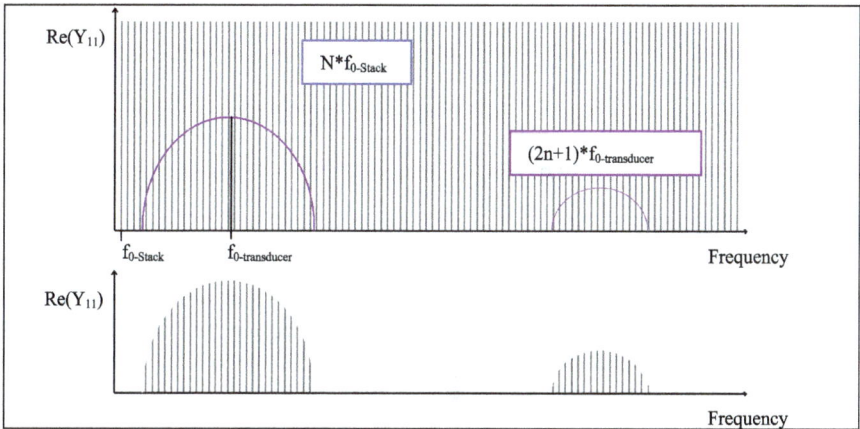

Figure 2. Schematic representation of the typical electrical response of HBARs.

Single port resonator structures can be easily achieved using HBARs, the use of two series devices being generally adopted to avoid etching the piezoelectric layer to reach the back electrode. Despite this favorable aspect, the exclusive use of single-port resonators limits HBAR applicability fields. Therefore, the possibility to fabricate four-port devices has been considered and experimentally tested (as shown in section 3.2). The leading idea consisted in transversely (or laterally) coupling acoustic waves between two adjacent resonators. The principle of such devices was inspired from the so-called monolithic filters based on coupled bulk waves in single crystals [10]. This is achieved by setting two resonators very close to one another. The gap between these resonators must be narrow enough to allow for the evanescent waves between the resonator electrodes to overlap and hence to yield mode coupling conditions. This system exhibits two eigenmodes with slightly different eigen-frequencies: a symmetric mode in which the coupled resonators vibrate in phase and an anti-symmetric mode in which they vibrate in phase opposition, as shown in figure 3. The gap between the two electrodes controls the spectral distance between the two coupled modes.

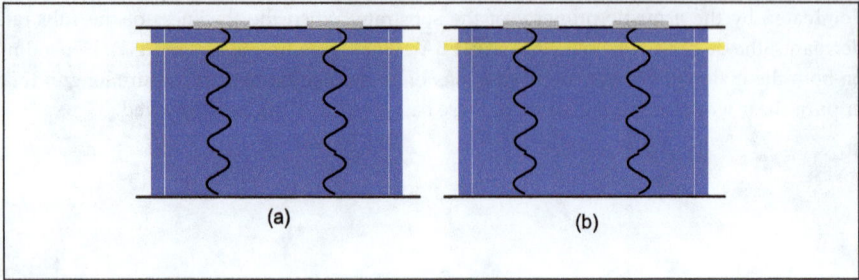

Figure 3. Principle scheme of the laterally-coupled-mode HBAR filter (a) symmetrical mode (b) anti-symmetrical mode

3. HBAR devices

3.1. Electrical and acoustic behavior

3.1.1. Single-port resonator

As explained above, the electrical response spectrum of such HBAR presents a large number of overtones. A large band representation allows for the observation of several envelopes themselves composed of several overtones. The central frequencies of these envelopes correspond to fundamental and even overtone resonances of the only transducer and therefore are mainly controlled by the transducer thickness.

Figure 4 shows the S_{11} response for the considered structure for different substrate thicknesses, illustrating the impact of this parameter on the overtone characteristics. The highest electromechanically-coupled overtone corresponds to the mode matching at best maximum energy location within the transducer, whereas the other overtones do exhibit smaller coupling factor proportionally to their spectral distance with the central overtone. The case of FBAR (t_s=0 μm in figure 1) is reported on this graph to effectively localize the central frequency of resonance and anti-resonance of the only transducer. In presence of a substrate, mode coupling between the two layers is made possible and several overtones appear for substrate thickness larger than the transducer one, as illustrated in figure 4. As suggested previously, the spectral distance between two overtones is mainly due to the substrate properties (velocity and thickness) when the later exhibits a thickness much larger than the other layers of the whole stack. Figure 4 also shows the evolution of the electromechanical coupling coefficient (generally noted k_s^2 for radio-frequency acousto-electric devices) when increasing the substrate thickness. The reduction of k_s^2 when increasing the substrate thickness is directly related to the energy ratio within the transducer and in the whole HBAR stack. Increasing the substrate thickness yields more energy in the whole HBAR structure and less energy within the transducer. Another interpretation consists in considering that the coupling of the transducer alone is spread on all the modes of the structure near the transducer resonance. Increasing the number of modes yields a reduction of the electromechanical of each mode coupled to the transducer mode. A trade-off between the mode density and the stack thickness therefore is mandatory to optimize the HBAR response. Increasing the number of modes experimentally tends to provide higher quality coefficients for modes close to the transducer one but also reduces the corresponding coupling and significantly impact the device spectral density, yielding more difficulty to exploit well defined resonance.

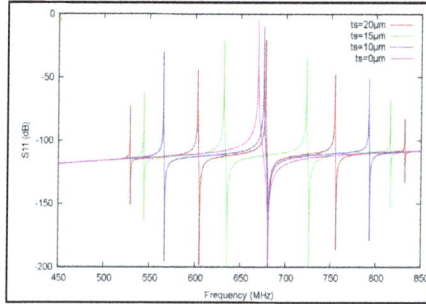

Figure 4. Impact of acoustic substrate. The reflection parameter-S11 with respect to a 50Ω load is measured for different substrate thicknesses. A material stack consisting of an acoustic substrate of LiNbO3 (YXl)/163°, an aluminum electrode of 10nm thick, a 10µm thin piezoelectric layer of LiNbO3 (YXl)/163° and a 10nm thick bottom aluminum electrode is considered here for a theoretical description of the HBAR characteristics. Electrode thickness are chosen extremely thin to neglect their acoustic influence. Acoustic and dielectric losses are only consider in LiNbO3 layers for scaling the maximum achievable quality factors. For all computations, an active electrode surface of 100x100µm² has been considered for normative purposes.

For a given stack, the coupling coefficient of each group of overtones (these groups being defined by fundamental and even overtones of the transducer alone) depends on the material coupling coefficient of the transducer and on the order of the considered group. Indeed, the third order group presents a coupling coefficient divided by 9 compared to the fundamental group (one third of the fundamental mode coupling at excitation times one third at detection), the fifth is divided by 25, and so on. LiNbO3 presents material coupling coefficient noticeably higher (3 to 7 times larger) than other material generally used for HBAR fabrication, such as AlN, ZnO. As a consequence, even transducer overtone groups can be effectively used with such a material and more especially when exciting pure shear waves as exposed further. Figure 5 shows the electrical response of a single-port HBAR built with (YXl)/163° LiNbO3 piezoelectric layer and substrate. Only shear waves are excited and all even group can be visible from the fundamental to the 11th harmonic of the layer alone near 2GHz.

Each overtone in a given group presents a specific coupling coefficient k_s^2 and a specific quality factor Q. Central overtones present the highest coupling coefficients within a group, but not always the highest quality factors. Indeed, the substrate (Sapphire for instance) is usually chosen with acoustic quality better than the transducer material (Aln, ZnO) as it is expected to act as the effective resonant cavity, whereas the transducer material is selected for its piezoelectric strength. As explained above, the energy ratio within the transducer and the substrate is higher for the central overtones than for the overtones located at the edge of the group.

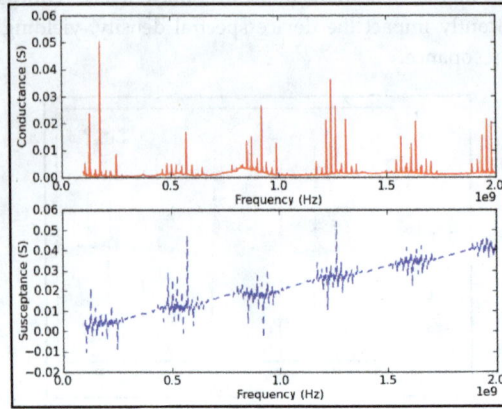

Figure 5. Single-port HBAR device built using LiNbO₃/LiNbO₃ (YX*l*)/163° cut.

In figure 6, a HBAR is constructed with LiNbO₃ material for the transducer. As shown further, LiNbO₃ presents better acoustic quality than Quartz, which is used for the HBAR substrate here to improve the device temperature stability (see section 4.2). In that example the overtone at 433.3MHz exhibits the best coupling coefficient k_s^2 as well as the best quality factor Q, due to the acoustic quality of LiNbO₃ compared to Quartz (see section 4.1).

According the above assumptions concerning material quality selection, the quality factor of overtones located at the edge of group is generally higher than the ones in the center of the group. Practically, it turns out that small-coupling overtones always exhibit better Q than the central overtones in a given group. One explanation of this objective result can be related to electrically generated losses (losses related to electrode resistivity and series resistance tends to increase with current).

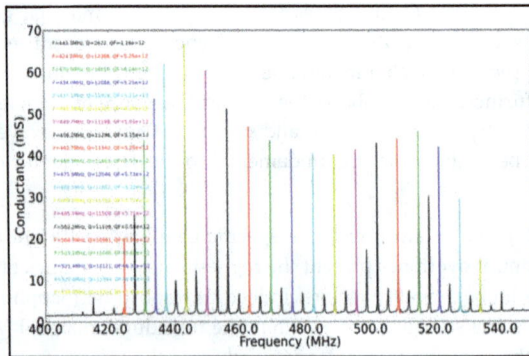

Figure 6. 5th Envelope of single-port HBAR device constituted by LiNbO₃ (YX*l*)/163°/Quartz.

3.1.2. Transversely-coupled HBAR

As explained above, the possibility to fabricate four-port devices has been considered and experimentally tested. Two HBAR resonators were fabricated on a LiNbO₃ (34μm) / Au (300nm) / LiNbO₃ (350μm) stack. Two 145x200μm² surface aluminum electrodes, were patterned upon the stack and separated by a gap of 10μm. Figure 7 shows a typical coupled-mode filter response for a device manufactured atop a LiNbO₃/LiNbO₃ structure. Rejection in excess of 20dB is demonstrated at 720MHz with a single filter cell. Insertion losses of about 15dB are emphasized and could be easily improved by impedance matching. The measured transfer function actually exhibits a double mode response, providing a first evidence of the device operation according to the above assumptions.

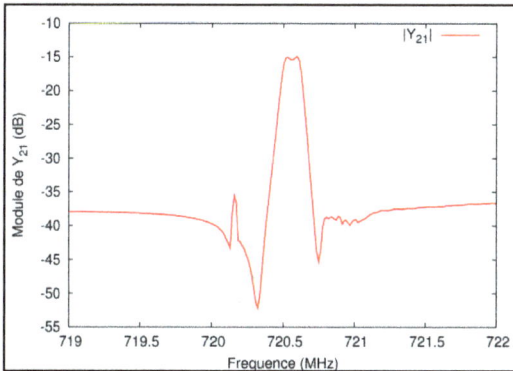

Figure 7. Four-port laterally coupled HBAR devices 0.1% band 720MHz LiNbO₃/LiNbO₃ filter.

Furthermore, the following experiment was applied for definitely validating the lateral mode coupling. The admittance of the first resonator was measured for two different loading conditions applied to the second resonator (Figure 8). For open circuit conditions, a main peak corresponding to the first resonator contribution is observed (Figure 8, left) together with a weaker contribution near the main resonance. For short-circuit conditions, the admittance measured on the first resonator shows two almost-balanced resonance peaks, (Figure 8, right). This behavior is explained by the fact that no current crosses the second resonator when in open condition, yielding a small contribution of the anti-symmetrical mode (due to poor boundary condition matching) whereas loaded electrical condition allows for an effective excitation of the later mode, yielding almost balanced contributions of symmetrical and anti-symmetrical modes as experimentally observed.

3.2. HBAR micro-fabrication

Two main approaches can be implemented to manufacture HBAR devices. The first approach consists in physical or chemical deposition of thin piezoelectric layers (such as ZnO, PZT, AlN and so on) onto the chosen substrate. The first HBAR was manufactured along this approach [11]. The main advantage of this kind of HBAR is the capability of the related techniques (sputtering, epitaxy, sol-gel spinning/firing, pulsed laser ablation, *etc.*) to deposit thin layers which allow for achieving device naturally operating at high frequency (for instance in the vicinity of the 2.45GHz ISM Band, or even more). This approach also did provide among the highest Q factor ever measured for an acoustic-based resonator at room temperature [3], with $Q.f$ product values in excess of 10^{14} at parallel resonance (7.10^{13} at series resonance).

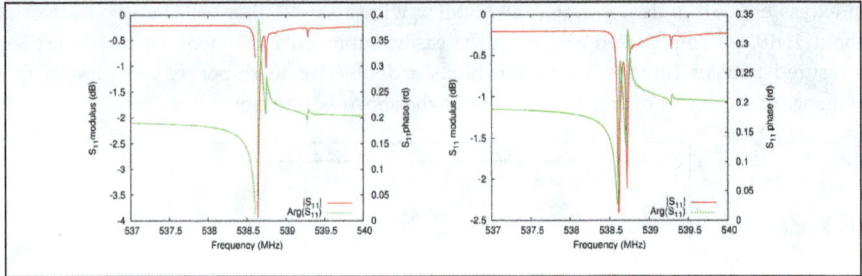

Figure 8. Admittances of one of the two resonators of the laterally-coupled structure as a function of the electrical conditions applied to the associated resonator (left) open circuit (right) 50Ω loading

Although this approach revealed efficient for operational device manufacturing, some drawbacks can be identified which limit the interest of the related resonators. Among these, one of the most problematic concerns the electromechanical coupling coefficient one can obtain particularly with AlN and ZnO, the most used thin piezoelectric layer for RF acoustic devices. As deposition techniques (principally reactive sputtering but also pulsed laser deposition (PLD)) generally allows for depositing well-controlled homogeneous C-oriented layers (i.e. with the C crystal axis oriented along the normal of the coated surface), the maximum accessible coupling remains much below 10%. Also the corresponding modes are purely longitudinal, with reduced degree-of-freedom for effectively controlled layer orientation suited for shear wave excitation/detection. Thin layers such as PZT can overcome this limitation but they generally exhibit notably high visco-elastic coefficients and significant dielectric loss which again limit their interest for high frequency (above 1GHz) applications. More generally, acoustic losses of most piezoelectric layers obtained by sputtering, sol-gel and techniques providing poly-crystalline materials always reveal larger than single-crystal ones. As explained before, the coupling coefficient of each high-overtone resonance depends on the number of overtones and on the intrinsic material electromechanical coupling coefficient. Poor material coupling coefficients prevent the use of overtones modulating the third (and therefore higher order) overtone of the piezoelectric transducer. Finally, compensating longitudinal modes thermal drift is particularly difficult as most of the high acoustic quality materials exhibit negative temperature coefficients of the corresponding phase velocity (as well as the transducer materials, ranging from -20 to -60ppm.K^{-1}). These negative aspects pushed to seek for other manufacturing approaches.

The opportunity to use single crystal layers for acoustic transduction therefore appears as an alternative solution. Assuming the possibility for manufacturing thin single-crystal films atop any material stack makes possible the use of specific crystal cut to select the polarization of the excited acoustic waves as well as its electromechanical coupling coefficient.

The development of the so-called Silicon On Insulator (SOI[TM]) wafers has demonstrated the huge opportunities offered by the Smart Cut[TM] approach [12]. Moreover, its application for transferring single crystal Lithium Niobate thin layer into silicon proved to be effective for SAW device development [13]. As this technology requires a severe know-how and complex technological facilities and environment, an alternative fabrication technique based on metal diffusion at the interface between the materials to be bonded together [14] has been developed together with a lapping/polishing technique for HBAR manufacturing [15].

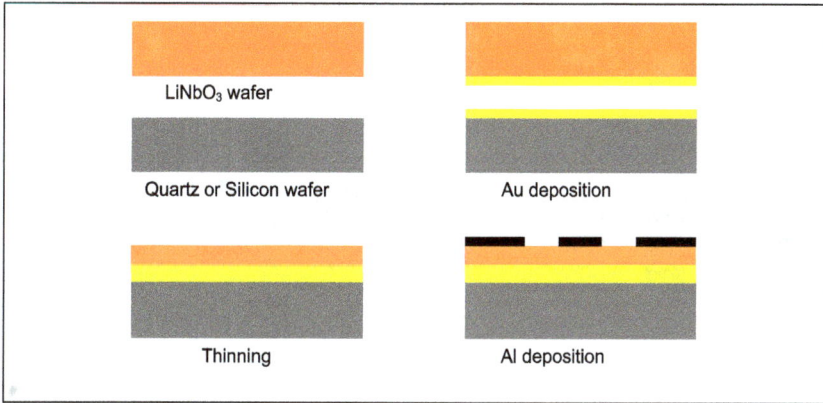

Figure 9. Process flow-chart for the fabrication of the HBAR based on bonding and lapping technology.

In this particular approach, contrary to the sputtering method, thermal process forbids to stack materials presenting notably differential thermal expansion coefficients. Smart Cut[TM] approach allows one to produce thin single crystal layers (such as LiNbO3 for instance, or even LiTaO3 [16]) with typical thickness below 1μm. Along this approach, embedded metal electrodes are fabricated using the Smart Cuttrade technology [16].

The above-mentioned bonding and lapping technology has specifically been developed to allow for material stacking at room temperature, for exploiting any material of any crystal orientation. The process flow-chart reported in figure 9 allows one for a collective manufacturing of HBAR devices. This process is based on the mechanical diffusion of sub-micron gold layers, providing an effective acoustic liaison of the chosen material as well as the HBAR back electrode at once. As the bonding operation is achieved at room temperature, no significant thermal differential effects are observed and the resulting wafer can be handled and further processed provided thermal budget remains smaller than 100°C (as experimentally observed).

Along the proposed approach, optical quality polished surfaces are preferred to favor the bonding of the wafers. A Chromium and Gold thin layer is first deposited by sputtering on both wafers to bond (LiNbO₃ and Quartz in the example of figure 6 and 9). The LiNbO₃ wafer is then bonded onto the substrate by mechanical compression of the 200nm thick gold layers into an EVG wafer bonding machine as shown in figure 10. During the bonding process, the material stack is kept at a temperature of 30°C and a pressure of 65N.cm⁻² is applied on the whole contact surface. The bonding can be particularly controlled by adjusting the process duration and various parameters such as the applied pressure, the process temperature, the quality of the vacuum during the process, *etc*. In the reported development, the process temperature is kept near a value close to the final thermal conditions seen by the device in operation. Since substrate and piezoelectric materials have different thermal expansion coefficients, one must account for differential thermo-elastic stresses when bonding both wafers and minimize them as much as possible.

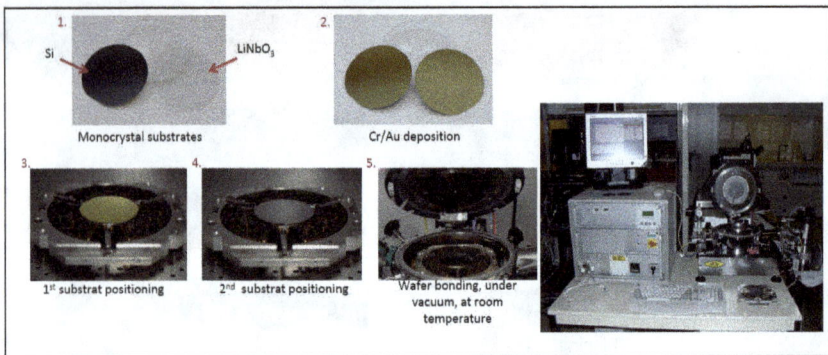

Figure 10. EVG wafer Bounder and illustration of Gold bonding process.

Once the bonding achieved, it is necessary to characterize the quality of the bonding. Due to the thickness of the wafers and the opacity of the stack (metal layers), optical measurements are poorly practicable. To avoid destructive controls of the material stack, ultrasonic techniques have been particularly considered here. The reliability of the bonding then is analyzed by ultrasonic transmission in a liquid environment. The bonded wafers are immersed in a water tank and the whole wafer stack surface is scanned. Figure 11 presents photography of the bench. Two focused transducers are used as acoustic emitter and receiver. They are manufactured by SONAXIS with a central frequency close to 50MHz, a 19mm active diameter and a 30mm focal length. The beam diameter at focal distance at -6dB is about 200μm.

Such a lateral resolution enables one to detect very small defects. The principle of the characterization method is based on the measurement of the received acoustic amplitude which depends on the variation of the acoustic impedance of the bonding area. If the bonding presents a defect at the interface between the two wafers, a dust or an air gap in most cases, the reflection coefficient of the incident wave is then nearly 1. The amplitude of the received wave is strongly reduced or even vanishes. Figure 12 shows a C-Scan of a Silicon/LiNbO₃ wafer bonding characterization. The blue color corresponds to bonded surfaces, whereas yellow and green regions indicate bonding defects.

Figure 11. Ultrasonic characterization bench dedicated to non destructive control of the bonding interface.

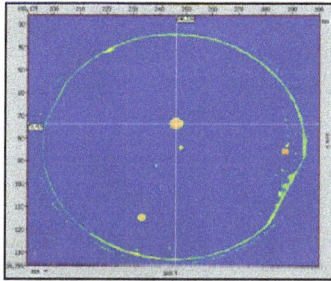

Figure 12. Characterization of a Silicon/LiNbO3 bonding – surfaces are bonded at 95%.

This method presents three major advantages:

- The control of the bonding can be made during the polishing steps without destruction; or the control can be done at the end of the process, indeed, the different layers obtained by sputtering do not disturb the measure.
- There is no constraint related to time resolution as in pulse-echo method, as the wafer thickness is not dramatically larger than the wavelength.
- The analysis of the ultrasonic transmitted signals is very simple because only the amplitude of the first detected signal contains the useful information.

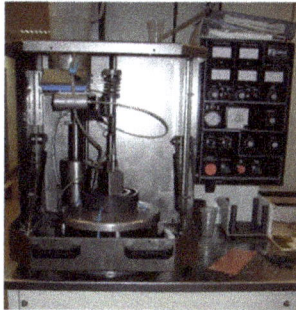

Figure 13. SOMOS lapping/polishing machine.

The piezoelectric wafer is subsequently thinned by lapping step to an overall thickness of 20µm. The lapping machine used in that purpose and shown in figure 13 is a SOMOS double side lapping/polishing machine based on a planetary motion of the wafers (up to 4" diameter) to promote abrasion homogeneity. An abrasive solution of silicon carbide is used here. The speed of the lapping is controlled by choosing the speed of rotation of the lapping machine stages, the load on the wafer, the rate of flow or the concentration of the abrasive. Once close to the expected thickness, the lapping process is followed by a micro-polishing step. This step uses similar equipments dedicated to polishing operation and hence using abrasive solution (colloidal silica) with smaller grain. This polishing step is applied until the average surface roughness r_a remains larger than 3nm. Afterward, the wafer is considered ready for surface processing.

The final step of the HBAR fabrication is the deposition and patterning of the top-side electrode. Aluminum electrodes are then deposited on the thinned LiNbO$_3$ plate surface with a lift-off process. This top electrode allows for connecting the HBAR-based sensor and for characterization operations.

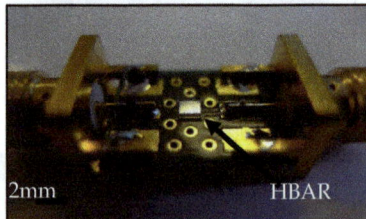

Figure 14. Flip chip of HBAR resonator on PCB substrate.

For all HBAR device, one technological problematic concerns packaging. Due to the HBAR operation, both sides must be kept free of any stress or absorbing condition. HBAR packaging therefore requires specific developments to meet such conditions. Experimental developments reveal that flip-chip techniques are the most appropriate approach in that purpose (as shown in figure 14).

4. HBAR optimization

4.1. Minimizing losses in HBARs

Since the 80's, HBAR devices have demonstrated high quality factor at high frequencies compared to other acoustic devices such as BAW, SAW. $Q.f$ products around 1.1×10^{14} have already been obtained for high overtones using aluminum nitride (AlN) thin films deposited onto sapphire [3]. Hongyu Yu *and al.* showed HBAR with a structure of 0.10µm Al /0.88µm ZnO /0.10µm Al /400µm Sapphire which was measured to have a loaded Q of respectively 15,000 and 19,000 for series and parallel resonant frequencies around 3.7GHz. The temperature coefficient of the resonant frequency is -28.5ppm/°C [17]. Resonators obtained by LiNbO$_3$ wafer as a transducer bonded on another LiNbO$_3$ wafer used as the HBAR substrate exhibit Q factors of 53,000 at 1.5GHz using the Gold bonding technique [5] and $Q.f$ product above 8.10^{13} with an 800nm thickness for the piezoelectric layer by Smart Cut approach [18]. Understanding losses phenomena helps to design high quality factor devices. Loss origins can be classified into two categories: material (intrinsic) and geometry (technology-related). Due to the architecture of HBAR, the quality factor of such devices depends on the crystalline losses and on the material isotropy, on the surfaces parallelism and any loading due to the electrodes.

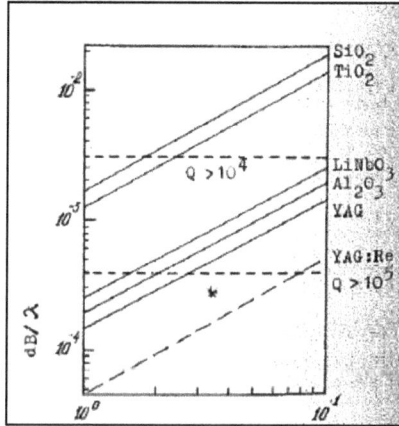

Figure 15. Losses per wave length and the resonator's quality Q as a function of frequency (GHz) for various materials [19].

As explained before, the quality factor is directly link to the acoustic quality of the substrate. Some works have already been done to compare and improve materials to favor high acoustic resonance quality [19], [11]. Figure 15 shows an example of these works [19].

The polishing process providing damaged-free ultra-smooth surfaces is essential, as well as checking the substrate quality by X-ray topography. To take into account current industrial needs, using the technology of material crystal growth is crucial to obtain large wafers. In this context, $LiNbO_3$, $LiTaO_3$ Sapphire, and YAG are the preferred candidates as they do present effective acoustic quality (*i.e.* reduced visco-elastic and dielectric damping properties) and available as 4 inch wafers, excepted for YAG substrates.

The defect of parallelism between two surfaces of HBAR devices dramatically limits the quality factor [11]. Figure 16 shows the quality factor of HBAR modes on Sapphire-base structures versus the plate tilt. As clearly highlighted by this graph, the parallelism must be perfect to prevent power flow yielding Q factor limitations. For example, a HBAR built on a 4 inch wafer with a total thickness variation (TTV) of $3\mu m$ (commercially accessible for Silicon) does not suffer from any parallelism defect and therefore the quality of its resonances is almost not limited by this effect ($Q>10^5$). However, one can see that a thickness variation of $3\mu m$ on 1cm yields effective limitation of the quality factor ($Q<10^4$).

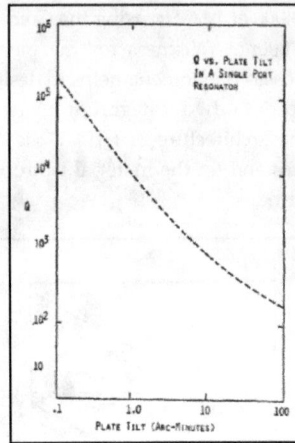

Figure 16. Parallelism-limited Q in a single-port resonator built on Z cut Sapphire substrates [11].

The shape, size and nature of the electrodes can be also important to manufacture high Q HBAR devices. Some works have been done on electrodes of HBAR devices [20], [21], [22], [23].

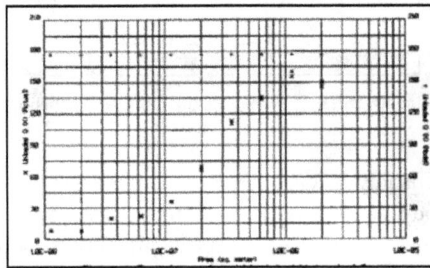

Figure 17. Experimental and Modeled Unloaded Q versus aperture [20].

D. S. Bailey *and al.* showed that the HBAR does not follow the one dimensional computer model [20]. Indeed, figure 17 shows the difference between the experimental Q and the theoretical Q versus the aperture of the electrode. The difference is due to the diffraction effect. The optimum electrode area can depend on two main parameters: the clamp capacitance C_0 and the geometry. This capacitance C_0 is proportional to the surface area and influences other parameters of resonator such as difference in impedance at series and resonance frequency. With a large active area, defects in transducer crystal or of geometry can happen more easily. The optimization of the area shape and surface to limit the diffraction effect and improve the quality factor is an area of ongoing works.

Furthermore, for ultra-high frequency HBAR devices, the electrodes are not thin compared to transducer layer. The thickness and the nature of electrodes have an influence on the quality factor and the other parameters such as the electromechanical coupling coefficient and the resonance frequency. Many works have been done on this subject [23], [24], [21].

The conditions of metal sputtering can influence the nature of the metallic electrode. Indeed, the conditions of metal sputtering for thin layers modify the density and the rate of impurity of the layer. The optimum must be found to have the highest metal density with the lowest impurities. Furthermore, some works compare the influence of different metallic layers (Al, Au, W, Ag) on the quality factor. If we consider the modified Butterworth-Van Dyke (MBVD) model, the best electrode is constituted with the lowest resistivity (Au), but experimentations also show the influence of other parameters. Thus, a Molybdenum layer used as an electrode shows better results due to better acoustic impedance [23].

Generally speaking, low losses applications also require a temperature compensation for the resonator. One solution is to have intrinsic compensation of temperature and it is the purpose of the next paragraph. Another solution is to control frequency by measuring temperature.

4.2. Temperature compensation

One challenge of the radio-frequency bulk acoustic devices is the temperature stability of their resonance frequency. A lot of work has been achieved exploiting thin piezoelectric films for developing temperature-compensated HBARs, with various successes. The possibility to use single crystal thinned films appears as an alternative to control the piezoelectric film properties (velocity, coupling, temperature sensitivity, and so on.) and to globally reconsider material association according to the technological assembly process previously presented.

The celebrated Campbell&Jones method [25] is used here for predicting the Temperature Coefficient of Frequency (TCF) of any mode of a given HBAR. As it has been reported hundred times in previous papers, only the main basic equation is reported below:

$$f = \frac{v}{2e} \rightarrow \frac{df}{f}(T) = \frac{dv}{v}(T) - \frac{de}{e}(T) \tag{1}$$

f, e, v and T are respectively frequency, thickness of resonator, wave velocity and temperature.

Which means that the frequency changes due to temperature variations is computed as the difference between the development of the velocity and of the stack thickness versus temperature. Theoretically using a standard anisotropic 1D model reveals that zero temperature coefficients of frequency (TCF) can be obtained and optimized along the mode order. It is well-known that Quartz and fused Silica (glass) do exhibit positive TCFs. So the use of the other temperature-compensated Quartz orientations, and hence of any other material sharing such property, has been checked theoretically and reveals applicable as well.

As example, Lithium of Niobate and Quartz have been associated for the fabrication of shear-wave based HBARs. LiNbO$_3$ provides crystal orientations for which very strongly coupled shear waves exist (k_s^2 in excess of 45%) whereas AT cut of Quartz allows for compensating second order frequency-temperature effects [WO2009156658 (A1)]. Although this idea was already proposed using other material combinations [US Patent #3401275A], no real design process was presented until now and therefore the possibility to actually determine structures allowing for high frequency operation with first order TCF smaller than 1ppm.K^{-1} was quite hypothetical, but improvement of numerical tools allows this design.

Nevertheless, some works show the possibility to have an intrinsic compensation of the temperature for HBAR devices [8], [26]. Figure 18 shows the temperature dependence for different configuration of HBAR devices constituted by LiNbO$_3$ and Quartz layers with different cut orientations. This work shows clearly that the choice of materials and the cut orientation of these materials have a direct impact on the frequency shift with temperature variations [26].

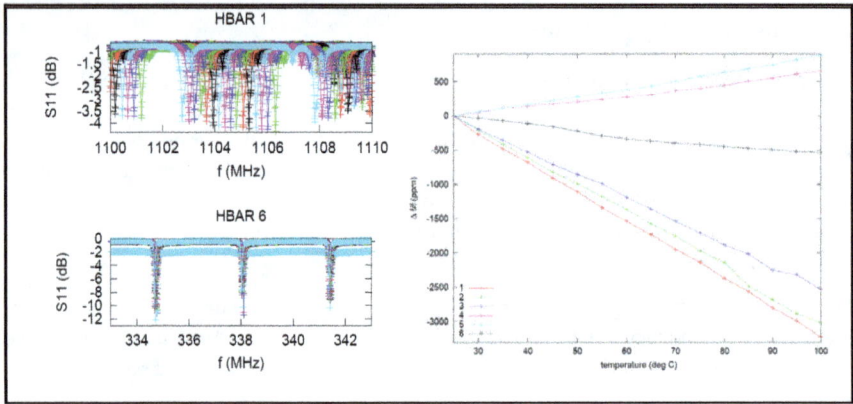

Figure 18. Electrical responses for two different configurations of HBAR (left), frequency variation versus temperature for six configurations of HBAR (right) [26].

Moreover, the frequency dependence on temperature is different for each overtone of HBAR devices. The thickness ratio between the transducer layer and the substrate also influences the frequency variations with different temperatures. Figure 19 shows the computation of the temperature coefficient of frequency (TCF) of a HBAR for various Lithium of Niobate / Quartz thickness ratios. This HBAR device is built on a (YXl)/163° LiNbO$_3$ thinned plate bonded on (YXlt)/36°/90° Quartz substrate of 50µm [27]. One can see that depending on the harmonic number, the TCF_1 changes from +1 to -14ppm.K^{-1}. Furthermore, depending on the harmonic of the transducer alone, the TCF_1 may notably change and thus it cannot be considered as a simple periodic function versus harmonic number. Therefore, it is mandatory to accurately consider all the actual features of the structure for an accurate design of a resonator, *i.e.* the operation frequency, the harmonic number and the thickness ratio for a given structure. To complete this, one should also account for the actual thickness of the device as this parameter will control the possibility to select one (frequency/harmonic number) couple. Finally, it clearly appears that the analysis of such HBAR TCF requires a numerical analysis and that if an intuitive approach allows for a first order definition of crystal orientations, the complicated distribution of energy within the stack versus all the structure parameters induces more intrication in the design process.

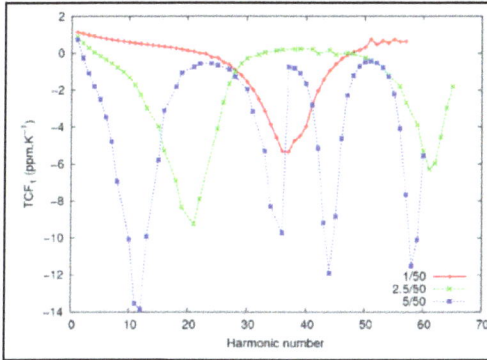

Figure 19. Plot of TCF of a HBAR built on a (YX*l*)/163° LiNbO3 thinned plate bonded on (YX*lt*)/36°/90° Quartz substrate for various Lithium of Niobate/Quartz thickness ratio (Quartz thickness arbitrary fixed to 50 μm) [27].

5. HBAR applications

5.1. HBAR sensors

Industrial acoustic-resonator-based sensors require adapted electronics to be efficiently operated. Two main approaches have been developed in that purpose:

-   The first way is to use the acoustic resonator in an oscillator loop. Compared to normal oscillator use in frequency/time applications, some specific operation regimes must be considered for sensors [28]. Particularly, the resonance frequency is assumed to drift along the measured parameter on a large frequency domain. The Q-value of the resonator may notably vary as well as other parameters like electromechanical coupling and the overall electrical conductance (connected one another) yielding the need for improved electrical robustness of the circuit. The electronics therefore must be able to adapt its operation parameters. Distance between electronic and sensor is often important compare to classic oscillator. Due to these reasons, resolution of the system is limiting to 1.10^{-8}. As example, sensors at 434MHz have resolution limitation of 5Hz due to oscillator loop measurement method. Finally, to use this method, we need resonator with low harmonic generation. BAW, SAW and FBAR can use this electronic.
-   The second way is to have electronic interrogation which finds frequency resonance in determined range of frequency. With classic method it is possible to obtain 100Hz of resolution for 434MHz sensors [28]. If electronics is improving, we can achieve 5Hz of resolution for 434MHz sensors [29]. In this case, clock of electronic is really important for performance. All kind of resonators sensors can be interrogated by this technique, especially HBAR device which present high overtone generation.

Acoustic sensor is passive sensor. Device combine with antenna could be having great interest. Indeed, electromagnetic waves can be changed on electrical waves on electrodes, which can excite acoustic waves by piezoelectric effect. Furthermore this phenomenon is linear and invertible. So, wireless interrogation is possible with acoustic sensors. Wireless communication presents great interest for all hard environments. In that way, acoustic sensors can be use in engine, close environment and more generally in all environments where wire can not be employed.

With wireless interrogation, antenna size, quality factor of resonator, frequency have a strong impact. With increasing of frequency, antenna size decrease. Indeed, the size of antenna is equal to the quarter of wavelength. When higher ISM band is used, quality factor need to be increase to give the same obstruction of the ISM bandwidth. At -3dB, the bandwidth of resonator is proportional to frequency divided by quality factor. And finally, the flight time is proportional to quality factor divided by frequency. They are two consequences of this flight time. Firstly, to have enough energy when frequency increase, the quality factor need to increase. As example, SAW resonators at 434MHz ISM band have quality factor of 10,000 and can be interrogated by wireless approach. To pass at 2.45GHz ISM band, a quality factor equal to 20,000 is required. HBAR devices achieve these characteristics. Secondly, refresh rate increases with frequency. With bandwidth of few kHz the refresh rate is around one millisecond. In this case, quality factor of sensor could not be too higher. So, quality factor of HBAR device need to be optimize for wireless sensor application.

HBAR devices present a great advantage for achieving sensors device. As previously discussed, frequency shift due to temperature effects can be minimized and even compensated, but also magnified as well. As a consequence, HBAR temperature sensors are considered first. Moreover, due to high number of overtones of such devices, it is also possible to develop sensors exhibiting different sensitivity to a given parametric effect at different frequencies. Acoustic devices can also be effectively exploited as stress sensor or pressure sensors. The fabrication of SAW pressure sensor based on thinned Quartz membrane (for instance) was strongly investigated due to the dependence of the wave velocity versus tensile stress at the surface of the membrane when bent by pressure. In the case of bulk wave propagating in such a membrane, the strain variations across the membrane thickness forbid the use of such an approach to develop pressure sensor applications. This can be easily demonstrated using for instance static finite element analysis with a very simple mesh. Indeed, the strain and hence the stress change their signs along the membrane thickness when submitted to pressure. As a consequence, the strain variation across the HBAR generates equilibration of the velocity variations. On the one hand, the strain below the membrane neutral line yields an increase of resonant frequency of the HBAR; on the other hand, the strain above the neutral line yields a decrease of this frequency. Consequently, the resulting frequency shift is negligible. One solution consists in the fabrication of a micro-cavity within the HBAR stack near the neutral line. If the transducer of the HBAR structure is straight above this micro-cavity, the emitted bulk waves are reflected by this micro-cavity and hence confined in this membrane location. The micro-cavity then plays the role of a mirror for the waves. The structure of such device is shown in figure 20. The surface of the cavity should at minimum coincide strictly to the surface of the transducer, but to ease the fabrication (particularly to manage alignment issues) the cavity largely overlaps the transducer aperture. The micro-cavity/micro-mirror could be placed at different deph into the HBAR stack. Its location will define the HBAR sensibility to stress [8].

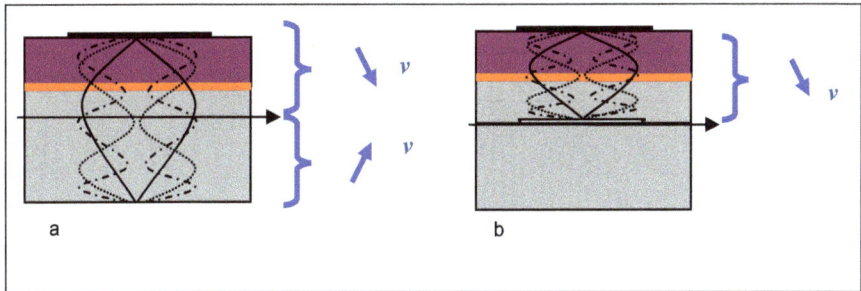

Figure 20. HBAR structures presented low frequency sensitive to stress (a), and highly frequency sensitive to stress with the realization of micro-mirror under the transducer aperture [8].

A lot of works has been done on liquid or gaz HBAR / FBAR sensor and as most representative example, gravimetric sensor. The basic principle of gravimetric acoustic wave sensors is the measurement of the phase velocity variations due to an adsorbed mass or a layer thickness change atop the device during a chemical reaction: this phase velocity is dependent on the boundary conditions of the propagating acoustic wave and is affected either by the layer properties or its thickness. The usual principle exploits bulk acoustic waves, yielding the well-known concept of Quartz Crystal Micro-balance (QCM). The gravimetric sensitivity of the QCM is directly related to its thickness and as a consequence to its fundamental frequency f_0. Adsorption on one side of the resonator modifies its resonance conditions and thus allows for a gravimetric detection. Furthermore, it is possible to functionalize the surface with specific reactants to provide information on the concentration of the adsorbed (target) species in the medium surrounding the sensor [31]. The case of HBAR is particularly attractive as one can expect probing the adsorbed material at various frequencies, providing frequency-dependent information such as viscosity for instance.

Copper electro-deposition on the back side of a HBAR has been used for calibrating the gravimetric sensitivity of its overtones. This approach was particularly implemented as it provides an independent estimate of the deposited metal mass through the measurement of the current. A negative current indicates copper reduction (deposition on the working electrode) whereas a positive current indicates oxidation (copper removal from the working electrode). Simultaneous to the current monitoring, the acoustic phase and magnitude at fixed frequency are recorded [30]. The figure 21.a shows four different overtone frequencies (red dot) recorded. Figure 21.b. shows relative frequency variations and clearly shows the sensitivity difference of the four HBAR overtones. Sensitivity of gravimetric HBAR directly depends on the stack thickness and more precisely on copper thickness versus transducer thickness. The best gravimetric HBAR sensor is constituted by the thinnest stack with metallic thickness equal to quarter of wavelength [31].

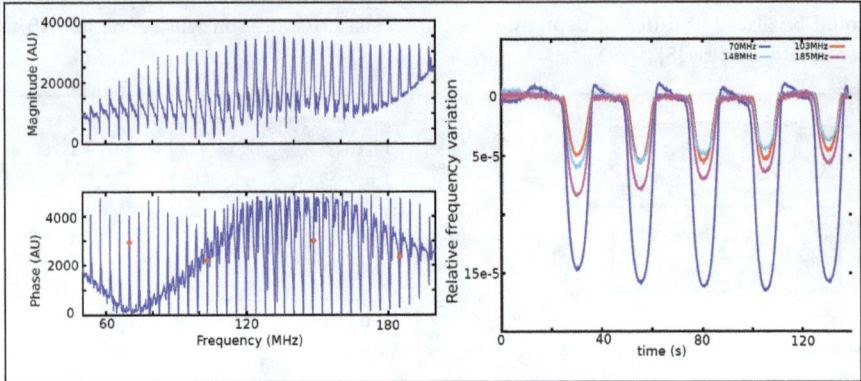

Figure 21. (a) Overtones of the fundamental transducer layer mode of a gravimetric HBAR sensor with four probed frequencies (red dot), (b) relative frequency variations of these four frequencies with their relative sensitivities [30].

5.2. HBAR-stabilized oscillators

Radio-Frequency oscillators can be stabilized by various resonating devices. Their stability is mainly conditioned by the spectral quality of the resonator even if the oscillator loop electronics must be optimized to lower the generated noise as much as possible. For mid-term stability, temperature compensation is a key point and allows for notably improving the corresponding figures of merit. The possibility to build temperature compensated HBARs has been shown in paragraph 4.2 and is a key-point for the fabrication of oscillator exhibiting short-term stability compatible with practical applications.

Moreover, the frequency stability of an oscillator can be characterized by its single-sideband phase noise, $L\{f_m\}$. Leeson's equation [33] shows that low phase noise operation can be achieved by increasing the loaded quality factor Q_{load} of the resonator. According to Leeson's model, a high resonator quality factor (Q) or circulating power level improves the phase noise and, therefore, the short-term stability of the oscillator. Considering these aspects, HBAR device features for the frequency stability reveal more favourable than FBAR or SAW device ones. Therefore HBAR should allow for notably improving oscillator performances. However, multi-overtone features of HBAR do not facilitate resonance lock for oscillator applications. Therefore SAW or FBAR device are generally used to filter the frequency of HBAR. Consequently, the compactness of HBAR is deteriorated due to the need for this filter. Optimizing HBAR spectral response then is still an open question and should receive more attention in future developments. One can note that using single port HBARs with optimized frequency separation between the overtones [32] may allow to get rid of this filtering operation.

HBAR then are capable to address high frequency source applications without requiring multiplication stages as usually achieved. The idea then is to evaluate the effective interest of HBAR for direct frequency synthesis, reducing the oscillator architecture complexity and potentially improving the corresponding operational features.

Hongyu Yu and *al.* have presented a local oscillator based on a HBAR resonator associated to an atomic clock [34]. Atomic clocks are used for embedded applications which need high stability performance such as GPS station. The atomic transition allows having long-term stability in this application, but it presents poor short-term stability. To success oscillator based on this atomic transition, local oscillator is needed. This local oscillator stabilizes the short-term variation of the global oscillator with its short-term performance. The local oscillator need to have good phase noise (better than -70dBc/Hz at 1kHz for instance) to prevent global degradation of clock stability. Figure 22 shows the phase noise measurement data of the 3.67GHz Pierce oscillator and the 1.2GHz Colpitts oscillator, and the Allan deviation of the free-running 3.67GHz oscillator that consumes only about 3mW [34]. Local oscillator based on HBAR resonator need frequency control to achieve the atomic transition frequency. With the modulation of the HBAR frequency with an external synthesizer and FBAR filters, the local oscillator locked to the coherent population trapping resonance.

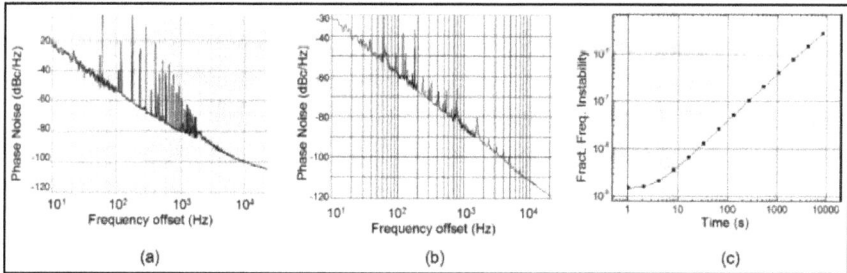

Figure 22. Phase noise measurement data of (a) the 3.67GHz Pierce oscillator and (b) the 1.2GHz Colpitts oscillator. (c) Allan deviation of the free-running 3.67GHz oscillator that consumes only about 3mW [34].

Other applications require low phase noise oscillator such as embedded RADAR. A radio-frequency oscillator operating near the 434MHz-centered ISM band validates the capability of the above-mentioned HBAR for such purposes. The composite substrates have been built using 3-inches (YXl)/163° LiNbO$_3$ cut wafer bonded and thinned down to 15µm onto a 350µm thick (YXlt)/34°/90° Quartz base. Single-port resonators operating near 434MHz (exploiting the third harmonic of the thinned Lithium of Niobate plate as the HBAR "motor") have been then manufactured. Electrical and thermoelectric characterizations have shown quality factor of the resonance in excess of 20,000, yielding a $Q.f$ product of about 10^{13} and a third order frequency-temperature behavior. A SAW filter was used to select the ISM band and to filter the high spectral density HBAR response (figure 23). The oscillator then has been measured using a phase noise automatic bench. A phase noise better than -- 160dBc/Hz at 100kHz has been measured as well as a -165dBc/Hz level at 1MHz from the career (figure 23). Short-term stability characterizations show that the resonator stability is better than 10^{-9} at room conditions (no temperature stabilization).

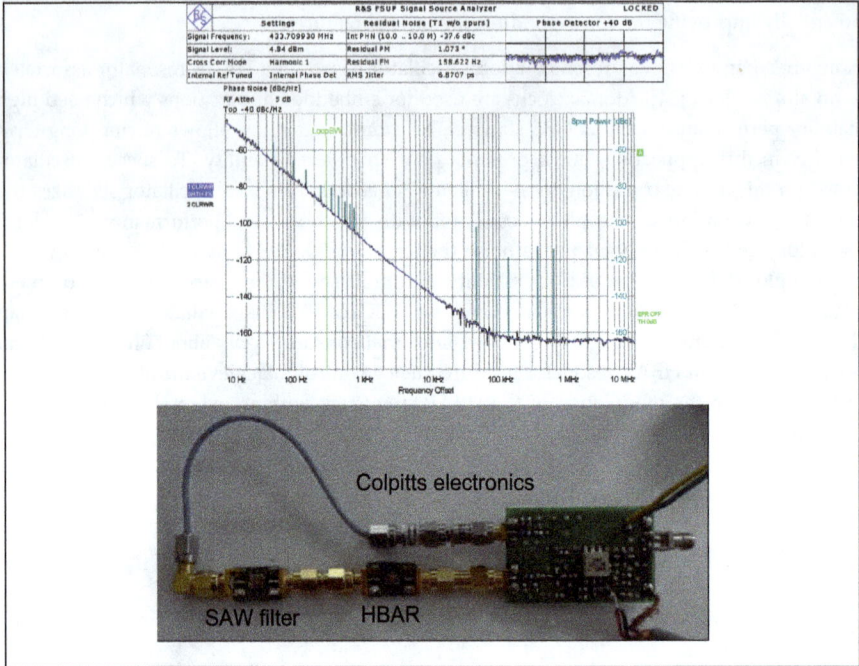

Figure 23. Phase noise curves for oscillators at 434MHz constituted by High Overtone Bulk Resonator, SAW filter and Colpitts electronics.

To achieve higher frequency (above 1GHz), four-port resonators are mandatory due to the difficulty to adjust the oscillator tuning elements. For this application, the temperature stability is required and therefore the resonator exploits a (YX*l*)/163° LiNbO$_3$ thinned layer atop a (YX*lt*)/36°/90° Quartz substrate. The electrodes defining the coupled transducers (four-port resonator) are two half-circles (300µm diameter) separated by a gap of 20µm, yielding favorable conditions for using the resonator to stabilize an oscillator loop at 935MHz and 1.636GHz. The device were cut and packaged, mounted in an oscillator loop, and measurements of phase noise were performed. Figure 24 shows the phase noise of the two corresponding oscillators compared with the phase noise of an oscillator stabilized by a classical BAW resonator at 100MHz, "octar 507X100" from AR-Electronics. The oscillator near 1.6GHz clearly shows better performances than the one at 935MHz, with a noise level lower than -130dBc/Hz at 10kHz from the carrier. In order to compare the 100MHz oscillator with the 1.6GHz one, the low frequency source has to be multiplied by 16, *i.e.* +12dB must be added to the noise level. It gives a level of -140dBc/Hz at 10kHz which is not far from HBAR solution.

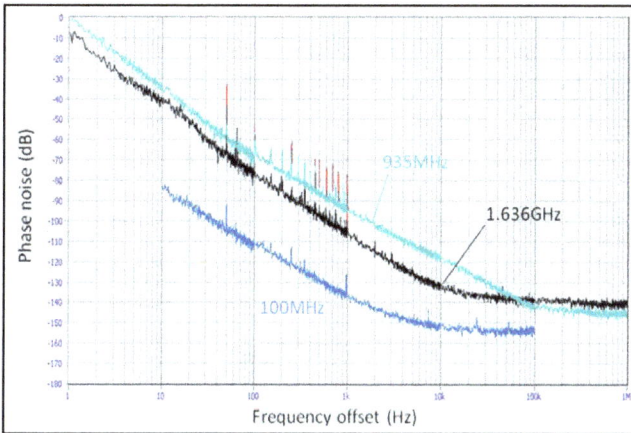

Figure 24. Phase noise curves for oscillators at 935MHz, 1,636MHz compared with phase noise of an oscillator stabilized by a classical BAW resonator at 100MHz [33].

6. Conclusions and perspectives

HBAR have been developed for the fabrication of passive radio-frequency devices capable to overcome standard SAW and BAW limitations considering the quality of the resonance, complexity of technological fabrication and operation frequencies.

These devices actually maximize the Q factor that can be obtained at room temperature using elastic waves, yielding quality factor times frequency products $(Q.f)$ close or slightly above 10^{14}, *i.e.* effective Q factors of about 10,000 at 10GHz in theory (practically, Q factors in excess of 50,000 between 1.5 and 2GHz were experimentally achieved). Single-port or four-port resonator has been described and the different approach of manufacturing as been explained. The choice of acoustic wave and acoustic substrate permits to address large range of application. First one is chose value of the quality factor and the electromechanical coupling coefficient of overtone frequencies. Second one is to minimize frequency shift due to temperature variation for chosen frequency. All these possibilities allow us to address different applications such as sensor or low phase noise applications.

Although HBAR device knows since several decades, HBAR device has not yet achieve is development maturity. Futures works will concern improvement of fabrication, frequency control, and wireless sensors. The large number of the parameters for optimizing HBAR in function of the applications requires well-control generic process of fabrication. The selection of the frequency resonance is also a key point for the emergence of HBAR devices, as the frequency tuning. And finally, wireless HBAR sensor need strong effort of development.

Some simple applications and large potentiality of HBAR conception has been presented. Two main approaches exist for the realization of HBAR devices. The first one based on piezoelectric deposition gives easily high quality factor and frequency device. The largest potentiality of conception to address different applications is obtained by the second approach based on mono-crystal wafer assembled. Further developments required to promote the industrial exploitation.

All previous technological industrial development for solidly mounted resonator for instance could be easily use for fabrication of HBAR based on piezoelectric deposition. More technological development is required to control thickness, repeatability and so on (see section 4.1) for the second approach. In both case, packaging aspect is a key-point. Both side of HBAR need to be free for acoustic reason. Flip-chip approach seems to give the best result for industrial needs.

In both cases, design tool needs to be developed to realize conception of all HBAR devices. SAW design tool could be a good base to develop such software.

More works also need to improve performances or to fixe limits of all different HBAR devices which are specialized for each application. These developments are the precondition for industrial actions.

Author details

T. Baron, E. Lebrasseur, F. Bassignot, G. Martin, V. Pétrini and S. Ballandras
FEMTO-ST, université de Franche-Comté, CNRS, ENSMM, UTBM, Département Temps-Fréquence, France

Acknowledgement

This work was partly supported by the Centre National d'Etudes Spatiales (CNES) under grant #04/CNES/1941/00-DCT094 and still supported by the CNES under grant #R-S08/TC-0001-026, and by the Direction Generale pour l'Armement (DGA) under grant #05.34.016.

7. References

[1] K. M. Lakin and J. S. Wang, UHF composite bulk wave resonators. *IEEE Ultrasonics Symposium Proceedings*, pp. 834-837, 1980
[2] K.M. Lakin, Thin film resonator technology. *IEEE Trans. on UFFC*, Vol.52, pp.707-716, 2005
[3] K.M. Lakin, G.R. Kline, K.T. McCarron, High Q microwave acoustic resonators and filters. *IEEE Trans. on Microwave Theory and Techniques, Vol. 41, n°12*, pp,2139-2146, 1993
[4] S.P. Caldwell, M.M. Driscoll, D. Stansberry, D.S. Bailey, H.L. Salvo, High overtone bulk acoustic resonator frequency stability improvements. *IEEE Trans. on FCS*, pp. 744-748, 1993
[5] D. Gachon, E. Courjon, G. Martin, L. Gauthier-Manuel, J.-C. Jeannot, W. Daniau and S. Ballandras, Fabrication of high frequency bulk acoustic wave resonator using thinned single-crystal Lithium Niobate layers. *Ferroelectrics, Vol. 362*, pp. 30-40, 2008

[6]   Curran Daniel R & Al, US Patent #3401275A, 1968-09-10
[7]   M. Pijolat, D. Mercier, A. Reinhardt, E. Defaÿ, C. Deguet, M. Aïd, J.S. Moulet, B. Ghyselen, S. Ballandras, Mode conversion in high overtone bulk acoustic wave resonators. *Proc. of the IEEE International Ultrasonics Symp.*, pp.201-204, 2008
[8]   T. Baron, E. Lebrasseur, J.P. Romand, S. Alzuaga, S. Queste, G. Martin, D. Gachon, T. Laroche, S. Ballandras, J. Masson, Temperature compensated radio-frequency harmonic bulk acoustic resonators pressure sensors. *Proc. of the IEEE International Ultrasonics Symposium*, pp. 2040-2043, 2010.
[9]   D. Gachon, T. Baron, G. Martin, E. Lebrasseur, E. Courjon, F. Bassignot, S. Ballandras, Laterally coupled narrow-band high overtone bulk wave filters using thinned single crystal lithium niobate layers. *Frequency Control and the European Frequency and Time Forum (FCS)*, 2011
[10] D. Royer & E. Dieulesaint. Elastic Waves in Solids II. *Springer*, 2000.
[11] R.A. Moore, J.T. Haynes, B.R. McAvoy. High Overtone Bulk Resonator Stabilized Microwave Sources. *1981 IEEE Ultrasonics Symposium*, pp.414-424, 1981
[12] B. Aspar, H. Moriceau, *et al.*. The generic nature of the Smart-Cut process for thin film transfer. *Journal of Electronic Materials*, Vol. 30, n°7, 2001, pp. 834-840
[13] T. Pastureaud, M. Solal, B. Biasse, B. Aspar, J.B. Briot, W. Daniau, W. Steichen, R. Lardat, V. Laude, A. Laëns, J.M. Frietd, S. Ballandras, High Frequency Surface Acoustic Waves Excited on Thin Oriented LiNbO3 Single Crystal Layers Transferred Onto Silicon. *IEEE Trans. on UFFC, Vol.54, n°4*, pp 870-876, 2007
[14] J.C. Ponçot, P. Nyeki, Ph. Defranould, J.P. Huignard. 3 GHz Bandwidth Bragg Cells. *IEEE 1987 Ultrasonics Symposium.* pp. 501-504, 1987
[15] T. Baron, E. Lebrasseur, *et al.*. Development of Composite Single Crystal Wafers for Sources and Sensor applications exploiting High-overtone Bulk Acoustic Resonators. *JNTE* 2010
[16] J.-S. Moulet, M. Pijolat, J. Dechamp, F. Mazen, A. Tauzin, F. Rieutord, A. Reinhardt, E. Defay, C. Deguet, B. Ghyselen, L. Clavelier, M. Aid, S. Ballandras, C. Mazure. High piezoelectric properties in LiNbO3 transferred layer by the Smart Cut™ technology for ultra wide band BAW filter applications. *Electron Devices Meeting, 2008. IEDM 2008. IEEE International.* pp. 1-4. 2008
[17] Hongyu Yu, Chuang-Yuan Lee, Wei Pang, Hao Zhang and Eun Sok Kim. Low Phase Noise, Low Power Consuming 3.7 GHz Oscillator Based on High-overtone Bulk Acoustic Resonator. *2007 IEEE Ultrasonics Symposium*, pp.1160-1163, 2007
[18] M. Pijolat, A. Reinhardt, *et al.*. Large Qxf Product for HBAR using Smart Cut (TM) transfer of LiNbO3 thin layers onto LiNbO3 substrate. *IEEE Ultrasonics Symposium 1-4 (2008)* pp. 201 – 204, 2008
[19] S. Ivanov, I. Koelyansky, G. Mansfeld and V. Veretin. Bulk Acoustic Wave High overtone resonator. *1990 1er congrès Français d'acoustique*, pp 599-601, 1990
[20] D. S. Bailey, M. M. Driscoll, and R. A. Jelen. Frequency Stability Of High Overtone Bulk Acous Tic Resonators. *1990 ultrasonics symposium.* pp. 509-512, 1990

[21] Lukas Baumgartel and Eun Sok Kim. Experimental Optimization of Electrodes for High Q, High Frequency HBAR. *2009 IEEE International Ultrasonics Symposium Proceedings.* pp. 2107-2110, 2009

[22] S.G.Alekseev, I.M.Kotelyanskii, G.D.Mansfeld, N.I.Polzikova. Energy Trapping in HBARs Based on Cubic Crystals. *2006 IEEE Ultrasonics Symposium.* pp. 1478-1480, 2006

[23] Hui Zhang, Zuoqing Wang, and Shu-Yi Zhang. Electrode Effects on Frequency Spectra and Electromechanical Coupling Factors of HBAR. *IEEE transactions on ultrasonics, ferroelectrics, and frequency control, vol. 52, no. 6,* june 2005

[24] Hui Zhang, Shu-Yi Zhang, Kai Zheng. Parameter characterization of high-overtone bulk acoustic resonators by resonant spectrum method. *Ultrasonics 43 (2005)* pp. 635–642. 2005

[25] J.J. Campbell, W.R. Jones, A method for estimating crystals cuts and propagation direction for excitation of piezoelectric surface waves. *IEEE Trans. On Sonics and Ultrasonics,* Vol. 15, pp. 209-217, 1968

[26] T. Baron, D. Gachon, *et al..* Temperature Compensated Radio-Frequency Harmonic Bulk Acoustic Resonators. Proc of the IEEE IFCS, pp 625-655, 2010

[27] T. Baron, G. Martin, E. Lebrasseur, B. Francois, S. Ballandras, P.-P. Lasagne, A. Reinhardt, L. Chomeloux, D. Gachon, J.-M. Lesage. RF oscillators stabilized by temperature compensated HBARs based on LiNbO3/Quartz combination. *Frequency Control and the European Frequency and Time Forum (FCS), 2011 Joint Conference of the IEEE International.* pp. 1-4, 2011

[28] J.-M Friedt, C. Droit, G. Martin, and S. Ballandras, "A wireless interrogation system exploiting narrowband acoustic resonator for remote physical quantity measurement", *Rev. Sci. Instrum. vol. 81, 014701 (2010)*

[29] C. Droit, G. Martin, S. Ballandras, J.-M Friedt, "A frequency modulated wireless interrogation system exploiting narrowband acoustic resonator for remote physical quantity measurement", *Rev. Sci Instrum. Vol 81, Issue 5, 056103 (2010)*

[30] D. Rabus, G. Martin, E. Carry and S. Ballandras. Eight channel embedded electronic open loop interrogation for multi sensor measurements. *European Frequency and Time Forum (EFTF),* 2012

[31] G.D. Mansfeld. THEORY OF HIGH OVERTONE BULK ACOUSTIC WAVE RESONATOR AS A GAS SENSOR. *13th International Conference on Microwaves, Radar and Wireless Communications.* vol.2, pp. 469 – 472, 2000

[32] Jérémy Masson. Étude de capteurs résonants acoustiques interrogeables à distance à base de films minces micro-usinés sur silicium. *Mémoire de thèse,* 2007

[33] E. Lebrasseur *et al..* A Feedback-Loop Oscillator Stabilized Using Laterally-coupled-mode Narrow-band HBAR Filters. *2011 IEEE International Ultrasonics Symposium Proceedings,* 2011

[34] Hongyu Yu, Chuang-yuan Lee, Wei Pang, Hao Zhang, Alan Brannon, John Kitching, and Eun Sok Kim. HBAR-Based 3.6 GHz Oscillator with Low Power Consumption and Low Phase Noise. *IEEE Transactions on Ultrasonics, Ferroelectrics, and Frequency Control, vol. 56, no. 2, February 2009.* pp.400-403, 2009

Electromagnetic and Acoustic Transformation of Surface Acoustic Waves and Its Application in Various Tasks

Sergey E. Babkin

Additional information is available at the end of the chapter

1. Introduction

1.1. Electromagnetic and acoustic transformation

Electromagnetic and acoustic transformation (EMAT) is transformation of high-pitched electromagnetic oscillations in the inductive sensor over a specimen into acoustic oscillations in the specimen. For the transformation to be performed the padding constant magnetic field is required. This process is referred to as a direct EMAT. Further oscillations are extended in the specimen in the shape of acoustic waves. Acoustic waves can be deduced outside by means of revertive EMAT when acoustic oscillations in the surface layer of a specimen will be transformed to electromagnetic oscillations in the receiving sensor. The overall process is as follows: a direct EMAT, a distribution of acoustic waves and a revertive EMAT, which is in practice referred to as a double EMAT or just EMAT [1, 2].

Materials in which EMAT is possible to occur:

- ferromagnetics, i.e. substances which possess electrical conductivity, magnetic properties, and magnetostriction (e.g. iron, nickel, steels);
- the conductors which have no essential magnetic properties (e.g. copper, aluminum);
- materials possessed at least one of following properties: magnetism, conduction, magnetostriction (ferrite, amorphous, rare-earth materials).

EMAT mechanisms

EMAT may proceed by three basic mechanisms.

1. A vortical current mechanism (Lorentz force mechanism). Electrons in a surface layer will fluctuate under the influence of Lorentz force:

$$F_L = f\left(J,H\right),$$

where J is a current density, H is a magnetic field.

In conductors this thin layer is called as a skin layer.

2. A magnetic mechanism. This mechanism defines power interaction of an alternative electromagnetic field of the sensor, h, and a constant magnetic field of a specimen, H.

$$F_m = f\left(h,\, H,\, \mu_{ij}\right),$$

where μ_{ij} is a tensor of magnetic conductivity of a material.

3. A magnetostriction mechanism. For the materials possessing magnetostriction, this mechanism is responsible for changing linear dimensions of microvolumes in the surface layer of a specimen under the sensor depending on the alternative field of the sensor.

$$F_\lambda = f\left(Q_{ij},\, M\right),$$

where Qij is a tensor of magnetoelastic communication, M is magnetization of a specimen.

The first and the second mechanisms are often considered as one which is referred to as an electromagnetic or electrodynamic mechanism of transformation.

Technical realization of EMAT

In technical realization terms EMAT is defined a non-contact method of generation and reception of acoustic waves (ultrasonic waves). The method of generation and reception of ultrasonic waves by means of piezoelectric transducers (PET) appears to be the nearest and best analog in this field. Therefore, the technical methods developed for PET methods generally are appropriate for EMAT methods as well.

There are two main techniques, i.e. a resonance technique and a pulse technique.

1. A resonance technique. A small specimen (a cylinder or a rectangular parallelepiped) is placed into solenoid in order to create a magnetic field. By means of a round wire EMA coil which is put on the specimen the loose oscillations are generated in it. A standing waves resonance is reached by changing the frequency of generation. Thus receiving EMA coil shows a signal maximum. A resonance technique is hard to realizing for a PET method, because the contact is required. These problems are easily solved by the non-contact EMAT method.

An amplitude of the received signal, a resonance frequency, a size of a magnetic field are the key information parameters. The attenuation and Q-factor of the system is possible to define.

2. A pulse technique. Short electric pulses of the generator are transformed to acoustic waves by means of EMAT. These pulses are of high-pitched filling. The generation

EMAT sensor can be to used as a revertive EMAT sensor, if the sensor receives a reflected signal.

Each technique developed for a PET method are suitable for pulse EMAT (an echo method, a shadow method etc.).

Key informational parameters of the pulse technique: amplitude of the received signal, speed of a wave, size of a magnetic field, wave attenuation.

2. The surface acoustic waves

Classically the term «the surface acoustic waves» (SAW) is considered to involve the extending of the waves along the surface of the solid and vacuum; waves poorly fade extending along the surface boundary, the waves quickly fade when moving away from surface boundary into a solid. It concerns only Rayleigh waves and the Guljaev-Blyushteyna waves but in piezocrystals.

There is an expanded treatment of the term «the surface acoustic waves» [3]. Firstly, surface waves at the boundary of the solid body and the gases are considered. It allows Stonly waves and the waves of leak being referred to SAW. They have vertical polarization and a quasyrayleigh structure. Secondly, the expanded treatment considers waves in a layer on the solid body surfaces, or waves with the inhomogeneous elastic properties which are observed in a closely to a surface layer. Loves waves and generalized Lamb waves are considered as well. Thirdly, if the wave extending along the surface is thought to be the main factor, the waves in plates, i.e. Lamb waves and horizontal polarization waves (SH-wave) can be referred to surface waves.

3. Electromagnetic and acoustic transformation of surface acoustic waves

The surface waves in a solid body can be generated in the different ways. The most widespread way is the way using PET. The main advantages of the EMA method of SAW generation in comparison with PET consist of a) not – contact generation and SAW receiving and b)availability to use different SAW which are hardly generated by PET. To reach this purpose it is enough to change a configuration of the wire EMA coil as well as its orientation regarding the magnetic field.

From the practical point of view Rayleigh and Lamb waves being waves vertical polarization as well as SH–waves and Love waves being waves of horizontal polarization are of great interest.

1.   Waves of vertical polarization.

The Rayleigh wave can be generated and received by means of meandr EMAT at the normal and tangential magnetic field, H_z and H_x (fig. 1). The double EMAT effectiveness is characterized by the amplitude of the received signal in the receiving coil (E). He is measured in volt usually.

On tangential magnetizing (H_x) the amplitude of the received signal according to the mechanisms of transformation is described by the following approximate formulas[1, 4].

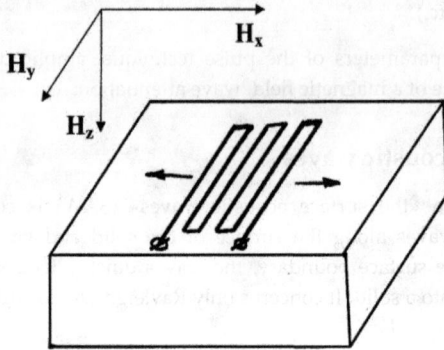

Figure 1. Position of the meandrovy coil on an specimen.

For the magnetostriction mechanism:

$$E^{ms} \approx \frac{1}{\mu_d}\left(\frac{d\lambda}{dH}\right)^2 P_1,\tag{1}$$

where μ_d is differential magnetic permeability, λ is the linear magnetostriction ($\lambda = \Delta l/l$ is specific elongation of the specimen), P_1 is constant factor which depends from parametres of the generating sensor, the reception sensor; physical constants of a material (for example, density, electroconductivity); strengthenings of a reception path, H is a magnetic intensity.

For the sum of vortical current and magnetic mechanisms (the electrodynamic mechanism):

$$E^i + E^m \approx (\mu_0 H)^2 P_2,\tag{2}$$

where μ_0 is a permeability of vacuum, P_2 is constant factor the similar P_1.

At normal magnetization (H_z).

For the magnetic mechanism:

$$E^m \approx (\mu M)^2 P_3,\tag{3}$$

where, μ M are magnetic conductivity and magnetization of the material in this orientation, P_3 is constant factor the similar P_1.

For the vortical current mechanism:

$$E^i \approx (B)^2 P_4, \tag{4}$$

where B is a magnetic induction, P_4 is constant factor the similar P_1.

The dependence of the amplitude of the received signal (E) on the magnetic field size, the so-called «field curve EMAT» is the total characteristic EMAT SAW [5]. Two sensor (such as in fig. 1) place on a surface of the sample. One sensor generates the surface wave, the second sensor accepts a wave. The accepted signal strengthen and measure in volt. Often E normalized in relation to any value E (for example, to the maximum value) and receive abs. units. This size postpone on axis Y. The magnetic field size in the sample is shown on an axis X (B (T) or H (A/cm), as B=$\mu\mu_0$H).

Fig. 2 shows a typical field curve for Rayleigh waves EMAT taking ARMCO iron as an example (curve 1). The curves for the transformation mechanisms calculated on the basis of equation (1) and (2) are presented in fig.2 as well (curve 2 and 3, accordingly). At calculation of the equations use known functions for a material: B=$\mu\mu_0$H, λ = f(H) [12].

The amplitude of the signal (E) is normalized concerning a maximum.

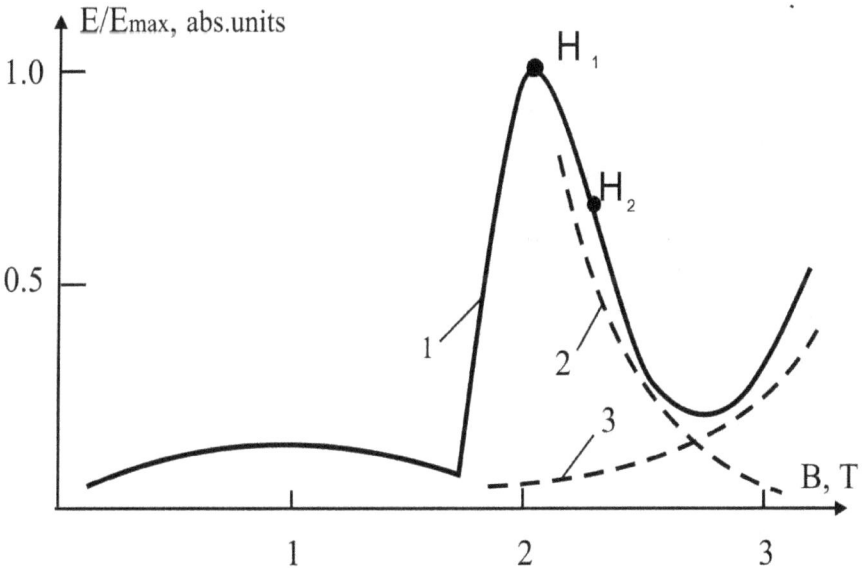

Figure 2. The experimental EMAT field curve of Rayleigh waves in a tangential field for iron (curve 1). The theoretical curve 2, 3 are presented equations (1) and (2).

An EMAT field curve of Rayleigh waves registered in a normal magnetic field(curve 1) and the curves 2, 3 calculated using (3) and (4) are shown in fig.3. As seen from fig.2 and fig.3 the experimental process is well described by the formulas.

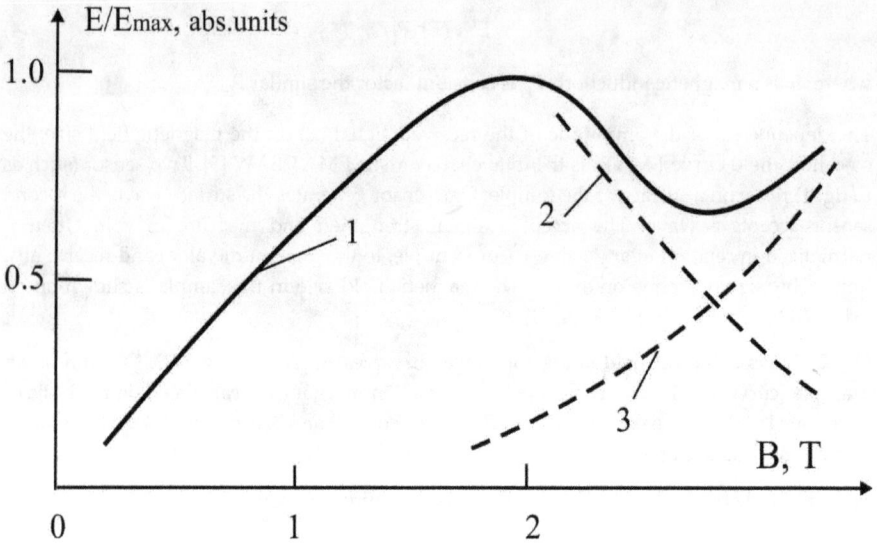

Figure 3. 1 – is the experimental EMAT field curve of Rayleigh waves in a normal field for iron. The theoretical curve 2, 3 are presented equations (3) and (4).

2.   Waves of horizontal polarization.

Waves of horizontal polarization are generated in a case when current elements of the meander coil are parallel to a magnetic field. In this case the elementary volumes of the specimen under the EMA coil are in the cross fields, i.e. under the big constant magnetize field and the small variable magnetic field of the coil. Such an arrangement results in shear modes of horizontal polarization which are synchronized under EMA current elements of the coil (fig. 1 field H_y).

The following simplified formula [6] is derived based on the theoretical calculation for SH waves in a **d-** thick specimen [6]:

$$E_{SH} \approx \frac{1}{d}\left(\frac{\lambda}{H}\right)^2 P_5,$$

(5)

where P_5 is constant factor the similar P_1.

A structurally similar formula is deduced for the waves of horizontal polarization in the layer (Love waves), but a layer thickness and some known restrictions concerning speeds ratio should be taken into consideration.

The experimental field curve of EMAT of SH waves in ARMCO iron is presented in fig. 4.

If I construct the chart of the equation (5) on this coordinates, I also will see good coincidence to experiment.

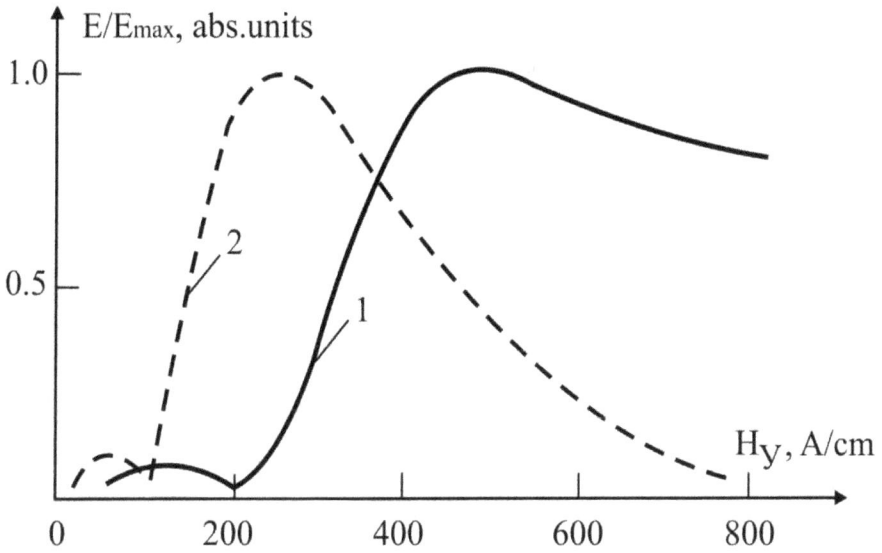

Figure 4. The experimental field curve of EMAT of SH waves in iron (1), in comparison with Rayleigh waves (2).

Love waves have been studied using 20-80 microns thick nickel films applied on the surface of an aluminum substrate. The field curve is presented on fig. 5.

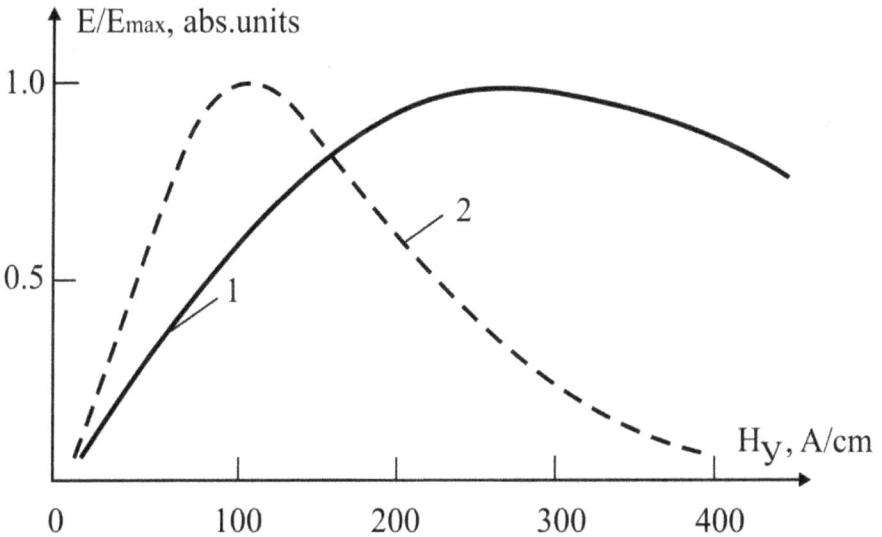

Figure 5. The experimental field curve of EMAT of Love waves in a nickel film on an aluminum substrate (1), in comparison with Rayleigh waves (2).

SAW identification

SAW identification is essential in studying EMAT of the surface wave experimentally.

SAW signals should be distinguished from the signals of other waves. E.g. from the volume waves signals starting from the surface of the specimen to its depth at different angles with interfering reflections in the receiving EMA sensor. It is also necessary for SAW to be recognized in between.

Basic identification methods

1. Pulp method. It involves simple damping by a finger of the surface of the specimen which strongly affects the SAW wave amplitudes but doesn't affect other waves. As long as it is a subjective method to make it being more objective a method of buttered drops is used. Using a pipette to drop the oil from the same height an identical damping is obtained. It allows distinguishing even the Rayleigh surface waves from SH waves. In the same conditions using a 6-drop method a 28 % decrease of Rayleigh wave amplitude and a 7 % decrease of SH waves are observed.
2. A Rayleigh wave C_R strongly differs from longitudinal waves C_l in speed, its speed being slightly differs from the speed of transverse waves C_t (C_R= 0, 87 -0, 96 C_t). A zero mode of a SH wave speed is equal to C_t. To be more sure in distinguishing from a Rayleigh wave the accuracy of speed determination is be within 0, 5 %.
3. A sensor driving method is based on the following: if a generating sensor moves regarding to the receiving one, the signals on the a receiving oscilloscope screen move synchronously.

Thus, we can tell the following.

-   We understand work electromagnetic and acoustic transformation of surface acoustic waves in the theory and in practice.
-   It is understood how to choose operating mode electromagnetic and acoustic transformation of surface acoustic waves for the decision various applied problems.

4. Using EMAT surface acoustic waves

4.1. Non-destructive testing

4.1.1. Defectoscopy of the surface defects

Using an ultrasonic method for testing materials and products EMAT incorporates all the advantages of ultrasonic testing methods. A piezoelectric transducers (PET) method is similar to EMAT. The main ultrasonic techniques developed for PET, are applied to EMAT.

In fig. 6 the design of EMA of the Rayleigh wave converter on the basis of П-shaped electromagnets [7] is shown.

The sensor consists of two identical half-cells: П-shaped electromagnet 3 and meandrovy coil 5, 6. Both converters are located in the case 2 and are filled in with a filler 4. One half-cell generates the surface wave in a specimen 1, another accepts the surface waves. It is a separate testing regime. It is possible to use the combined testing regime when one half-cell generates a wave and accepts the signal reflected from defect.

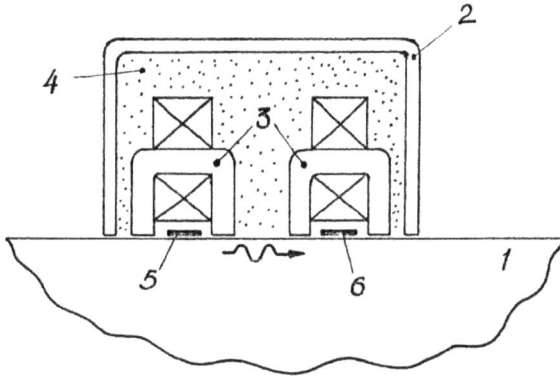

Figure 6. Design of the EMAT sensor of waves of Reley. 1-an specimen, 2 - a housing, 3 - electromagnets, 4 - filler, 5, 6 - generating and receiving EMA coils.

A magnetic field can be created the strong permanent magnets, but in this case it will be uncontrollable. Electromagnets permits operating a magnetic field, choosing the working point on field curve EMAT.

EMAT loses to a way PET in sensitivity, but in this case a contact is not required. It is meant it that using the EMAT method an immediate contact between EMA of the sensor and a specimen is not needed, and the sensor can operates through air or other gaps. For example, the layer of scale or paint isn't an absolute obstacle for EMAT. The amplitude of the received signal has been recognized to be strongly dependent on the gap size. It is due to two factors: 1) an electromagnetic field of EMA of the sensor in a gap called "the sensor – an specimen surface" decreases according to the exponential curve law, 2) a magnetic field of the specimen decreases as a gap occurs. For ferromagnetic materials the field curve of Rayleigh waves EMAT effectiveness has a characteristic maximum defined by the magnetostriction mechanism (fig. 2). The amplitude of the received signal depends on a working point choice in a field curve.

Usually the working point at non-destructive monitoring is chosen at top of the main maximum of the field curve (point H_1 in fiq.2). The maximal receiving signal is obtained in this way. For most ferromagnetic materials this maximum lies within 100 – 300A/cm magnetic fields. The same effectiveness using the vortical current mechanism is reached in magnetic fields about 5000A/cm in size.

If the working point of monitoring is chosen at the maximum top (H1 in fig. 2), the emergence of the gap in a magnetic circuit will result in decreasing the magneticfield in size and falling transformation effectiveness, i.e., the amplitude of the received signal. However, if the working point is chosen on the slope of the field curve (H2), a magnetic field decreases with the gap increasing which leads to the transformation effectiveness. Thus, both factors, i.e., the removal of EMA of the coil from a surface of the specimen and a magnetic field decrease in an specimen compensate each other. [8].

If the gap between the coil and the surface is introduced before the measurements are carried out the measurement results are more reliable. In this case the uncontrollable gap less influences on the received signal. If the working point is chosen in the slope of the field curve the signal received does not depend on the gap (fig. 7).

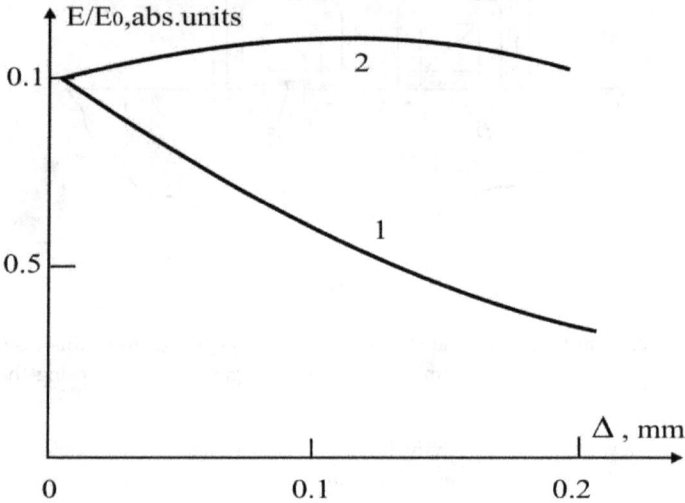

Figure 7. Dependence of the received signal from dimension of the gap between sensor and specimen at the working point H₁(1), at the working point H₂ and preliminary gap (2).

The experiments on detection of model defects showed that by means of EMA of the converter of Rayleigh waves presented in fig. 6, it is possible to reveal the surface defects such as cracks 5мм in length, 0, 5мм in depth. Such defects come to light both at reflection from the defect, and when weakening a signal passing the defect (weakening of 20 %).

4.1.2. Structurescopy of metals

Structural changes are well recognized by EMAT in ferromagnetic materials taking place in technological processes. This results from the fact that at EMAT four subsystems are involved in a material: magnetic, electric, magnetoelastic and elastic. All these factors influence informational parameters of the EMAT method.

Thus, if a dependence between the parameters of the technological process and the information parameters of the method of one subsystem is not a one- direction dependence the EMAT method can be used.

So the dependence between the amplitude of the received signal of Rayleigh waves sensor (fig. 6) and the hardness of the steel 38XTC specimens being treated at different temperatures is practically linear. To increase the specimen hardness it was treated at high temperatures. The coefficient of correlation was equal to 0, 92. It allows EMAT controlling.

Another structural component is anisotropy of the material properties which can be also measured by the sensor described above. The sensor is rotated on the specimen surface showing the information parameters change. However, the data can be distorted due to different factors.

To overcome this trouble a meander -curve sensor was suggested to be curved into a ring (fig. 8) [9].

In this case each element of the dl coil generates SAW in K1 and K2 directions. In general two circular waves are observed: a coverging to the center wave and a diverging from the center one. A converging to the O point wave then turns into a diverging wave. It is received by the same circular EMA coil when the wave moves under it.

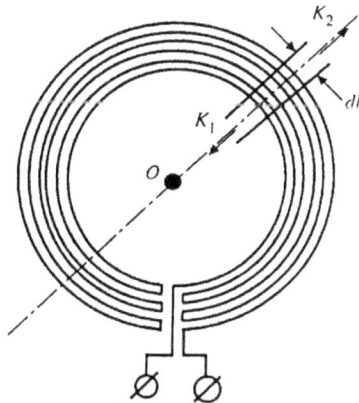

Figure 8. Circular EMAT sensor.

Two case of magnetizing are likely to observed for the circular EMA sensor. In case of normal magnetizing a rod electromagnet is used (fig. 9a), and in case of a tangential magnetizing it is realized by means of a cylindrical m-shaped electromagnet (fig. 9b). Cracks are revealed by such a sensor.

But the most interesting area for the sensor to be applied is in controlling anisotropy of sheet ferromagnetic materials. It allows revealing three types of anisotropy: elastic, magnetoelastic and magnetic.

At generation and reception of a circular surface wave in a completely isotropic metal we have the individual fine-bored received pulse which is a superposition of the signals from all directions (fig. 10a). In case of elastic anisotropy (anisotropy of sound speed) signals in the different directions take different time. And the received pulse extends, or splits into a number of pulses. An elastic anisotropy of the material is characterized by the pulse duration or the distance between extreme pulses (fig. 10b).

The amplitude of the obtained pulses (A1, A2) demonstrates the effectiveness of EMAT in different directions which is defined by magnetoelastic properties of a material. Magnetoelastic anisotropy of materials is defined by the amplitude distinction of pulses.

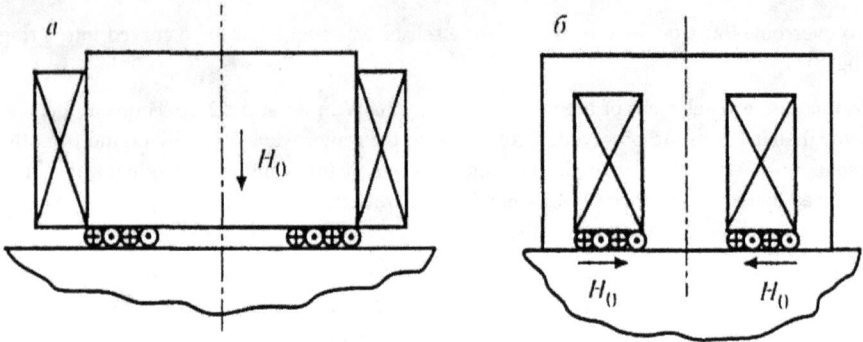

Figure 9. Tangential (a) and normal (в) magnetization of EMAT sensor.

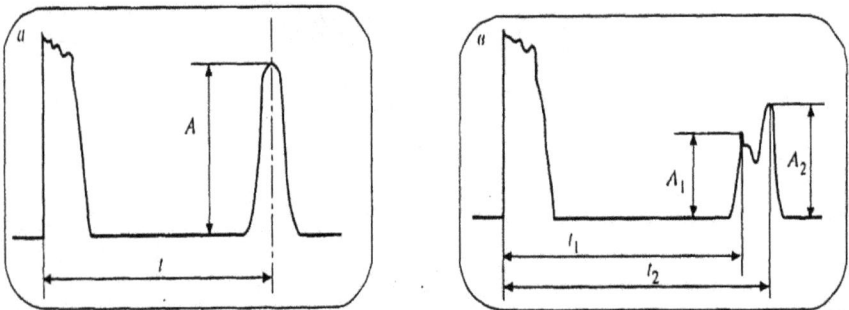

Figure 10. Oscillograms of signals, isotropic material (a), non-isotropic material (b), which are received by means circular EMAT sensor.

The magnetic anisotropy of a sheet is shown as follows. Effectiveness of EMAT from a magnetize field has a maximum. In case of a magnetic anisotropy the maximum of EMAT signal is reached in different directions at different currents of magnetization. Thus, on changing a magnetization current for an isotropic material the range of optimal currents is narrow, and for a magnetic anisotropic material the range is wider the one mentioned above.

The developed technique was tested on various materials and showed high effectiveness.

4.1.3. Tension controlling

EMAT SAW enables tension and operational loadings controlling. For this purpose the sensor (fig.6) was used. The sensor platform area is to be 30 x 60мм. Adjusting to obtain a magnetostriction maximum the amplitude of the received signal is an information parameter.

At stretching the specimen made of steel 25 in the elastic area it has been shown that the amplitude behaves depending on the positions of the sensor i.e. lengthwise and crosswise of the loading. Thus dependence is almost linear.

On measuring the sound speed, it has been found out that in the elastic area it changes less than 0, 5 %. Hence, at controlling tension concerning the sound speed more precise equipment and computer processing of signals should be applied.

The studying of the field curve of Rayleigh waves EMAT under «the sensor is along loading» shows that at specimen stretching in the elastic area and at the beginning of the area plastic deformation both the amplitude of the received signal and an optimal field of magnetizing (a field of the maximal signal in the EMAT curve) for a magnetostriction maximum have changed. It depends on whether magnetostriction is positive or negative [10].

For materials with the negative magnetostriction of saturation, λ_s, (nickel, constructional steel) the dependence is as follows.

Fig. 11 shows the dependence of the maximum point of the field curve of EMAT on loading, the maximum point being normalized in relation with its initial value. At the initial stage of a loading the amplitude of the maximum grows, and this maximum is displaced to the area of the bigger fields. Then, in the area of the inflection point a decrease of a maximum value is observed. The inflection point lies in the area of the elasticity limit (the conditional point where the size of permanent deformations makes 0, 05 %).

It can be explained from the physical point of view. In [4] it is noted that at λ_s $\sigma<0$ pulling stresses of σ are considered to be positive, two factors being competitive: on the one hand the energy of a magnetic field, , on the other hand the crystallographic anisotropy and magnetoelastic energies. Tension in a metal makes a magnetization vector turn perpendicular to the magnetic field. As a result, it is more difficult to obtain an amplitude maximum of EMAT.

The amplitude of the EMA signal is described by formula (1). As the loading in the elastic area increases magnetic conductivity, μ, decreases, and the parameter $\dfrac{d\lambda}{dH_0}$ grows. It results in the received signal growth. At larger loadings there is tension which leads to the signal decrease. So a maximum occurs in the elasticity limit.

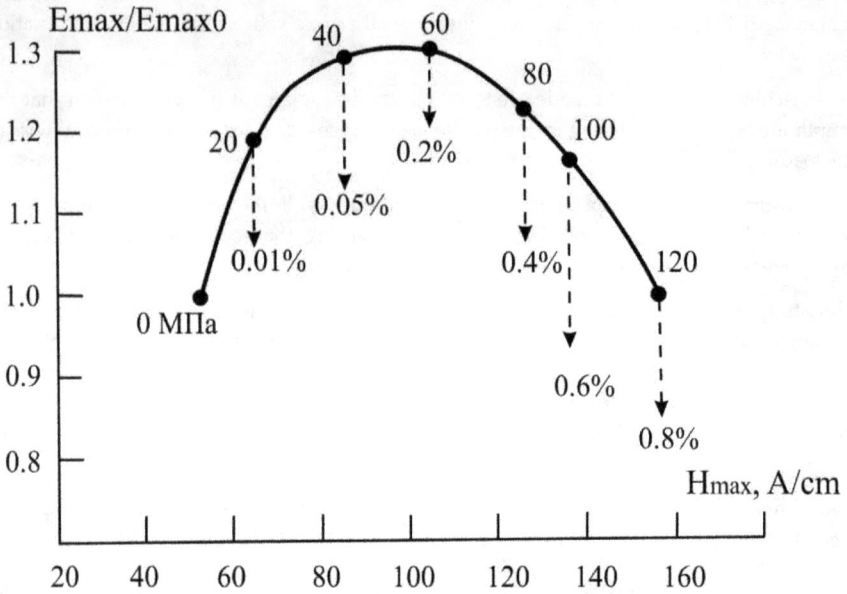

Figure 11. The shift of the maximum point of a field curve of Rayleigh waves EMAT in nickel from the loading (the loading is given in MPa, and the length change after loading is given in percent).

If materials have a positive magnetostriction of saturation, the behaviour of the maximum of the field curve for Rayleigh waves EMAT is essentially different. Using alloy H18 (18% Ni, 82 % Fe) λ$_s$> 0 as an example it is shown that a received EMA signal unequivocally falls (fig. 12). As to a magnetic field first an optimal magnetic field decreases and then it becomes larger, the change takes place at the point close to a limit of material elasticity. It can be also explained from the physical point of view.

Thus, it has been shown that using Rayleigh waves EMAT at least two information parameters are obtained which are used for the applied loadings to be characterized.

It can be carried out either by calibration curves at monitoring or without calibration curves if the tendency of change of parameters EMA is used.

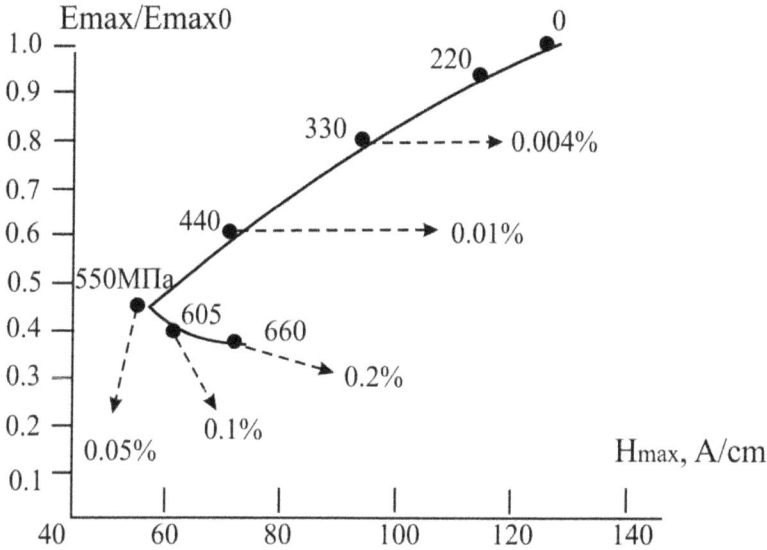

Figure 12. The shift of the maximum point of a field curve of Rayleigh waves EMAT in H18 alloy from the loading (the loading is given in MPa, and the length change after loading is given in percent).

4.2. Physical study of materials

It has been shown that the EMAT SAW method are defined by different subsystems of the material. Information on the parameters of the material related to these subsystems is revealed by the information parameters of the EMA method.

4.2.1. Magnetostriction assessment of a material

Magnetostriction is one of the important characteristics of a material. Due to the wide use of these materials in science and industry the assessment of magnetostriction properties of a material is of great importance.

The main characteristic of magnetostriction properties of a material is the field curve of magnetostriction, i.e., dependence of the linear magnetostriction, λ, on a magnetic field. Though from the practical point of view it is enough to know magnetostriction of saturation, λ_s.

At magnetostriction measurement the method based on tension resistivity of sensors stuck by glue on an specimen is the most extensively used. There are some shortcomings in this method. Alternatively a quick test based on non-contact EMAT SAW [11] may be suggested.

So for Rayleigh waves EMAT formula (1) is used. The task is as following: based on the known curve of EMAT we are to obtain a curve of magnetostriction, $\lambda = f(H)$. It is clear that from curve (1) a field curve of magnetostriction can be obtained by the integration method only. This process is difficult and inaccurate.

Horizontal polarization waves EMAT (SH wave) is more preferable to be used to obtain λ =f(H). A field curve of SH waves EMAT is described by the simplified formula:

$$E_{SH} \approx \frac{1}{\mu d}\left(\frac{\lambda}{H}\right)^2 P \qquad (7)$$

where - $P = \rho C_t / \sigma$ is a material parameter, μ is a static magnetic conductivity, d is a sheet thickness, ρ is material density, C_t is a speed of shift waves, σ is an electrical conductivity.

Thus we obtain :

$$\lambda = \sqrt{\frac{E_{SH}d}{P}}H\sqrt{\mu} \qquad (8)$$

Knowing B=f (H) it is easy to calculate μ=f (H), and then knowing the material parameters it quite easy to calculate a curve λ=f (H).

The calculations above have been used to obtain λ=f (H) for four materials: iron, cobalt, nickel, permendur (brand 49КФ2). Fig. 1 presents the orientation scheme of the magnetic Hy field for EMAT of SH waves. Fig. 13 shows field characteristics of EMAT of the materials mentioned above. In fig. 14á field characteristics of the magnetostriction module calculated on the basis of formula (8) are presented. Fig. 14б shows the same curves based on data [12]. High curves coincidence is observed.

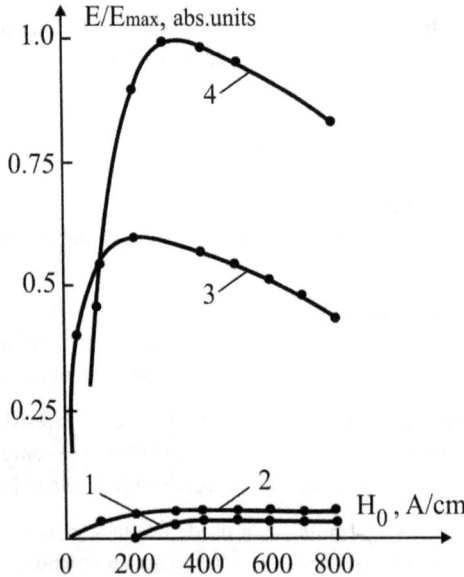

Figure 13. The experimental EMAT field curve of SH- waves (1-iron, 2 - cobalt, 3 – alloy 49КФ2, 4 – nickel).

In 800A/cm fields magnetostriction of materials approaches saturation. To obtain the quantitative value of saturation magnetostriction the required data were obtained by a method of tension resistors. The comparative table on the magnetostriction of saturation received by an EMAT method and a method of resistance strain gages is given in table 1.

As shown in the table, the relative ratios for the EMA method are obtained with an accuracy of 5-7 %.

It should be noted that EMA method allows defining the magnetostriction module λ only. A sign of magnetostriction is complicated to be defined, but it is quite possible. To carry it out it is necessary to observe a phase of the received signal and then to compare it with a signal phase of the known material.

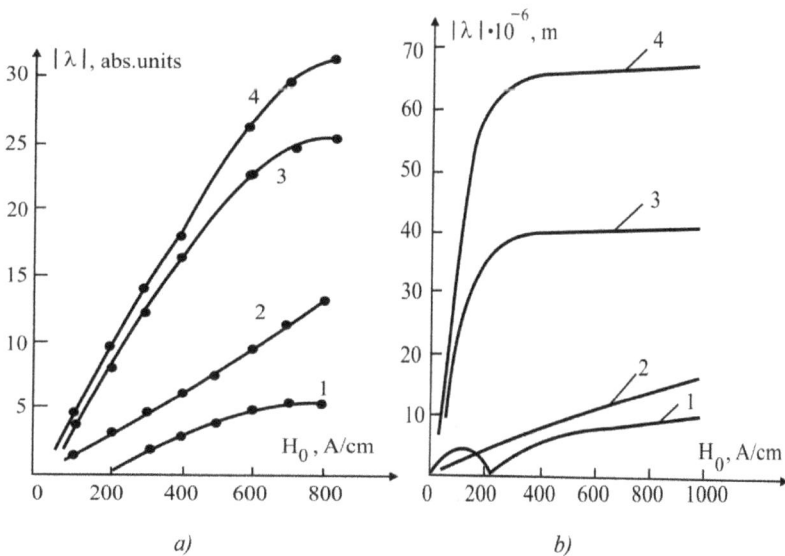

Figure 14. Field curves of the magnetostriction module calculated based on the experimental data (a) the data are given according to literature data [12] (b) (1-iron, 2 - cobalt, 3 - alloy 49КФ2 (Permendur), 4 – nickel).

Material	$\lambda_{max}/\lambda_{max\ Fe}$ EMA method	$\lambda_{max}/\lambda_{max\ Fe}$ Tenzometod
Fe	1	1
Co	2, 6	2, 8
Ni	5, 2	5, 0
49KΦ2	6, 2	6, 5

Table 1. The comparative table on the magnetostriction of saturation received by an EMAT method and a method of resistance strain gages.

4.2.2. Assessment of other characteristics of a material

Electric properties of ferromagnetic materials can be estimated using the amplitude of the received signal in large magnetic fields (more than 800A/cm). EMAT effectiveness depends on the electrical conductivity of a material under the same conditions.

Magnetic properties of a material can be estimated by the location of the magnetostriction maximum.

Two parameters are used to estimate elastic properties of a material:

Firstly, a sound speed in a material. To measure the speed of the Rayleigh surface wave, C_R, and the speed of horizontal polarization waves, C_{SH}, the techniques described are employed. Here, it is should be considered that C_{SH} for a zero mode is equal to the speed of shift waves of the material, C_t. These measurements enables estimating the elastic constants of a material such as shift modulus and a Poisson's ratio. The constants defined allows us to determine the remained elastic modules of a material.

Secondly, sound attenuation in a material is estimated by EMAT SAW methods. In this case two receiving EMA coils are used.

Having defined the attenuation parameter it is easy to find such important production characteristics as good quality of the system and an internal friction parameter of a material.

4.3. Prospects of use of EMAT SAW

The main examples of using the surface waves EMAT are presented in the paper only. The use of EMAT SAW in non-destructive monitoring as well as in studying new materials and the phenomena is possible in the nearest future.

Love waves EMAT may be used for non-destructive monitoring of coatings and thin films and for their thickness measurements as well. SH waves EMAT of can be applied in cylindrical specimens where such waves are known to be generated and received very efficiently.

One can find papers on the use of surface waves EMAT to study magnetic phenomena in rare-earth metals. At the increased or decreased temperatures of the specimens it is easier to apply the EMAT method than the method of piezoelements.

When studying various acoustic waves the EMAT method is more perspective due to easy generation of acoustic waves.

There are combined methods of generating and receiving acoustic waves where EMAT is used together with other methods, for example, with laser methods and with methods of an optical interferometry.

All we said above allows us to draw a conclusion that the EMAT methods are very promising in the future.

Author details

Sergey E. Babkin

*Physical-Technical Institute, Ural Branch of Russian Academy of Sciences,
Izhevsk, Russia*

5. References

[1] Комаров В.А. Квазистационарное электромагнитно-акустическое преобразование в металлах. – Свердловск: УНЦ АН СССР, 1986, 235с.

[2] Васильев А.В., Бучельников В.Д., Гуревич С.Ю., Каганов М.И., Гайдуков Ю.П. Электромагнитное возбуждение звука в металлах. – Челябинск. – М.: Издательство ЮурГУ. – 2001. –339 с.

[3] Викторов И.А. Звуковые поверхностные волны в твердых телах. М.: Наука, 1981.

[4] Ильясов Р.С., Бабкин С.Э., Комаров В.А. О механизмах ЭМА преобразования волн Рэлея при различных частотах. - Дефектоскопия, 1988, №10, с.77 – 82.

[5] Thompson R.B. A Model for the Electromagnetic Generation and Detection of Rayleigh and Lamb Waves. – IEEE Trans. Sonics and Ultras., 1973, v.20, N 4, p.340-346.

[6] Комаров В.А., Бабкин С.Э., Ильясов Р.С. ЭМА преобразование волн горизонта льной поляризации в магнитоупругих материалах. - Дефектоскопия, 1993, №2, с.11 – 17.

[7] Бабкин С.Э., Ильясов Р.С., Комаров В.А., Рубцов В.И. Устройство для бесконтактного возбуждения и приема волн Рэлея в ферромагнетиках. – Дефектоскопия, 1989, №6, с.93-94.

[8] Бабкин С.Э., Ильясов Р.С., Зверев Н.Н. Отстройка от зазора при электромагнитно-акустическом способе контроля. - Дефектоскопия, 1998, №4, с.19-22.

[9] Бабкин С.Э., Ильясов Р.С. Кольцевой электромагнитно-акустический преобразователь поверхностных волн. - Дефектоскопия, 2002, №1, с.78-82.

[10] Бабкин С.Э., Ильясов Р.С. О возможности использования параметров ЭМАП для оценки предела упругости и остаточных деформаций ферромагнитных материалов. - Дефектоскопия, 2010, №1, с.83-89.

[11] Бабкин С.Э., Ильясов Р.С., Комаров В.А. Бесконтактный метод оценки магнитострикции материалов. - Дефектоскопия, 1996, №1, с.41-45.

[12] Bozort R.M., Ferromagnetism, D.Van Nostrand Co., Princeton, 1951.

Chapter 4

Surface Acoustic Wave Based Magnetic Sensors

Bodong Li, Hommood Al Rowais and Jürgen Kosel

Additional information is available at the end of the chapter

1. Introduction

Since the radar system was invented in 1922, the development of devices communicating by means of reflected power has experienced a continuously growing interest. In 1948, Harry Stockman published a paper [1] in which he laid the basis for the idea of radio frequency identification (RFID), and the first patent had been filed in 1973 by Charles Walton. After decades of research and commercialization, RFID products became a part of everyday life (e.g. logistics, access control, security). With the growing interest in remote and battery-free devices, researchers are pushing the boundaries of RFID technology to find solutions in new fields like sensing applications.

For many sensors such as those operated in remote or harsh environments, the sensitivity is not the only evaluation criteria. The lifetime, especially of the power source, and the complexity added by wiring often demand wireless and passive operation. Batteries have limited lifetime and also add to the size and mass of the sensors. Alternatively, energy harvesting or an RF-based wireless power supply can be employed [2, 3]. The former method depends on environmental conditions such as solar radiation, temperature change, chemical reagents, vibration etc., which are often not constantly or not sufficiently available. RF power sources, on the other hand, transmit power wirelessly and with full control over amount and timing.

Passive and remote sensors utilizing SAW transponders are devices, which are powered by an RF source. These systems require an interrogation device that requests the sensor signal, a SAW transponder plus a sensing element and two antennas. The basic idea is that an RF signal of certain frequencies generated by the interrogator is received by the SAW transponder, which reflects back a signal modified by the sensing element. This signal contains the environmental information in an amplitude and phase change, which is converted into the physical parameters by the interrogator. In most cases, SAW sensors are coded by having different reflector designs in order to have multiple measurement capabilities from sensors located in the same interrogation area. A great amount of research has been carried out in the past decades in this field, and, as a result, different wireless SAW sensors have been developed to measure a variety of physical and chemical parameters including temperature, stress, torque, pressure, humidity, magnetic field, chemical vapor etc. [4-9]. Several devices are already commercialized [10-12].

SAW-based magnetic sensors have, so far, not been studied in detail. Magnetic sensors are one of the most pervasive kinds of sensors for a large number of applications and are employed in different fields like automotive, biomedical or consumer electronics. Integrating magnetic sensors with SAW transponders enables remote and passive operation, thereby, opens a door for further applications. Early in 1975, a magnetically tuned SAW phase shifter was proposed by Ganguly et al [13]. A thin film of magnetostrictive material was fabricated on the delay line of a SAW device. A phase shift was observed due to the dependence of the wave propagation velocity on the external magnetic field. Recently, a new concept of a magnetic sensor based on a SAW resonator has been published [8]. A magnetostrictive material was used to fabricate the interdigital transducers (IDT) of the SAW device. The resonant frequency of the device changes with an external magnetic field. A different idea was put forth by Hauser and Steindl [14-16] combining a SAW transducer with a giant magnetoimpedance (GMI) microwire sensor. The GMI sensor has a magnetic sensitivity, at least, one order higher than contemporary giant magnetoresistance (GMR) sensors and can be used to measure very low magnetic fields such as those generated by the human heart or muscles. A GMI sensor is operated by an ac current and the impedance changes upon changes of a magnetic field. This makes it a suitable load for a SAW transponder, which converts this impedance change into a magnitude and phase change of the reflected acoustic waves. In order to reduce the size and improve the level of integration of the senor, a new design of an integrated SAW transponder and thin film GMI sensor has been proposed and developed recently by the authors [17, 18]. The SAW transponder and GMI thin film were integrated on the same chip using standard micro-fabrication technology suitable for mass fabrication.

The ideal SAW-based magnetic sensor is small and highly integrated, inexpensive, passive, remotely controlled and have a high magnetic sensitivity together with a large linear range. With regard to these criterions, an SAW-GMI sensor is a very promising candidate.

SAW-based magnetic sensors have been studied for several years. However, this topic has yet not been comprehensively summarized, and the aim of this chapter is to provide a systematic review of the past research as well as the latest results. The performance of the devices crucially depends on different design parameters in a complex fashion. This will be shown by a detailed description and analysis for a device consisting of a SAW transponder and GMI thin film sensor.

2. SAW based passive sensors

A basic SAW device consists of an input interdigital transducer (IDT) and an output (or reflector) IDT, which are fabricated on a piezoelectric substrate. The area between the input IDT and output IDT is called the delay line. The IDT is made of two metallic, comb-like structures arranged in an interdigital fashion, whereby the distance between two fingers of a comb defines the periodicity (p) (Fig. 1). Upon application of a voltage, charges accumulate at the fingers of the IDT depending on the capacitance of the structure. The resulting electric field generates stress in the substrate due to the piezoelectric effect. If an ac input voltage is applied, the continuously changing polarity of the charges will excite an SAW (Rayleigh wave) traveling through the substrate. At the operating (resonant) frequency of the SAW device, the value of p equals the wavelength of the SAW, and the SAW amplitude shows a maximum value due to constructive superposition.

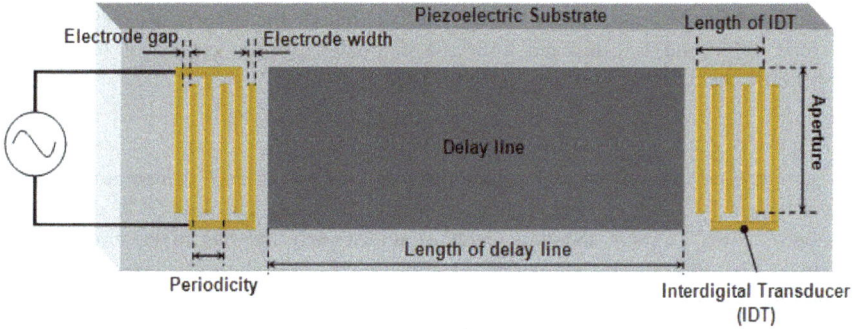

Figure 1. Schematic of a SAW device.

2.1. Basics of SAW devices

SAW Resonant Frequency: The resonant motion of an acoustic resonator is caused by the coupling between the transducer (IDT) and the acoustic medium. The resonant frequency, or operating frequency, is determined by the periodicity and the acoustic wave velocity (v)

$$f = \frac{v}{p}. \tag{1}$$

The value of v mainly depends on the substrate's material. A typical SAW velocity for piezoelectric materials is several thousand meters per second. Due to the intrinsic anisotropy of piezoelectric materials, v is dependent on the direction of propagation. Since different acoustic modes have different wave velocities, a device can resonate at different frequencies. The SAWs are Rayleigh waves, which have a longitudinal and a vertical shear component that can couple with any media in contact with the surface. This coupling strongly affects the amplitude and velocity of the wave allowing SAW sensors to directly sense, e.g., mass loads.

Electro-Mechanical Coupling Coefficient: The electro-mechanical coupling coefficient (κ) defines the conversion efficiency of the piezoelectric material between the electrical and mechanical energies, determined by

$$\kappa^{-2} = \frac{input\ energy}{converted\ energy}. \tag{2}$$

A high coupling coefficient reduces the insertion loss caused by the energy conversion, which results in smaller energy consumption as well as larger effective readout distance of a SAW-based wireless sensor.

SAW Delay Line: The SAW delay line refers to the area between the input IDT and output IDT on the substrate (Fig. 1). It creates a time delay between the input signal and the output signal depending on the SAW velocity and the length of the delay line. Due to this feature, SAW devices are widely used in RF electronics. It is also used in sensing applications, where the measurand causes, e.g., a change in the SAW.

Temperature Coefficient of Delay (TCD): The TCD reflects the temperature dependence of the time delay and is connected with the thermal expansion coefficient (α) and the temperature coefficient of the phase velocity (TCV) by

$$TCD = \alpha - TCV .\tag{3}$$

The temperature dependence of the time delay is the basis of SAW temperature sensors, where higher TCD values yield higher sensitivity. However, for other SAW devices, the influence of the TCD on the time delay is undesirable and has to be minimized or eliminated. For this purpose, temperature compensated cuts of the crystalline substrates are employed, where the TCD is minimized over certain temperature ranges [19-20]. Piezoelectric bi-layers are another concept that has been utilized in order to compensate the TCD in sensing applications [21, 22].

2.2. Basic design concepts of passive SAW sensors

Passive SAW sensors typically operate as resonators, delay lines or loaded transponders. In case of resonators, the reflection of the interrogation signal from the SAW device is a function of the SAW device's resonant frequency, which depends on the measurand. In case of delay lines, the request signal is separated from the response signal by a time difference, whereby this time difference depends on the measurand. Similarly, the request signal and response signal are separated by a time difference in case of a loaded transponder. However, the time difference is constant and the measurand affects the signal amplitude. Intrinsic SAW sensors utilize a change of the substrate's properties. For example, intrinsic temperature sensors were realized by detecting the resonant frequency or phase change of the SAW in materials with large TCD [23]. Intrinsic stress sensors utilize the length change of the delay line caused by mechanical strain applied to the substrate. The stress can be evaluated by measuring the SAW phase shift [24]. Extrinsic SAW sensors can be realized by integrating a SAW device and an additional sensing element. A common extrinsic sensor concept utilizes selective thin films on top of the delay line leading to a change in mass by the measurand [13, 25]. This can be, for example, a thin film with high CO_2 solubility and selectivity [26]. As CO_2 dissolves into the film, the additional mass load causes a detectable phase shift in the SAW. Another extrinsic concept utilizes a sensitive IDT. For example, in case of a magnetostrictive IDT, a magnetic field applied to the sensor causes a change of the resonant frequency [8]. A loaded transponder is another extrinsic design, where the output

IDT is connected to a sensor, which changes the IDT's electrical characteristics as a function of the measurand. An example for a load sensor is a pair of conducting rods placed in the earth with a certain distance from each other. As the water level changes, the resistance between the rods changes, which can be detected as a magnitude and phase change of the signal reflected from the load IDT [27]. Another example for a load is a giant magnetoimpedance sensor [16, 17]. A change in the magnetic field yields a change in the sensor's impedance. Consequently, changes the reflectivity of the output IDT.

Some SAW sensors, their classification and method of detection are presented in Table 1.

Sensor Type	Commercialization	Year	Intrinsic/ Extrinsic	Design	Detection Method	Access	Paper
Temperature	Yes	1990	Intrinsic	Resonator	Frequency	None	[23]
		2003	Intrinsic	Delay line	Phase velocity	None	[28]
Pressure	Yes	2001	Extrinsic	Loaded Transponder	Phase	Capacitive pressure sensor	[25]
		2007	Intrinsic	Delay line	Phase	None	[24]
Bio/Chem	No/Yes	2006	Extrinsic	Resonator	Frequency	Thin film	[29]
		2011	Extrinsic	Delay line	Phase	Thin film	[26]
		2001	Extrinsic	Loaded Transponder	Amplitude/ Phase	Conducting rods	[30]
Magnetic	No	1975	Extrinsic	Delay line	Phase	Thin film	[13]
		2011	Extrinsic	Resonator	Frequency	Magnetostrictive IDTs	[8]
		2006/11	Extrinsic	Loaded Transponder	Amplitude	GMI wire/ thin film	[16,17]
Sound	No	2005	Extrinsic	Loaded Transponder	Phase	Capacitive pressure sensor	[31]
Torque	Yes	1996	Intrinsic	Delay line	Phase	None	[32]

Table 1. SAW-based passive sensors.

3. Passive and remote SAW-based magnetic sensors

Magnetic sensors are one of the most versatile sensors employed not only for the task of measuring magnetic fields but for a large number of different applications, thereby detecting the measurand indirectly, e.g., via a change of material parameters in construction monitoring or a change of distance in position monitoring. A passive and remote operation of magnetic sensors can be advantageous in many cases and considerably increase their applicability.

A SAW-based passive magnetic sensor can be realized either by adding an additional material layer, which is sensitive to magnetic fields, or by loading the output IDT with a magnetic sensor. In the first case, the magnetic layer changes the delay line or the resonant frequency of the SAW device. While in the second case, the sensor changes the reflection signal of the output IDT. Since SAW devices are operated by RF power, the sensor element has to work at the operation frequency of the SAW device. Among the available magnetic sensors, GMI sensors are the most suitable candidates as they have a high magnetic sensitivity as well as a high operating frequency.

3.1. Magnetostrictive SAW devices

Magnetostriction defines the relationship between the strain and the magnetization states of a material. It is an important property of ferromagnetic materials and was first observed by James Joule in 1842 in nickel samples. For a positive/negative magnetostrictive material, an applied magnetic field causes the material to expand/shrink in the field direction. Inversely, when a stress is applied to the magnetostrictive material, its magnetic anisotropy will change accordingly.

A magnetostrictive-piezoelectric resonator consists of amorphous magnetostrictive material layers as the electrodes sandwiching a piezoelectric core (Fig. 2). An ac signal applied to the electrodes causes the quartz layer to oscillate. The resonant frequency of this oscillation depends on the thickness of the piezoelectric material, the crystal orientation, temperature and mechanical stress, etc.

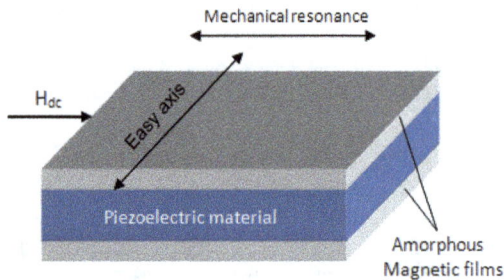

Figure 2. Structure of a composite magnetostrictive/piezoelectric resonator. The magnetic anisotropy is perpendicular to the external magnetic field H_{dc}.

When a magnetic field H_{dc} is applied, the length change induced in the magnetostrictive film exerts stress to the piezoelectric material and, consequently, shifts the resonant frequency of the device. Utilizing this concept, a magnetic sensitivity high enough to detect the terrestrial field has been achieved [33]. In a similar work, a magnetostrictive-piezoelectric tri-layer structure has been embedded in a coil. The dc magnetic field sensitivity was as high as 10^{-8} T [34].

3.1.1. Passive resonator

A SAW-based, passive resonator for magnetic field detection was developed recently by Kadota et al [8]. Nickel, which is a negative magnetostrictive material, was used to fabricate the sensing IDT on a quartz substrate (Fig. 3). Upon the application of a magnetic field, stress will be induced to the substrate by the IDT change causing a change in the resonant frequency. This sensor showed a frequency change of 200 ppm for a magnetic field of 100 mT applied perpendicularly to the direction of SAW propagation.

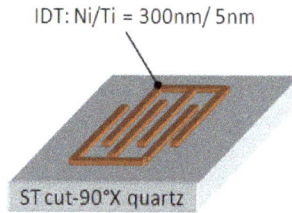

Figure 3. Schematic of a magnetic sensor device using a magnetostrictive IDT on a SAW substrate.

3.1.2. Passive phase shifter

A magnetically tuned SAW phase shifter is a one-port SAW structure with a magnetic sensing functionality achieved through a delay line sensitive to magnetic fields. This idea was first introduced by Ganguly et al in 1975 [13]. In their device, the acoustic velocity is varied by an external magnetic field. This functionality is facilitated by a magnetostrictive thin film deposited on top of the delay line (Fig. 4). The propagation velocity of the SAW in the film region depends on the magnetic field. Hence, there is a correlation between the time shift of the reflected signal and the magnetic field.

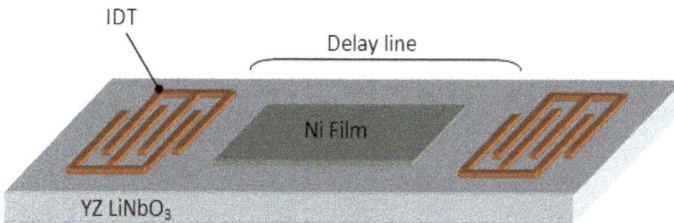

Figure 4. Schematic of a magnetically tuned SAW wave phase shifter.

Later, research efforts focused on different magnetostrictive materials and measurement methods [9, 35, 36], and a magnetic sensitivity of 10^{-4} %/kOe was achieved.

3.2. SAW-GMI sensors

A SAW-based, magnetic and passive sensors comprises a two-port SAW transponder and a magnetic sensor acting as a load at the output IDT. Among the available magnetic field sensors, giant magnetoimpedance (GMI) sensors offer favorable characteristics like high sensitivity to magnetic fields and high operation frequency (compatible with SAW transponders) making them a very suitable load. SAW-GMI sensors have been fabricated by combining SAW transponders with GMI wire sensors as well as thin film GMI sensors. Both of these methods have shown a higher magnetic sensitivity than direct designs.

3.2.1. Basics of GMI sensors

The GMI effect was first observed in Co-based amorphous wires by Panina and Mohri in 1994 [37] and has since attracted strong interest due to its sensitivity enabling magnetic field measurement with a nT resolution. The GMI effect is the impedance change of an ac-powered ferromagnetic conductor upon the change of a magnetic field. The relative impedance change, also called GMI ratio, is expressed as

$$\text{GMI Ratio (\%)} = 100\% \times \frac{Z(H) - Z(H_0)}{Z(H_0)} \quad \text{or} \quad \text{GMI Ratio (\%)} = 100\% \times \frac{Z(H) - Z(H_{max})}{Z(H_{max})}, \qquad (4)$$

where $Z(H_0)$ is the impedance at zero magnetic field and $Z(H_{max})$ is the impedance at saturation field. Both definitions have particular aspects that should be considered. In case of the first expression, $Z(H_0)$ depends on the remanent state of the magnetic material while, in the second case, $Z(H_{max})$ is not always achievable and equipment dependent.

The GMI effect is explained by classical electromagnetism. The change of the complex impedance mainly originates from the skin effect in conjunction with a change of the complex permeability. Analytically, the complex impedance (Z) of a conductor is defined by

$$Z = \frac{U_{ac}}{I_{ac}} = \frac{\int_L \frac{1}{\sigma} J_z(S) dz}{\iint_q J_z dq}, \qquad (5)$$

where U_{ac} is the applied ac voltage, I_{ac} is the current, L is the length and σ the conductivity. S and q refer to the surface and the cross section of the conductor, respectively. J_z is the current density in the longitudinal direction obtained by solving Maxwell's equations. In ferromagnetic materials, by neglecting displacement currents ($\dot{D} = 0$), Maxwell's equations can be written as follows:

$$\nabla \times H = J, \qquad (6)$$

$$\nabla \times J = -\frac{\mu_0}{\rho_f} (\dot{H} + \dot{M}), \qquad (7)$$

$$\nabla \cdot (H + M) = 0, \qquad (8)$$

J is the current density, H is the applied magnetic field, M is the magnetization of the ferromagnetic material, ρ_f is the free charge density and μ_0 is the permeability of vacuum. From Equ. (6) to (8), the expression

$$\nabla^2 H - \frac{\mu_0}{\rho_f} \dot{H} = \frac{\mu_0}{\rho_f} \dot{M} - \nabla(\nabla \cdot M), \qquad (9)$$

can be derived. Equ. (9) can be solved using the Landau-Lifshitz equation, which relates M and H

$$\dot{M} = \gamma M \times H_{eff} - \frac{\alpha}{M_s} M \times \dot{M}, \tag{10}$$

where γ is the gyromagnetic ratio, M_s is the saturation magnetization, α is the damping parameter and H_{eff} is the effective magnetic field expressed as [30]

$$H_{eff} = H + H_a + \frac{2A}{\mu_0 M_s} \nabla^2 M \tag{11}$$

where H is the internal magnetic field that includes the applied field and demagnetizing field, H_a is the anisotropy field and A is the exchange stiffness constant..

By combining Equ. (5) to (11), a theoretical impedance model can be evaluated for GMI sensors with different geometries [37-40].

Although the experimentally obtained GMI effect shows a large sensitivity compared to other effects exploited for magnetic sensors, the theoretically estimated values have not been achieved yet. Therefore, a lot of effort has been put into improving the magnetic properties of GMI materials [41-44]. At the same time, GMI sensors of different structures have been developed such as glass-coated wires, thin films, multi layer thin films, meander structures, ribbons, etc. [45-47]

3.2.2. Wire GMI sensors

As the first discovered GMI sensor structures, GMI wire sensors have been extensively studied. Based on the classical electromagnetism, the theoretical model of the GMI wire is (Panina et al, 1994) [37]

$$Z = \frac{R_{dc} kr \zeta_0(kr)}{2\zeta_1(kr)}, \tag{12}$$

where

$$k = \frac{1+j}{\delta_m} \tag{13}$$

and

$$\delta_m = \frac{c}{\sqrt{4\pi^2 f \sigma \mu_\emptyset}}. \tag{14}$$

R_{dc} is the dc resistance of the wire, ζ_0, ζ_1 are the Bessel functions, r is the radius of the wire, j is the imaginary unit, δ_m is the penetration depth, c is the speed of light, f is the frequency of the ac current, $\mu_\emptyset$ is the circumferential magnetic anisotropy. The origin of the GMI effect lies in the dependence of $\mu_\emptyset$ on an axial magnetic field resulting in a change of δ_m. In order to obtain a high GMI ratio, the value of δ_m has to be close to the thickness of the conductor. Hence, the thinner a ferromagnetic conductor and the lower its permeability, the higher the operation frequency required. A well-defined circumferential magnetic anisotropy in combination with a soft magnetic behavior is desirable, since it will provide a large permeability change for small magnetic fields.

Different amorphous and ferromagnetic materials were used to fabricate GMI wires [48], and various fabrication methods were developed such as melt spinning, in-rotating water spinning, glass-coated melt spinning etc. [45, 49, 50]. Glass-coated micro-wires (Fig. 5) present outstanding properties in terms of the magnetic anisotropy distribution, which is reinforced by the strong mechanical stress induced by the coating. $(Co_xFe_{1-x})_{72.5}Si_{12.5}B_{15}$ is one of the most typical materials. By adjusting x from 0 to 1, the magnetostriction of the material changes from positive at high Fe content to negative at high Co content. Negative magnetostrictive compositions in combination with the compressive, radial stress induced by quenching and the glass coating provide the best results, since it supports a strong circumferential anisotropy.

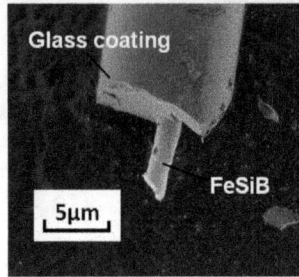

Figure 5. SEM image of a glass-coated amorphous micro-wire (Courtesy of M. Vazquez, Inst. Materials Science of Madrid, CSIC).

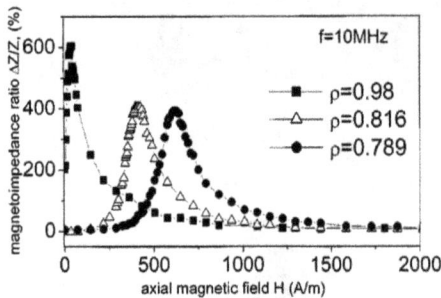

Figure 6. GMI ratio of $Co_{67}Fe_{3.85}Ni_{1.45}B_{11.5}Si_{14.5}Mo_{1.7}$ glass-coated wires with different geometric ratio ρ (the metallic nucleus diameter to the total microwire diameter) at 10 MHz.

Wire-type GMI sensors provide the best performance in terms of the GMI ratio with values as high as 615% (Fig. 6) achieved with optimized glass coated microwires (Zhukova et al, 2002) [43]. The value of the magnetic field at which the maximum GMI ratio is obtained increases as the diameter of the magnetic nucleus decreases compared to the diameter of the glass coating. This is attributed to the different anisotropies induced by the stress from the coating. Due to the high sensitivity provided by GMI wires they have been commercialized despite the facts that fabrication is not silicon based, does not use standard microfabrication methods and, as a consequence, integration with electronics is complex.

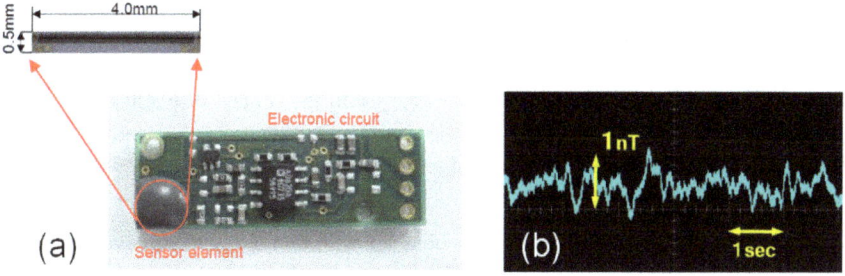

Figure 7. (a) Layout of the commercialized GMI sensor from Aichi Steel Co. (b) Noise output of the GMI sensor.

Fig. 7 shows a GMI sensor developed by Aichi Steel Co., which has a very high sensitivity of 1 V/µT and a noise level of 1 nT [51].

3.2.3. Ribbon GMI sensors

Magnetic ribbons discussed in this section are planar structures of rectangular shape with a thickness of a few tens of micrometers and a length and width from several millimeters to centimeters. Similar to the micro-wires, magnetostriction is utilized in order to create certain anisotropies during the ribbon's fabrication. Magnetic ribbons that exhibit a strong GMI effect have a high permeability as well as a transversal magnetic anisotropy.

For a planar film of infinite width, the impedance is given by

$$Z = R_{dc} \cdot \frac{jka}{2} coth(\frac{jka}{2}),$$ (15)

where R_{dc} is the dc resistance, a is the thickness of the ribbon, k and δ_m can be obtained from Equ. (14) with the only difference that μ_θ represents the transversal permeability instead of the circumferential one [52].

Again, Fe- and Co-based amorphous alloys are preferably used as the magnetic material. The standard fabrication method for the ribbons is melt spinning, where a rotating copper wheel is used to rapidly solidify the liquid alloy. This method produces magnetic ribbons with a thickness of about 25 µm and a width of several mm. With this thickness, ribbon GMI sensors operate at comparably low frequencies of hundred kHz up to a few MHz. A GMI ratio of, e.g., 640% has been obtained with a GMI ribbon made of $Fe_{71}Al_2Si_{14}B_{8.5}Cu_1Nb_{3.5}$ at 5 MHz [53].

3.2.4. Thin film GMI sensors

In theory, a single layer magnetic thin film is similar to a magnetic ribbon, and the same analytical expressions are applied for modeling the GMI effect. Practically, the main difference is the fabrication method. Thin film fabrication is a standard micro-fabrication technology producing a film thickness of some nanometers up to a few micrometers. Thin film GMI sensors are of great interest due to the advantages arising from the fabrication in terms of the flexibility in design and integration. They can easily be fabricated on the same substrate as the electronic circuit and other devices. In the context of passive and remote sensors, this is particularly relevant, since the GMI element can be easily integrated with an SAW device. For this reason, GMI thin film sensors will be discussed in more detail and our recent results will be presented.

Compared to wires and ribbons, the results obtained with thin film sensors have not been as good, and the highest GMI ratios reported are around 250% [42]. This may be due to the differences in the magnetic softness as well as the magnetic anisotropy, which is very well established in circumferential and transversal direction in wires and ribbons, respectively, and is difficult to control in thin films. In thin films transverse anisotropy is mainly realized through magnetic field deposition or field annealing, Fig. 8 (a) and (b) show the magnetization curve and domain structure of a $Ni_{80}Fe_{20}$ thin film (100nm thick) fabricated under a magnetic field of 200 Oe during deposition. A magnetic easy axis and domain structures in transverse direction are observed. Due to the small thickness, thin film GMI sensors normally operate at a higher frequency from hundred MHz to several GHz where the penetration depth is in the range of the film thickness.

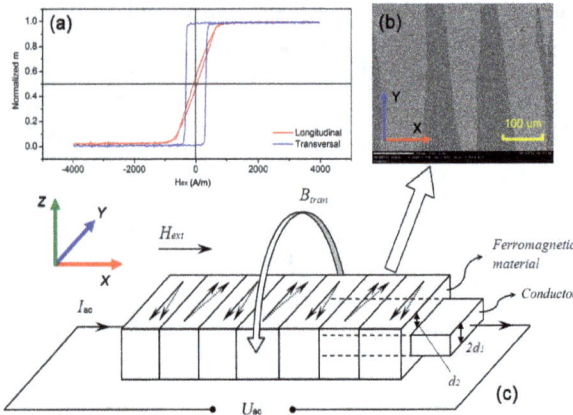

Figure 8. (a) Magnetization curves obtained by vibrating sample magnetometry of a magnetic thin film (100nm of $Ni_{80}Fe_{20}$) in transversal and longitudinal directions. (b) Domain pattern of the magnetic layer. (c) Schematic of a typical multi layer GMI structure. The arrows in the ferromagnetic material indicate the magnetization of individual domains (simplified). Upon application of an external magnetic field H_{ext}, the magnetization rotates into the direction of H_{ext} (dotted arrows).

In general, a GMI sensor with high sensitivity consists of a stack of several material layers. In case of a tri-layer element, one conducting layer is sandwiched between two magnetic layers as shown in Fig. 8 (c). The conducting layer ensures a high conductivity and, in combination with the highly permeable magnetic layers, a large skin effect is obtained [51,

54]. An alternating current I_{ac} mainly flowing through the conductor generates a transversal flux B_{tran}, which magnetizes the magnetic layers. Upon the application of an external field H_{ext} in longitudinal direction, the magnetization caused by I_{ac} will be changed. This is equivalent to a change of the transversal permeability of the magnetic layers and is reflected by an impedance change.

The analytical model of the impedance for a magnetic/conducting/magnetic tri-layer structure is given by

$$Z = R_{dc}\left(1 - 2j\mu\frac{d_1 d_2}{\delta_c^2}\right),\tag{16}$$

where R_{dc} is the dc resistance of the inner conductor, $2d_1$ is the thickness of the conductor, d_2 is the thickness of the magnetic layers as shown in Fig. 8 (c) and δ_c is the penetration depth of the conducting layer [39].

Analytical solutions for the impedance of thin film GMI sensors can only be found for rather simple structures. In order to calculate the impedance of more complicated geometries, for example, a sandwich structure with isolation layers between the conductor and the magnetic layers [41], a meander structure multilayer [46] or to take into account edge effects, the finite element method (FEM) provides a viable solution [55].

Fig. 9 shows the comparison of the GMI ratios simulated for a single magnetic layer, a tri-layer structure made of a magnetic/conducting/magnetic stack and a five-layer structure with isolation layers between the conducting and magnetic layers using the FEM. The simulated GMI sensors have a width of w = 50 μm and length of l = 200μm. The magnetic layers have a thickness of t_{mag} = 1 μm and the conducting layer has a thickness of t_{met} = 4 μm. The material of the isolation layer is SiO$_2$ with a thickness of 1μm. The conductivity of the ferromagnetic and conducting layers are 7.69×10^5 S/m ((CoFe)$_{80}$B$_{20}$) and 4.56×10^7 S/m (Gold), respectively. All parameters including M_s = 5.6×10^5 A/m, γ = 2.2×10^5 m/(A·s), α = 0.3, H_a= 1890 A/m and are taken from literature [56].

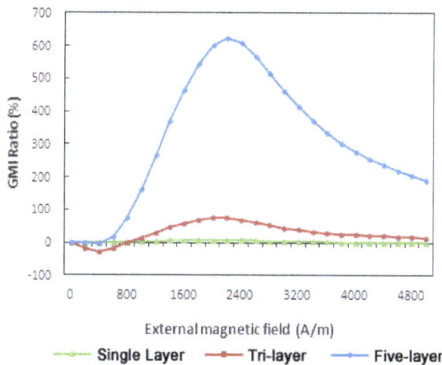

Figure 9. Simulated GMI ratios of single layer, sandwiched multilayer and isolated sandwiched multilayer structures.

The results clearly show the performance increase achieved with the multilayer structures. Specifically, the isolated sandwich structure has a superior performance, which is due to preventing the current from flowing in the magnetic layer.

For the fabrication of thin film GMI sensors, Co-based and Fe-based amorphous magnetic alloys were used in earlier studies. Recently, permalloy, which is a NiFe compound, became popular as it provides very high permeability, zero magnetostriction and simple fabrication. Meander shaped multilayers and different stacks of magnetic and conductive layers using permalloy were developed. Some results are summarized in Table 2.

Year	Material	Frequency	GMI Ratio (%)	Sensitivity (%/Oe)	Reference
1999	FeNiCrSiB/Cu/FeNiCrSiB	13MHz	77	2.8	[57]
2000	FeSiBCuNb/Cu/FeSiBCuNb	13MHz	80	2.8	[58]
2004	$Ni_{81}Fe_{19}/Au/Ni_{81}Fe_{19}$	300MHz	150	30	[59]
2004	$(Ni_{81}Fe_{19}/Ag)n$	1.8GHz	250	9.3	[42]
2005	$FeCuNbSiB/SiO_2/Cu/SiO_2/FeCuNbSiB$	5.45MHz	33	1.5	[60]
2011	NiFe/Ag/NiFe	1.8GHz	55	1.2	[61]
2011	NiFe/Cu/NiFe	20MHz	166	8.3	[62]

Table 2. Recent results on thin film GMI sensors.

GMI thin film sensors not only offer the advantages of standard microfabrication and straight-forward integration with SAW devices, but, as can be seen from Table 2, the operation frequency of GMI thin film sensors is also compatible with the one of SAW devices (usually from hundred MHz to several GHz and can be adjusted within a wide range.

3.2.5. Integrated SAW-GMI sensor

In the first studies, SAW transponders and GMI wire sensors were combined to form remote devices [14-16]. GMI wires were selected for their high sensitivity, and they were bonded to the output IDT of the SAW device, which operated as a reflector, in order to act as load impedance. The strong dependence of the impedance on magnetic fields causes a considerable amplitude dependence of the reflected signal on magnetic fields. Even though these studies provided good results for passive and remote magnetic field sensors, the fabrication method for the GMI wires, which is not compatible with standard microfabrication, is a considerable problem with respect to reproducibility and costs, hence, hindering commercial success of such devices. In order to conquer this problem, a fully integrated SAW-GMI design utilizing standard microfabrication processes is required. The most viable option is a thin film GMI sensors for the following reasons:

1. Thin film GMI sensors can be produced by the same metallization processes as the SAW transponders and on the same substrate.

2. Standard photolithography technique guarantees an accurate and reproducible alignment of the two devices.
3. Thin film GMI sensors provide a wide range of working frequencies up to GHz, which matches the high frequency requirement of the SAW transponders.
4. Thin film GMI sensor can have a minimized and flexible design as well as large magnetic field sensitivity.

In this section, a detailed description of our recent work on the design, fabrication and testing of an integrated SAW-GMI sensor is presented.

Design

Fig. 10 shows a schematic of a GMI thin film sensor integrated with a SAW transponder. A wireless signal applied to the source IDT (IDT1) is converted to an SAW and propagates towards the other end of the substrate, where it is reflected from the reference IDT (IDT2) and the load IDT (IDT3). The reflected waves containing the reference and load information are received by IDT1 at different time instants and reconverted to a wireless electrical signal sent out via the antenna.

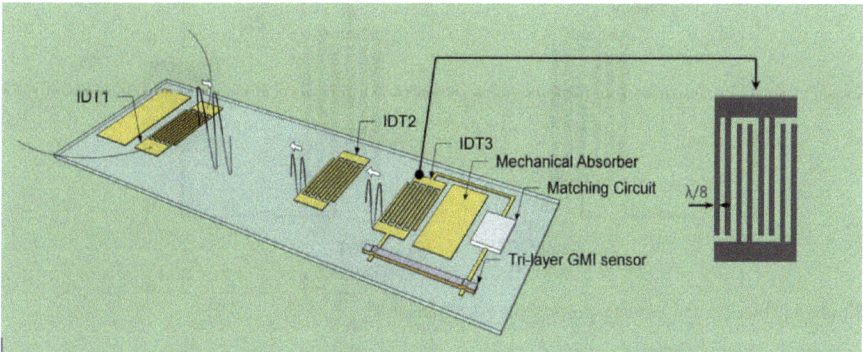

Figure 10. Schematic of an integrated passive and remote magnetic field sensor consisting of a SAW transponder and thin film GMI sensor.

In order to obtain high magnetic field sensitivity, the GMI sensor is matched to the output port (IDT3) at the working frequency of the SAW device. As the impedance of the GMI sensor changes with an applied magnetic field, the matching deteriorates, which causes the amplitude of the signal reflected from IDT3 to change. Since the piezoelectric material is sensitive to environmental changes, e.g. temperature, a reference IDT is used to provide a signal that enables the suppression of such noise by means of signal processing. Two metallic pads next to the input and output IDTs act as mechanical absorbers and suppress reflections from other structures on the substrate or the edge of the substrate.

Matching the sensor load to the optimal working point of IDT3 is a crucial aspect in the device design. Therefore, the influence of the load on the signal reflected from IDT3 is simulated. The interaction of a SAW with an IDT can be described by the P-matrix model introduced by Tobolka [63]. As shown in Fig. 11, P_{11} is the acoustic wave reflection at the output IDT [24]. Specifically, the dependence of P_{11} on the load impedance $Z = Z(H_{ext}) + Z_m$, where $Z(H_{ext})$ is the impedance of the GMI element and Z_m is the matching impedance, is expressed as

$$P_{11}(Z) = P_{11,sc} + \frac{2 \cdot P_{13}^2}{P_{33} + \frac{1}{Z}},\tag{17}$$

where $P_{11,sc}$ is the short circuit reflection coefficient, P_{13} is the electro-acoustic transfer coefficient and P_{33} is the input admittance of the transducer. In order to have a large change of P_{11}, which is equivalent to the sensitivity of the SAW device loaded by an impedance sensor, the influence of Z in equation (17) needs to be large. Therefore, a SAW transducer with a small $P_{11,sc}$ and large P_{13} will provide a large sensitivity. $P_{11,sc}$ can be minimized by using a double electrode IDT design as shown in Fig. 10, which provides cancelation of the internal mechanical reflections of the IDT.

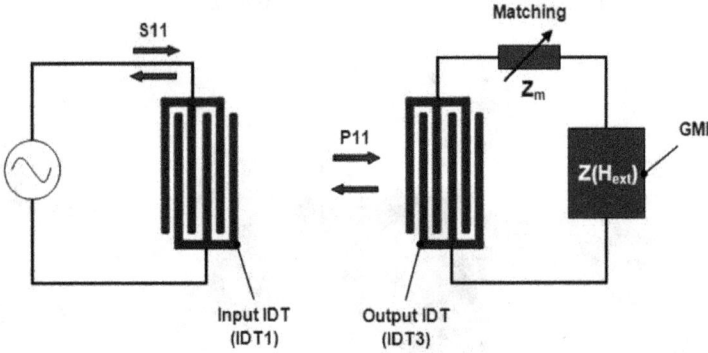

Figure 11. Electric and acoustic ports of the SAW sensor

The electro-acoustic transfer coefficient P_{13} and input admittance P_{33} can be obtained by,

$$P_{13} = \frac{1}{2\sqrt{2Z_a}} r_m (1 - e^{-j\varphi_m})\tag{18}$$

$$P_{33} = j\omega C_{IDT} + \frac{r_m^2}{Z_a}(1 - e^{-j\varphi_m})\tag{19}$$

Where r_m is the ratio of the electrical to acoustical transformer, C_{IDT} is the capacitance of the IDT, Z_a is the acoustic impedance and φ_m is the transit angle [63].

Since the GMI sensor is an inductive element, matching is accomplished by a series capacitance resulting a load impedance

$$Z = 1/j\omega C_m + R + j\omega L(H_{ext}),\tag{20}$$

where C_m is the matching capacitance, R is the average resistance (over the considered magnetic field range) of the GMI sensor and $L(H_{ext})$ the inductance of the GMI sensor.

Fig. 12 (a) shows the simulation result of the IDT's reflectivity as a function of the load. The slope of this plot corresponds to the magnetic field sensitivity. Therefore, the optimum matching capacitance can be determined. Fig. 12 (b) presents the rate of change of P_{11} for 1nH inductance changes (corresponding to a field change of approximately 50A/m). The result shows that with the optimum matching capacitance a maximum reflectivity change rate of 0.3dB/nH can be achieved. As the fabricated GMI sensor has an inductance change from 5nH to 15nH, a reflectivity change of 3dB can be expected.

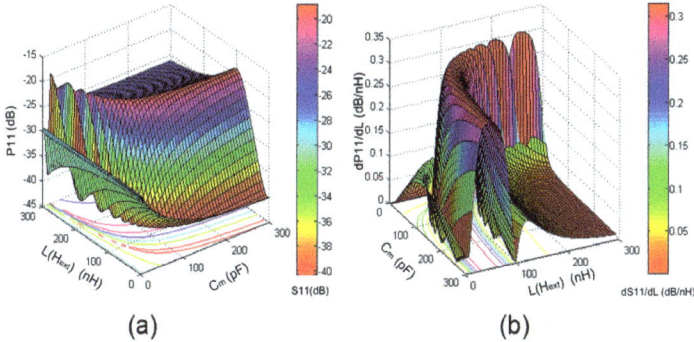

(a)   (b)

Figure 12. (a) Magnitude P_{11} as a function of the matching capacitance and sensor inductance. (b) Rate of change (absolute value) of P_{11} for 1nH load inductance change.

The piezoelectric substrate chosen for this application is LiNbO$_3$ as it provides a strong electromechanical coupling corresponding to a high value of P_{13}. The detailed design parameters of the SAW transponder are shown in Tab. 3. The working frequency of the device is 80MHz, resulting in a periodicity p of 50µm (Equ. (1). The value of p determines the electrode width and gap. The distances between the IDTs yield a 1.25µs delay between IDT1 and IDT2 and a 0.625µs delay between IDT2 and IDT3.

Design parameter		Design parameter	
Substrate material	LiNbO$_3$ (128 deg. Y-X cut)	Electrode material	Gold
Center frequency	80MHz	Aperture	30λ
Periodicity	50µm	Electrode/gap width	6.25µm
Electrodes per segment	4	IDT segment number	30

Table 3. Design parameters for the SAW device.

The GMI sensor consists of a tri-layer structure with two ferromagnetic layers of 100nm in thickness made of Ni$_{80}$Fe$_{20}$ and a conducting copper layer with a thickness of 200nm. The sensor has a rectangular geometry of 100 μm× 4000 μm. The conducting layer is connected to the IDT3 [18].

Fabrication

The fabrication of the combined device is accomplished in several steps as shown in Fig. 13. On a LiNbO₃ wafer, a 40 nm Ti adhesion layer and 200 nm gold layer are sputter deposited and patterned by ion milling into individual SAW devices. The leads and SMD footprints are designed together with the SAW device to facilitate an on-chip impedance matching circuit, which was accomplished by a 150pF capacitor connected in series with the GMI element. The GMI element comprises a tri-layer structure (Ni₈₀Fe₂₀(100nm) / Cu(200nm) / Ni₈₀Fe₂₀(100nm)) deposited at room temperature with a uniaxial magnetic field of 200 Oe applied in the transversal direction.

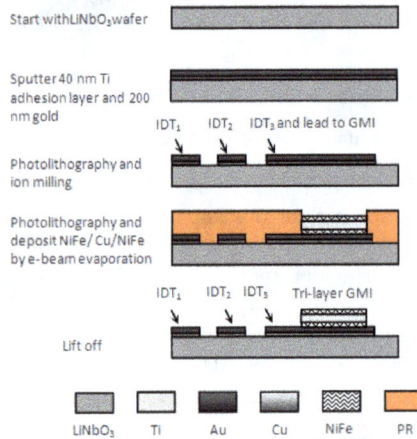

Figure 13. Fabrication flow chart of the integrated SAW-GMI device.

Results

A network analyzer (Agilent E8363C) is used to apply an RF signal to IDT1 and measure the electric reflection coefficient (S_{11}) of the input IDT, which is related to the admittance matrix of the whole device and P_{11}. The time domain signal of S_{11} is converted from the frequency domain using fast Fourier transform. As shown in Fig. 14 (a), two reflection peaks at 2.45μs and 3.55μs are observed indicating the reflections from the reference IDT and the load IDT accordingly. The magnetic response of the integrated device is determined by applying a variable magnetic field in longitudinal direction to the device. A 2.4dB amplitude change of the reflection signal can be observed. A comparison of the simulated and experimental results together with the measured GMI ratio curve is shown in Fig. 14 (b).

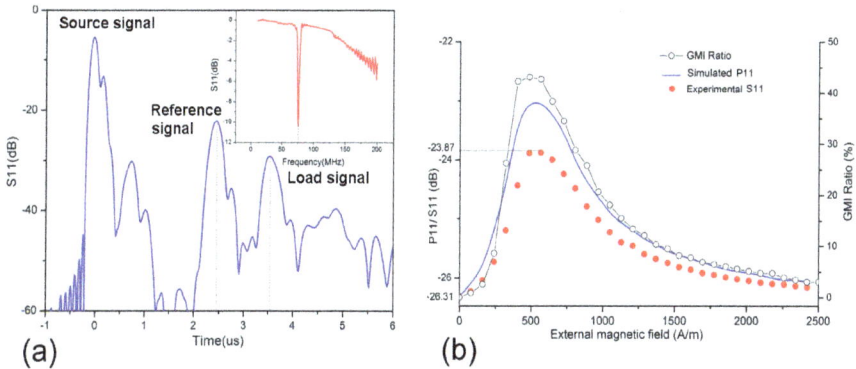

Figure 14. (a) Time domain measurement of the SAW-GMI device. Inset: Frequency domain measurement. (b) Comparison of the simulated and experimental device response together with the measured GMI ratio curve.

4. Potential applications

Magnetic field sensors are one of the most widely used sensors and employed for many different applications. Current commercial magnetic sensors are wire connected to a circuit providing power and readout. These wire connections prevent the sensors from being used for certain applications. In addition, as the complexity and the number of devices, increases in modern systems such as automobiles, wire connections are becoming an increasing problem due to limited space. For those and other reasons, wireless solutions are being much sought after.

As pointed out in the previous sections, SAW-based sensors have been developed for different applications, and this technology also provides a platform for realizing wireless and passive magnetic sensors. They can provide solutions for various applications, for example, where the sensors have to withstand harsh environmental conditions or are required to have a long lifetime without maintenance.

Out of the countless applications for SAW-based passive and remote magnetic field sensors, a few will be highlighted in the following.

Nanotechnology and miniaturized systems are becoming increasingly popular in the biomedical field. Technologies based on magnetic effects are of particular interest since they can be controlled remotely via magnetic fields. For example, NVE Corporation recently developed a battery operated magnetic sensor to be used as a magnetic switch for implantable devices. When a magnetic field is applied, the sensor turns on triggering a specified action. It turns off when the field is removed. The sensor works at a stable operating point of 15 Oe [64]. Magnetic beads have been extensively used in many biomedical applications. These magnetic beads are made of ferromagnetic material ranging in size between 5 nm to 500 um. A new application of such particles promises benefits in cancer therapy by employing the particles either as drug carriers or heat sources (hyperthermia) [65]. In order to have better control of the treatment, magnetic sensors are considered to measure and detect the concentration of these magnetic particles.

The automotive industry extensively uses magnetic sensors for different purposes, for example, to measure current in electric vehicles [66] or the rotation speed of gears [67]. Another application employs magnetic sensors to detect passing vehicles using lane markers [68]. Such a system could also be used to detect vehicle speed by measuring the time between two markers of a fixed distance. In yet another application, developed by Stendl et al [69], the wear of a vehicle's tire is detected by measuring the field of magnetic beads embedded in the rubber of tire treads. As the tread size decreases, the magnetic field also decreases. A wireless magnetic sensor is placed just below the threads.

Construction monitoring is an upcoming application for wireless sensor. Long-term monitoring of metallic reinforcements in, e.g., bridges or buildings requires passive and remote sensors, which are capable of detecting changes of the metal. Similarly, the detection of internal defects or corrosion of pipelines is of great interest. Gloria et al [70] developed an Internal Corrosion Sensor (ICS) consisting of a magnet and a Hall sensor. A disturbance in the magnetic field caused by changes of the metal changes the sensor readout. This information is used for both to detect and size the defects.

5. Conclusion

In this chapter we discussed different types of SAW-based, magnetic sensors including resonators, phase shifters and loaded transponders. Sensitivity to magnetic fields can be achieved by either changing the properties of the IDT or delay line utilizing magnetostrictive materials or loading the output IDT with a magnetic field sensor. GMI sensors feature a very high sensitivity and wide range of operating frequencies and, therefore, constitute an especially suitable load. The principle of GMI sensors is described in detail and different GMI structures are discussed. While the highest sensitivity has been obtained with GMI microwires, thin film GMI sensor are advantageous because they can be produced using standard microfabrication methods, and they can be easily integrated with a SAW transponder on the same substrate. These features are crucial with respect to production complexity and costs.

A SAW transponder combined with a GMI element connected to the output IDT is a passive and remote magnetic field sensor, which responds to an interrogation signal with a delayed response signal. The design of such a device needs to take into account different aspects like operation frequency, dimensions of IDTs and delay line or matching the load with the output IDT. In order to obtain a high sensitivity, an impedance change of the GMI element caused by a magnetic field, has to yield a large change in the SAW reflected from the output IDT. A model is presented to simulate the electro-acoustic interaction of the output IDT with the GMI sensor's impedance and the impedance matching capacitance. The simulation results provide information regarding the matching parameters and are invaluable for obtaining an optimized performance. A detailed description of the fabrication of an integrated SAW-GMI sensor is provided using standard microfabrication technologies. The GMI ratio of the fabricated sensor is 45 % The SAW-GMI sensor provides a sensitivity of 3 dB/mT, and its output corresponds well with the simulation results.

Magnetic field sensors have countless applications and are widely used in many different fields. The trend towards wireless operation, which is generally observed nowadays, drives the development of passive and remote magnetic field sensors. Several concepts of such sensors employing SAW devices have been presented in this chapter. The most promising one is a SAW-GMI sensor, which has been discussed in detail and which features wireless and battery less operation as well as durability and the ability to withstand harsh environments. This kind of sensor is considerable not only for providing existing applications with a wireless mode; it also largely expands the potential applications of magnetic field sensors.

Author details

Bodong Li and Jürgen Kosel
Electrical Engineering Department, King Abdullah University of Science and Technology, Thuwal, Saudi Arabia

Hommood Al Rowais
Electrical Engineering Department, Georgia Institute of Technology, Atlanta, Georgia, USA

6. References

[1] Stockman H (1948) Communication by means of reflected power, Proceedings of IRE. 36: 1196-1204

[2] Plath F, Schmeckebier O, Rusko M, Vandahl T, Luck H, Moller F, Malocha D C (1994) Remote sensor system using passive SAW sensors. ULTRASON. 1: 585-588

[3] Reindl L, Scholl G, Ostertag T, Scherr H, Wolff U, Schmidt F (1998) Theory and application of passive SAW radio transponders as sensors. IEEE Trans. Ultrason., Ferroelectr., Freq. Control. 45: 1281-1292

[4] Reeder T M and Cullen D E (1976) Surface-acoustic-wave pressure and temperature sensors. Proceedings of the IEEE. 64: 754-756

[5] Lee K, Wang W, Kim G and Yang S (2006) Surface Acoustic Wave Based Pressure Sensor with Ground Shielding over Cavity on 41° YX LiNbO₃. Jpn. J. Appl. Phys. 45: 5974–5980

[6] Pohl A (1997) Wirelessly interrogable surface acoustic wave sensors for vehicular applications. IEEE T Instrum Meas. 46: 1031-1038

[7]   Calabrese G S, Wohltjen H, Roy M K (1987) Surface acoustic wave devices as chemical sensors in liquids: Evidence disputing the importance of Rayleigh wave propagation, Anal. Chem. 59: 833–837

[8]   Kadota M, Ito S, Ito Y, Hada T, And Okaguchi K (2011) Magnetic Sensor Based on Surface Acoustic Wave Resonators. Jpn. J. Appl. Phys. 50: 07HD07

[9]   Hanna S M (1987) Magnetic Field Sensors Based on SAW Propagation in Magnetic Films. IEEE Trans. Ultrason., Ferroelectr., Freq. Control. UFFC-34: 191-194

[10] http://www.asrdcorp.com

[11] http://www.senseor.com

[12] http://www.transense.co.uk

[13] Ganguly A K, Davis K L, Webb D C, Vittoria C, Forester D W (1975) Magnetically tuned surface-acoustic-wave phase shifter. *Electronics Letters*. 11: 610-611

[14] Hauser H, Steindl R, Hausleitner Ch, Pohl A, Nicolics J (2000) Wirelessly Interrogable Magnetic Field Sensor Utilizing Giant Magneto-Impedance Effect and Surface Acoustic Wave Devices. IEEE T INSTRUM MEAS. 49: 648-652

[15] Steindl R, Hausleitner Ch, Pohl A, Hauser H, Nicolics J (2000) Passive wirelessly requestable sensors for magnetic field measurements. Sens Actuators A. 85: 169-174

[16] Hauser H, Steurer J, Nicolics J, Musiejovsky L, Giouroudi I (2006) Wireless Magnetic Field Sensor. Journal of Electrical Engineering 57: 9-14

[17] Al Rowais H, Li B, Liang C, Green S, Gianchandani Y, Kosel J (2011) Development of a Passive and Remote Magnetic Microsensor with Thin-Film Giant Magnetoimpedance Element and Surface Acoustic Wave Transponder. J. Appl. Phys. 109: 07E524

[18] Li B, M H. Salem N P, Giouroudi I, Kosel J (2011) Integration of Thin Film Giant Magneto Impedance Sensor and Surface Acoustic Wave Transponder. J. Appl. Phys. 111: 07E514

[19] Sinha B K, Tiersten H F (1979) Zero temperature coefficient of delay for surface waves in quartz. Appl. Phys. Lett. 34: 817

[20] Ebata Y, Suzuki H, Matsumura S and Toshiba K F (1982) SAW Propagation Characteristics on $Li_2B_4O_7$. Jpn. J. Appl. Phys. 22: 160-162

[21] Tsubouchi K.Sugai K, Mikoshiba N (1982) Zero Temperature Coefficient Surface-Acoustic-Wave Devices Using Epitaxial AlN Films. Ultrasonics Symposium, 340 – 345

[22] Dewan N, Tomar M, Gupta V, and Sreenivas K (2005) Temperature stable LiNbO3 surface acoustic wave device with diode sputtered amorphous TeO2 over-layer. Appl. Phys. Lett. 86: 223508

[23] Viens M, Cheeke J D N (1990) Highly sensitive temperaturesensor using SAW resonator oscillator. Sensors and Actuators A: Physical 24: 209-211

[24] Wang W, Lee K, Woo I, Park I, Yang S (2007) Optimal design on SAW sensor for wireless pressure measurement based on reflective delay line. Sensors and Actuators A: Physical. 139: 2-6

[25] Schimetta G, Dollinger F, Scholl G, Weigel R (2001) Optimized design and fabrication of a wireless pressure and temperature sensor unit based on SAW transponder technology. Microwave Symposium Digest, 2001 IEEE MTT-S International. 1: 355-358

[26] Lim C, Wang W, Yang S, Lee K (2011) Development of SAW-based multi-gas sensor for simultaneous detection of CO_2 and NO_2, Sensors and Actuators B: Chemical. 154: 9-16

[27] Reindl L, Ruppel C C W, Kirmayr A, Stockhausen N, Hilhorst M A, and Balendonck J (2001) Radio-Requestable Passive SAW Water-Content Sensor, IEEE Transactions on Microwave Theory And Techniques. 49: 803

[28] Wang S Q, Harada J and Uda S (2003) A Wireless Surface Acoustic Wave Temperature Sensor Using Langasite as Substrate Material for High-Temperature Applications. Jpn. J. Appl. Phys. 42: 6124

[29] Wu C (2006) Fabrication of Surface Acoustic Wave Sensors for Early Cancer Detection, Electrical Engineering, University of California, Los Angeles

[30] Knobel M, Vazquez M, and Kraus L, Buschow ed. K.H.J. (2003) Giant Magnetoimpedance, Handbook of magnetic materials. 15: 497-563

[31] Sezen A S, Sivaramakrishnan S, Hur S, Rajamani R, Robbins W, and Nelson B J (2005) Passive Wireless MEMS Microphones for Biomedical Applications, J. Biomech. Eng. 127: 1030

[32] Wolff U, Schmidt F, Scholl G, Magori V (1996) Radio accessible SAW sensors for non-contact measurement of torque and temperature. Ultrasonics Symposium Proceedings. 1: 359-362

[33] Yoshizawa N, Yamamoto I, and Shimada Y (2005) Magnetic Field Sensing by an Electrostrictive/Magnetostrictive Composite Resonator. IEEE TRANSACTIONS ON MAGNETICS 41: 11

[34] Dong S, Zhai J, Li J, and Viehland D (2006) Small Dc Magnetic Field Response Of Magnetoelectric Laminate Composites, Applied Physics Letters 88: 082907

[35] Koeninger V, Matsumura Y, Uchida H H, Uchida H (1994) Surface acoustic waves on thin films of giant magnetostrictive alloys. J ALLOY COMPD 211/212: 581-584

[36] Uchida H, Wada M, Koike K, Uchida H H, Koeninger V, Matsumura Y, Kaneko H, Kurino T (1994) Giant magnetostrictive materials: thin film formation and application to magnetic surface acoustic wave devices. J ALLOY COMPD 211/212: 576-580

[37] Panina L V and Mohri K (1994) Magneto-impedance effect in amorphous wires. Appl.Phys.Lett. 65: 1189-1191

[38] Machado F L A, Rezende S M (1996) A theoretical model for the giant magnetoimpedance in ribbons of amorphous soft-ferromagnetic alloys. Journal of Applied Physics. 79: 6558 - 6560

[39] Hika K, Panina L V, Mohri K (1996) Magneto-Impedance in Sandwich Film for Magnetic Sensor Heads. IEEE Trans Magn. 32: 4594-4596.

[40] Panina L V, Makhnovskiy D P, Mapps D J, and Zarechnyuk D S (2001) Two-dimensional analysis of magnetoimpedance in magnetic/metallic multilayers. J. Appl. Phys. 89: 7221

[41] Morikawa T, Nishibe Y, Yamadera H, Nonomura Y, Takeuchi M, Taga Y (1997). Giantmagneto-Impedance Effect in Layered Thin Films. IEEE Trans Magn. 33: 4367-4372

[42] de Andrade A M H, da Silva R B, Correa M A, Viegas A D C, Severino A M, Sommer R L (2004) Magnetoimpedance of NiFe/Ag multilayers in the 100 kHz–1.8 GHz range. Journal of Magnetism and Magnetic Materials 272–276: 1846–1847

[43] Zhukova V, Chizhik A, Zhukov A, Torcunov A, Larin V and Gonzalez J (2002) Optimization of giant magnetoimpedance in Co-rich amorphous microwires. IEEE Trans. Magn. 38: 3090-3092

[44] Le A T, Phan M H, Kim C O, Vázquez M, Lee H, Hoa N Q and Yu S C (2007) Influences of annealing and wire geometry on the giant magnetoimpedance effect in a glass-coated microwire LC-resonator, J. Phys. D: Appl. Phys. 40: 4582–4585

[45] Vázquez M, Adenot-Engelvin A L (2009) Glass-coated amorphous ferromagnetic microwires at microwave frequencies. Journal of Magnetism and Magnetic Materials. 321: 2066–2073

[46] Zhou Z, Zhou Y, Chen L (2008) Perpendicular GMI Effect in Meander NiFe and NiFe/Cu/NiFe Film. IEEE Transactions on Magnetics. 44: 2252 - 2254

[47] Pompéia F, Gusmão L A P, Hall Barbosa C R, Costa Monteiro E, Gonçalves L A P and Machado F L A (2008) Ring shaped magnetic field transducer based on the GMI effect. Meas. Sci. Technol. 19: 025801

[48] Phan M H , Peng H X (2008) Giant magnetoimpedance materials: Fundamentals and applications. Progress in Materials Science. 53: 323–420

[49] Squire PT, Atkinson D, Gibbs M R J, Atalay S J (1994) Amorphous wires and their applications. J Magn MagnMater. 132: 10–21

[50] Ohnaka I, Fukusako T, Matui T (1981) Preparation of amorphous wires. J. Jpn. Inst. Met. 45: 751–62.

[51] http://www.aichi-steel.co.jp/ENGLISH/pro_info/pro_intro/elect_3.html

[52] Panina L V, Mohri K, Uchiyama T, Noda M (1995) Giant magneto-impedance in Co-rich amorphous wires and films. IEEE Trans Magn. 31:1249–60

[53] Phan MH, Peng HX, Yu SC, Vazquez M (2006) Optimized giant magnetoimpedance effect in amorphous and nanocrystalline materials. J Appl Phys. 99: 08C505

[54] Sukstanskii A L and Korenivski V (2001) Impedance and surface impedance of ferromagnetic multilayers: the role of exchange interaction J. Phys. D 34: 3337

[55] Li B, Kosel J (2001) 3d Simulation of GMI Effect In Thin Film Based Sensors. J. Appl. Phys. 109: 07E519

[56] Dong C, Chen S, Hsu T Y (Xu Zuyao) (2003) A modified model of GMI effect in amorphous films with transverse magnetic anisotropy. J. Magn. Magn. Mater. 263: 78-82

[57] Xiao S, Liu Y, Dai Y, Zhang L, Zhou S, and Liu G (1999) Giant Magnetoimpedance Effect in Sandwiched Films. J. Appl. Phys. 85: 4127

[58] Xiao S, Liu Y, Yan S, Dai Y, Zhang L, and Mei L (2000) Giant Magnetoimpedance and Domain Structure in FeCuNbSiB Films and Sandwiched Films. Phys. Rev. B 61: 5734–5739

[59] de Cos D, Panina L V, Fry N, Orue I, Garcia-Arribas A, Barandiaran J M (2005) Magnetoimpedance in narrow NiFe/Au/NiFe multilayer film systems. IEEE Transactions on Magnetic. 41: 3697 - 3699

[60] Li X, Yuan W, Zhao Z, Ruan J and Yang X (2005) The GMI effect in Nanocrystalline FeCuNbSiB Multilayered Films with a SiO2 Outer Layer. J. Phys. D: Appl. Phys. 38: 1351–1354

[61] Corrêa M A, Bohn F, Escobar V M, Marques M S, Viegas A D C, Schelp L F, and Sommer R L (2011) Wide Frequency Range Magnetoimpedance in Tri-layered Thin NiFe/Ag/NiFe Films: Experiment and Numerical Calculation. J. Appl. Phys. 110: 093914

[62] Zhou Z, Zhou Y, Chen L and Lei C (2011) Transverse, Longitudinal and Perpendicular Giant Magnetoimpedance Effects in a Compact Multiturn Meander NiFe/Cu/NiFe Trilayer Film Sensor. Meas. Sci. Technol. 22: 035202

[63] Tobolka G (1979) Mixed matrix representation of SAW transducers. IEEE Trans. Sonics Ultrason. SU-26

[64] http://www.medicalelectronicsdesign.com/products/nanopower-magnetic-sensors-fit-implantable-devices

[65] Corchero J L, Villaverde A (2009) Biomedical Applications of Distally Controlled Magnetic Nanoparticles. Trends in Biotechnology 27: 468-476

[66] Ripka P (2008) Sensors based on bulk soft magnetic materials: Advances and challenges. Journal of Magnetism and Magnetic Materials. 320: 2466-2473

[67] Lenz J, Edelstein S (2006) Magnetic sensors and their applications. IEEE Sensors Journal. 6: 631-649

[68] Nishibe Y, Ohta N, Tsukada K, Yamadera H, Nonomura Y, Mohri K, Uchiyama T (2004) Sensing of passing vehicles using a lane marker on a road with built-in thin-film MI sensor and power source. IEEE Transactions on Vehicular Technology. 53: 1827- 1834

[69] Steindi R, Hausleitner C, Hauser H, Bulst W (2000) "Wireless magnetic field sensor employing SAW-transponder," Applications of Ferroelectrics. Proceedings of the 2000 12th IEEE International Symposium on. 2: 855-858

[70] Gloria N B S, Areiza M C L, Miranda I V J, Rebello J M A (2009) Development Of A Magnetic Sensor for Detection And Sizing of Internal Pipeline Corrosion Defects, NDT & E International. 42: 669-677

Chapter 5

Acoustics in Optical Fiber

Abhilash Mandloi and Vivekanand Mishra

Additional information is available at the end of the chapter

1. Introduction

Optical filters are the heart of optical networks; without the wavelength selective device wavelength division multiplexing and dense wavelength division multiplexing network will not exist. As the networks are progressing towards closer wavelength spacing, performance requirement for filters are becoming more demanding. Currently, the popular filters include gratings, thin-film filters, and Fabry-Perot filters and acoustoi optic tunable filters (AOTFs).

Acousto-optic (AO) effect in fibers has been studied to produce tunable filters, gain flatteners, modulators, frequency shifters, and optical switches reported. Most AO devices work on coupling from the fundamental mode (LP$_{11}$) of light to a higher order asymmetrical (LP$_{ll}$, LP$_{12}$ .... LP$_{1n}$) modes. Acousto-optics is defined as the discipline devoted to the interactions between the acoustic waves and the light waves in a material medium. Acoustic (vibrational) waves can be made to modulate, deflect and focus light waves by causing a variation in the refractive index. Acousto optic tunable filters are a promising technology for dynamic gain equalization of optical fiber amplifiers [1]. By launching an acoustic wave directly on the fiber, the device combines the merits of fiber and AOTF devices namely the low insertion loss, low polarization dependence loss, wide tunability, fast tuning speed and ease of packaging. When a flexural acoustic wave is applied to a tapered single mode fiber, coupling takes place between the core mode and the cladding mode. The coupled energy in the cladding mode is essentially absorbed by the fiber jacket as reported so that the device is a notch filter. It means the centre frequency and the rejection efficiency can be tuned by adjustment of the frequency and voltage being applied. Varying the amplitudes and frequency of a RF generator can change the spectral profile of these filters.

To improve the rejection efficiency of the filters, the thickness of the fiber can be reduced. This is achieved through the heating and the acid-etching method. In the heating method, the ratio of cladding to core size is maintained while in the acid etching-method, the ratio between the cladding and core can be changed.

2. Acousto-optic tunable filter

2.1. Device design

The fiber used in our experiment is a Corning SMF-28, standard telecommunication single mode fiber. A region of SMF is etched by dipping the fiber in a hydrofluoric acid solution, which has a concentration of 40%. Etching rate controls the thickness of the SMF and the diameter reduction is observed using a CCD camera.

When the optical signal enters the fiber and interacts with the acoustic energy in a jacket stripped segment of the fiber, the core mode of the light is converted to a higher order cladding mode producing a notch filter like characteristics in the transmission spectrum. Core mode converting to various cladding modes will produce a few notch filters, with each having its peak notch at a separate wavelength [2-4]. A vibrating PZT transducer driven by a RF generator produces the acoustic energy as stated by Yun, Hwang and Kim, (1996). The acoustic energy is further amplified and concentrated to the fiber by a machined aluminium horn.

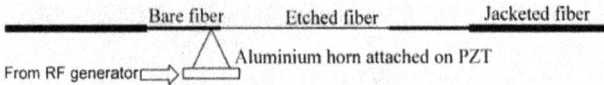

Figure 1. The setup to study AO interaction inside a fiber.

2.2. Horn design

An acoustic horn functions to transfer and amplify the surface acoustic wave to the fiber. All horns made were conical in shape, where the tip is narrow and the base is broad as described by Lee, Kim Hwang and Yun, (2003). All the horns fabricated for the AOTF experiment have a ratio of length to outer diameter ratio of 2. Length is defined as the length from the tip of the horn to the base of the horn. Horns taken are 1cm in length. Outer diameter of the horn is defined at the diameter at the base. The inside of the horns are made hollow. When the horn is made considerably small, the frequency dependence on the acoustic generator is low. In the experiments done, no transduction is observed when the fiber is not etched. Potential problems can be attributed to the size of the transducer and the adhesive used to bond the tip of the horn to the fiber. Solder acts as a strong, metallic, thermally stable, and acoustically transmitting joint. In these experiments however glue was chosen as the bonding agent. Of particular interest will be the horn tip size. Acoustic impedance at the horn tip is given by:

$$Z_r = c_a v_a A_r \tag{1}$$

where c_a is the longitudinal velocity inside aluminium, v_a is the density aluminium and A_r is the cross section of the horn tip. Acoustic impedance at the bond junction along the fiber is given by:

$$Z_s = 2c_s v_s A_s \tag{2}$$

where c_s is the longitudinal velocity inside silica, v_s is the density silica and A_s is the cross section of fiber. Acoustic impedance inside the fiber is counted twice because of bidirectional acoustic movement along the fiber. Optimum transduction occurs when $Z_r = Z_s$ and since acoustic impedance of silica is almost matching that of aluminium, according to engan et. al maximum acoustic wave transfer occurs when horn tip diameter is almost matching that of the fiber.

2.3. Tuning of peak wavelength

By driving a piezoelectric (PZT) device at an ultrasonic frequency the periodic perturbations can be created inside the fiber. In a phase-matched condition, where the momentum and energy conservation requirement $(L_B = \wedge)$ are met, the resonant frequency of an acoustic wave according to Birks, Russel and Culverhouse (1992) is given by

$$f = \frac{\pi b C_{ext}}{L_B^2} = \frac{\pi b C_{ext}}{\wedge^2} \qquad (3)$$

where b is the radius of the fiber, C_{ext} is the speed of fundamental acoustic mode, which for silica is 5760 ms^{-1}, $\wedge$ is the period of the microbend[1].

Assuming a phase-matched condition, the frequency needed to transfer the modes from core to cladding mode for various thickness of the fiber is given in Fig 2 and Fig 3. As the fiber diameter is reduced, the values of $df/d\lambda$ get smaller. For unetched fibres, the frequencies used to create the micro bends and thus, convert the modes are from 1.75 MHz to 2.25 MHz. For thin diameter fibers (20 μ m, 30 μ m, 40 μ m), the frequencies are from 800 kHz to 1.1 MHz.

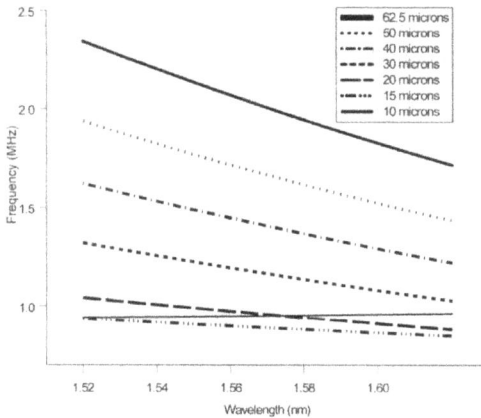

Figure 2. Calculated RF frequency to convert the LP01 mode to LP11 mode plotted against wavelength (for various thickness of fibre diameter).

[1] A microbend is the physical deformation of fiber achieved mechanically or chemically done to perturbing the optical modes to study mode coupling between core and cladding mode.

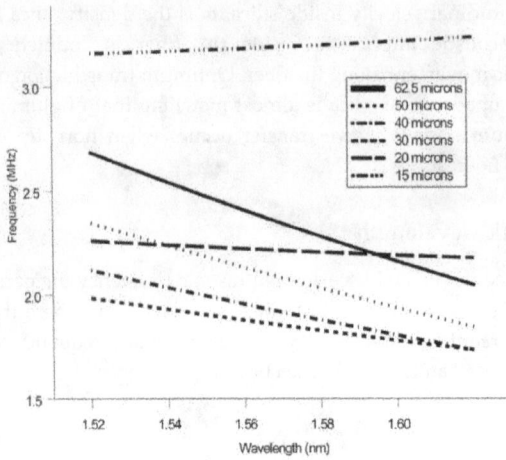

Figure 3. Calculated RF frequency to convert the LP0l mode to LP12 mode plotted against wavelength (for various thickness of fibre diameter).

Frequency from the RF generator can be used to control the peak wavelength tuning of the notch filters (Fig.4). The fiber used in the experiment has a diameter of 30 μm, and length of 17 cm. Higher frequencies of the RF generator will blue shift the peak wavelength of the filter [3-7]. The tuning range of the filter is slightly less than 300 nrn. From Eq.1.3, we deduce that, micro bend's period is inversely proportional to the frequency of the RF generator. For a larger value of period, the filter's peak is red shifted. Thin fibers have lower period values, thus etching the fibres will blue shift the peak wavelength of the notch filters. Frequency used to tune the peak wavelength as in for thin fibres is from 800 kHz to 1.1 MHz, which is in excellent agreement with the theory as in Fig.3 and Fig.4.

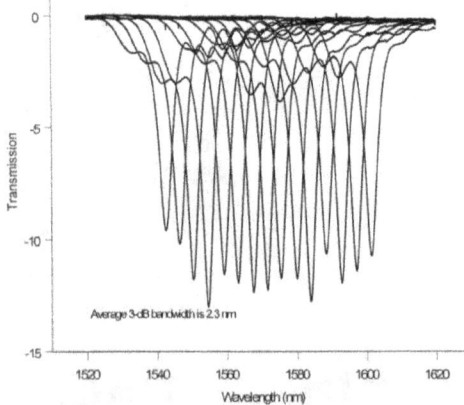

Figure 4. Measured peak wavelength tuning of the filter by changing the RF frequency. Frequency used is from 970 kHz to 1045 kHz. The fiber used has a thickness of 30 μm and length of 17 cm.

3. Tuning of attenuation depth

The RF generator's V_{P-P} level will be used to control the attenuation depth of the filter. V_{P-P} level is actually referring to the acoustic power transferred to the fiber. Increasing the V_{P-P} level will generally increase the bottom level of the filter as seen from Fig. 5. However in some cases increasing the V_{P-P} level will only distort the shape of the filter without increasing the notch's depth. For this strong over-coupled phenomenon, side lobes of the filter is actually increasing. One way to eliminate the problem is by limiting the interaction length of light inside the etched region. Here the power means RF generator's Vp-p level which will be used to control the attenuation depth of the filter [8-10]. Vp-p level is actually referring to the acoustic power transferred to the fiber. Increasing the Vp-p level will generally increase the bottom level of the filter as seen from Fig. 5. Here acoustic power supplied to PZT is 1.6 W to allow mode conversion.

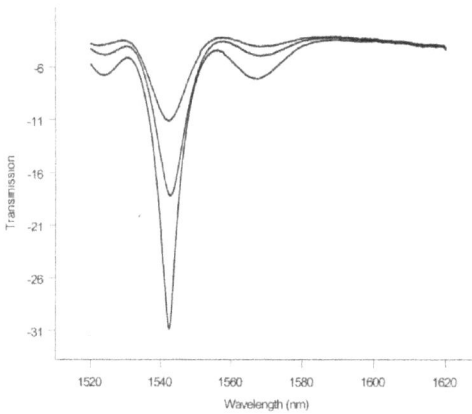

Figure 5. Measured attenuation variation of filter when the power of RF generator (V_{P-P}) is increased.

An effort to reduce the acoustic power fed into the fiber is by reducing the thickness of the fiber. The minimum acoustic power required by the device to operate or to allow mode conversion, is given by

$$P = 2\pi^3 \rho v_g (fR)^2 (u_t)^2 \qquad (4)$$

where ρ is the mass density of the fiber (ρ =2200kg/m3 for fused silica), V_g is the group velocity of the wave and R is is curvature of fiber, and u_t is the transverse acoustic amplitude which is given by:

$$u_t = \frac{\pi}{2} L_B \frac{a}{L} \frac{1}{0.908} \qquad (5)$$

where L is referring to the interaction length of acoustic and light inside the fiber and L_B is the optical beat length. Fig. 6 shows the calculated power required for mode conversion is lower for etched fibers. When the fiber is unetched the power required will be 287 mW. For a 20 μ m fiber, the power required for conversion is only 1.17 mW.

Experimentally, as seen from Fig. 7 for a 37 μm thick fiber, the acoustic power supplied to PZT is 1.6 W to allow mode conversion. Mode conversion was confirmed using far-field radiation pattern as reported by Doma, and Blake (1992). However, in a 26 μ m fiber, power requirement for mode conversion is reduced to a mere value of 42 mW. The difference in the power reduction with the calculated value, suggests that the loss at the point of contact is high[10]. It is believed that the horn design is still not optimized; nevertheless, this transduction is sufficient to demonstrate conversion between two modes. Typically, only the lowest order flexural acoustic mode should be made to travel inside the fiber, and this can be achieved by ensuring the horn tip's thickness is matching that of the fiber.

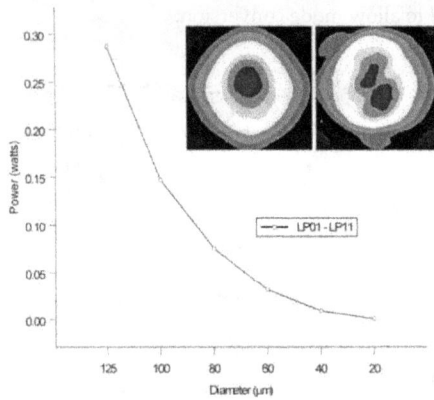

Figure 6. Calculated acoustic power required to allow mode conversion. Interaction length was set to 13 cm. Inset: Far field radiation pattern of modes involved in conversion. Left- LP01, Right- LP11.

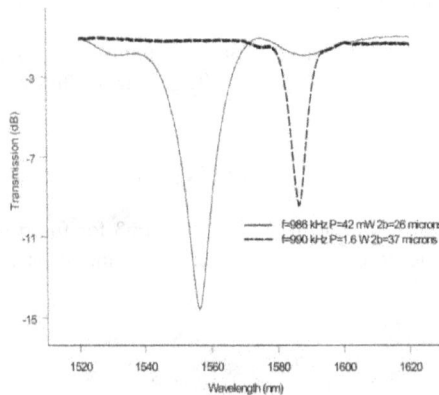

Figure 7. Measured transmission spectrum when fiber is etched. Significant reduction in acoustic power is observed.

4. Tuning of 3-dB bandwidth

The 3-dB bandwidth of the notch filter is given by the equation below as reported by D. Ostling, H.E. Engan (1995):

$$\Delta\lambda = \frac{0.8}{L}\left[\frac{\partial L_B(\lambda)}{\partial\lambda}\right]^{-1}\left[L_B(\lambda)\right]^2 \tag{6}$$

Where λ is the wavelength of the light, L is the length of the coupling interaction, and L_B is the optical beat length [11-14]. For a broadband filter, a short coupling length, a long beat length and small beat length dispersion is required. Without making the device short, only by etching the fiber to that thickness a broad filter can be obtained as reported. However this bandwidth is not tuneable and so is not suitable for spectral shaping. In this section, a similar achievement by only using a SMF to tune the 3-dB bandwidth of the filter is demonstrated. In this device, the notch filter's attenuation, peak wavelength tuning and 3-dB bandwidth can be simultaneously controlled in a single device.

To achieve this, a tunable acoustic absorber is added to the original AO setup as shown in Fig. 8. By moving the acoustic absorber along the etched region of the fiber, the interaction of light inside the acoustic region can be controlled. From Eq.6, we know that by controlling the coupling interaction length the 3-dB bandwidth of the filter can be controlled [15]. A strong acoustic absorbing material such as cotton or polystyrene can be used as the acoustic absorber [16]. The absorbing material functions to ensure no surface acoustic wave beyond the absorber's position are present. Since the interaction length of light inside the acoustic region can be controlled, over-coupling phenomenon can be monitored, to reduce the effects of undesirable side lobes. Broad filters require higher power to operate when the attenuation level is maintained the same as a narrow filter.

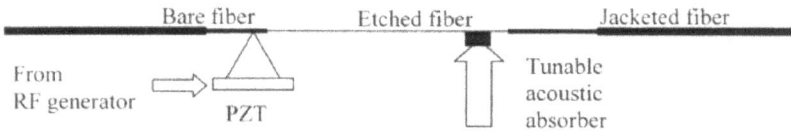

Figure 8. Setup to study the bandwidth variation using the AO interaction inside fiber.

From Fig. 9, the narrowest filter with a 3-dB bandwidth of 13 nm is obtained when maintaining the interaction length at 14 cm while the broadest filter with a 3-dB bandwidth of 28 nm is obtained when the interaction length is reduced to 7 cm. For an interaction length of 11.5 cm the spectral width is 16 nm and for 9 cm the spectral width will be 21 nm. The frequency used for wavelength tuning was from 960 kHz to 995 kHz and was sufficient to cover a wavelength span of 100 nm (1520 om-1620 nm).

For the broadest filter, the RF generator's V_{P-P} needed to generate coupling between the modes, seem to be the highest at 14 V. Meanwhile, for the narrowest filter, the V_{P-P} needed is only 6 V. Thus, we need a higher V_{P-P} to generate filters for shorter interaction length of light inside the grating region [17]. The introduction of tuneable acoustic absorber will change the strain dependency on the device. To limit the strain change introduced in the device only the tip of the absorber is allowed to touch the fiber, in our case, the resonant frequency change corresponding to the strain change was maintained at +/- 0.7kHz. Throughout the experiment the total IL was maintained less than 0.1 dB and the PDL was less than 0.4 dB.

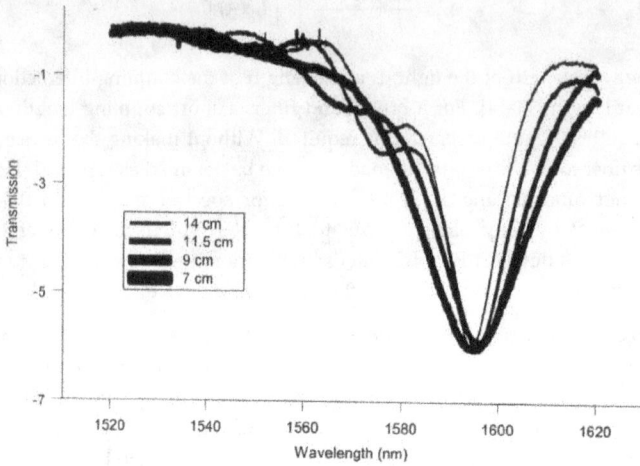

a)

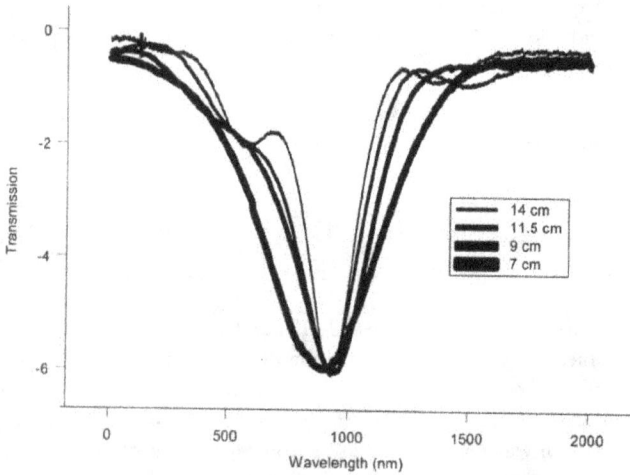

b)

a) 1596 nm b) 1566 nm

Figure 9. Measured bandwidth variation of filters at different peak wavelengths

5. Double-pass configuration

One of the key problems in fiber-based AOTF is the low attenuation level of the notch filter. Superposing two or more filters according to Yun, Lee, Kim and Kim (1999) produced by multiple transducers can increase the attenuation level. But this method introduces a very high crosstalk in the device especially when the filter's peak: wavelengths are very near to each another and prove [18] highly impractical. Alternatively to improve the attenuation level of the filter, a double pass AO setup reported by Satorious, Dimmick, Burdge (2002) and Culverhouse, Yun, Richardson, Birks, Farwell, Russell (1997) can be used. In the new setup as in Fig 1.10, a 3-port circulator is added before and after the AO device. Light comes in from port 1 of circulator 1 and goes through the acoustic region and experiences mode conversion. The LP 11 coupled mode is converted back to the fundamental mode at the jacket of the fiber. The light rounds circulator 2 and goes through the AO device and experiences mode conversion again. The produced notch filter is observed using the OSA connected to the port 3 of circulator 1. Since the period of the acoustic inside the fiber is not changed, the light going through this region experiences mode conversion at the same wavelength of the incident and returning light.

The insertion loss (IL) of a double-pass is increased to less than 3 dB and the Polarization Dependent Loss (PDL) was less than 0.6 dB. IL was not intentionally increased to a high value here, because 2 FC/FC connectors were introduced in the setup to connect port 2 of both circulators to the AO device. Splicing the ports to the device will reduce the IL loss to values less than 1 dB. Using higher quality circulators can further reduce PDL of the notch filters. The filters however will be more expensive to fabricate.

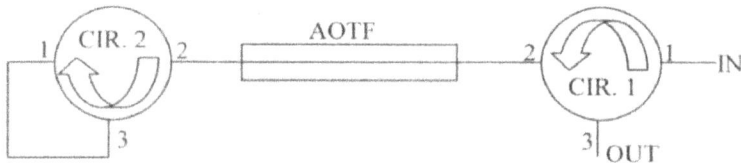

Figure 10. Double pass configuration to increase the maximum attenuation of the notch filter.

The AO band pass filter by Satorius *et. al.*, mentions side lobe suppression and maximum attenuation suppression using the double-pass configuration. Unlike the band pass filter, the notch filter will increase the side lobe level and maximum attenuation level using the double-pass configuration. The side lobe increment is not significant to exceed the bottom level of the main lobe of the notch filter.

From Fig. 11 the maximum attenuation of the notch filter was -28 dB for the double pass configuration and -12 dB for the single pass configuration. The maximum attenuation of the filter was increased to more than two times. The 3-dB bandwidth of the single pass AO device is 6.14 nm and the 3-dB bandwidth of the double pass AO device is 2.383 nm. This technique will be useful in producing narrow filters with high attenuation suitable in switching applications. However, there will be a frequency shift of 7 nm introduced using this setup.

The optical signal coupled from the slow mode (LP01) to the fast mode (LP11) will be downshifted in frequency when the acoustic wave is in the same direction as the optical signal. Frequency is shifted up when the fast mode is coupled to the slow mode for the same acoustic wave [19]. The frequency shift direction is reversed when the acoustic wave is in opposite direction with the optical signal as reported by Kim, Blake, Engan, and Shaw, (1986). In a double-pass setup, the optical signal is both the same and opposite direction to the acoustic wave, while in a single-pass setup, optical signal is maintained in the same direction as the acoustic wave. Thus, a frequency shift is observed in a double-pass setup.

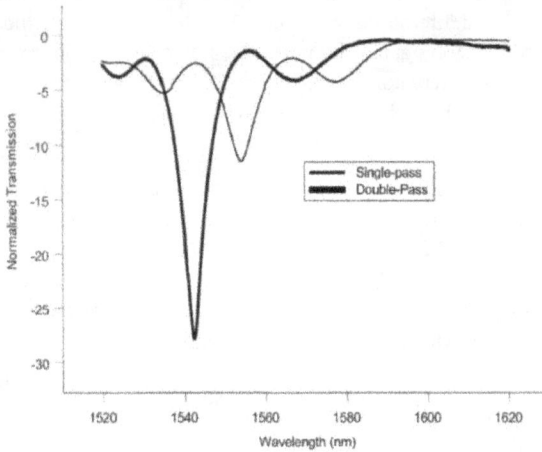

Figure 11. Measured normalized transmission spectrum using the double pass configuration. The result is compared with single pass configuration (refer to Fig4.1). There is a frequency shift of7 nm to the left using the double-pass setup.

6. Gain flattening filter

The technique to vary the 3-dB bandwidth of filter inside SMF is then extended as a dynamic gain equalizer for the gain profile of an Erbidium Doped Fiber Amplifier (EDFA). This is just one of the possible applications of AO interaction as efficient spectral shaping devices. Various efforts to dynamically control the gain flatness of the ASE spectrum using acousto-optic tuneable filters (AOTF) were well demonstrated. Passive gain equalization as reported by Vengsarkar, Pedrazzani, Judkins, Lemaire, Bergano, and Davidson (1996) is unable to encounter gain variations due to different input optical power of Wavelength Division Multiplexing (WDM) channels. Meanwhile, integrated AOTF as gain flattening filters have a serious limitation of high insertion loss and crosstalk problems. Fiber-based AOTF by H.S. Kim, Park, and B.Y. Kim (1998) and the setup by Feced, Alegria, and Zervas

(1999) uses two transducers with six synthesizers to obtain the desired spectral filters. In this technique, to shape the gain, the AOTF setup is using only one transducer and a single-taper. This is possible because the 3-dB [21-24] bandwidth of the filter we demonstrated can be varied on the same device. In our setup to flatten the gain profile of the Amplified Stimulated Emission (ASE) spectrum, an AOTF device with two frequency generators and a double-branched power combiner is used as in Fig 12. The power combiner typically introduces a 3-dB loss to the system, thus higher Vp-p from the RF generator is needed to produce the filters for spectral shaping. Total insertion loss of the setup is less than 0.2 dB. For the measurement, the EDF A was used as the ASE source and the output spectrum measured [20] using an Optical Spectrum Analyser (OSA).

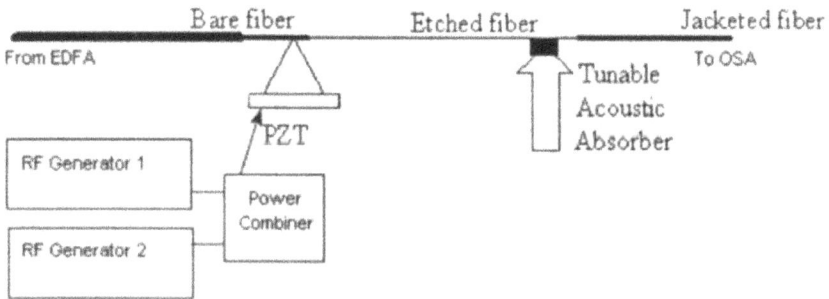

Figure 12. AOTF setup to flatten the gain of ASE spectrum.

The gain was flattened by changing the V_{p-p} level of the RF generator, and moving the tuneable acoustic absorber along the etched region of the SMF. The degree of freedom to shape the filter is very high, thus the necessity of cascading another AOTF to the setup is not needed. Fig. 13 shows the effect of shaping the filter on the Amplified Stimulated Emission spectrum of EDFA. Typically has it Amplified Stimulated Emission s peaks at 1532 run and 1550 run. For low gain, however there is a single broad peak at 1560 run. By using this method we show that, the [26-28] ASE spectrum can be flattened regardless of the peak's position and bandwidth using the same device. Since the tuning range is about 300 run, any Amplified Stimulated Emission spectrum that is lying from 1350 run to 1630 run can be successfully flattened using the same device.

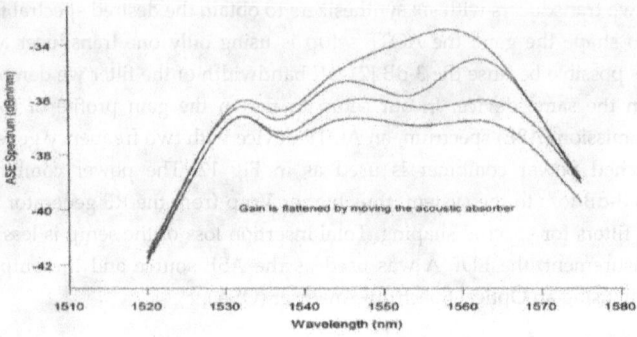

a)

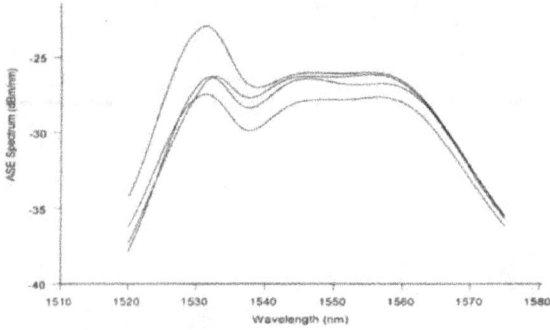

b)

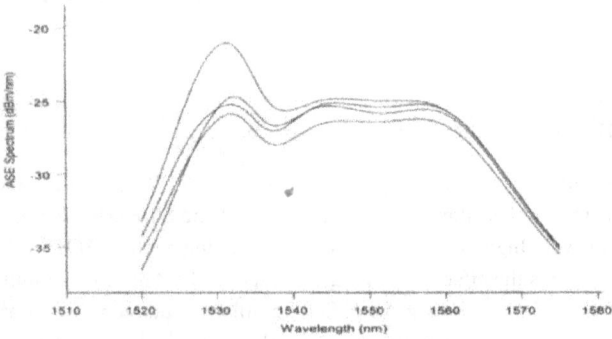

c)

Figure 13. The effect of moving the tuneable acoustic absorber on the Amplified Stimulated Emission spectrum at various gain levels: a) low gain single peak at 1560 nm b) and c) high gain two peaks at 1532 nm and 1550 nm.

Fig. 14 shows the flattened gain of ASE spectrum at various gain levels using this technique. For the lowest gain, at -30 dBm, which is achieved with a pump power of 96 mA, a broad filter is needed at 1545 nm; to obtain this; the tuneable acoustic absorber is positioned 14 cm after the AOTF device. The required resonant frequency to produce the coupling will be 990 kHz. A deeper notch is needed at 1532 nm; which is produced through the second frequency generator that is set at 993 kHz. The flattened gain is less than 0.8 dB. For gain at -25 dBm, which is achieved with a pump power of 150 mA, a filter is needed at 1556 nm; and a narrow deep notch is needed at 1532 nm; the required resonant frequency to produce the coupling respectively will be 986 kHz and 993 kHz. To obtain this, the tuneable acoustic absorber is positioned 16 cm after the AOTF device. The flattened gain is less than 0.9 dB.

Similarly for gain at -22 dBm, which is achieved with a pump power of 220 mA, a very deep filter is needed at 1532 nm and a small filter at 1545 nm. The resonant frequencies corresponding to these wavelengths are 990 kHz and 993 kHz respectively. To obtain the narrow filter, the tuneable acoustic absorber was set 17 cm after the AOTF device. And the measured flattened gain is less than 0.9 dB. Fig. 5 represents the notch filters obtained to flatten the gain of the Amplified Stimulated Emission (ASE) spectrum.

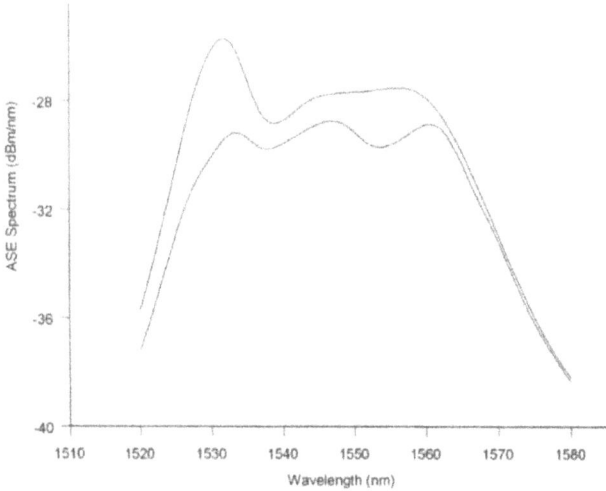

a)

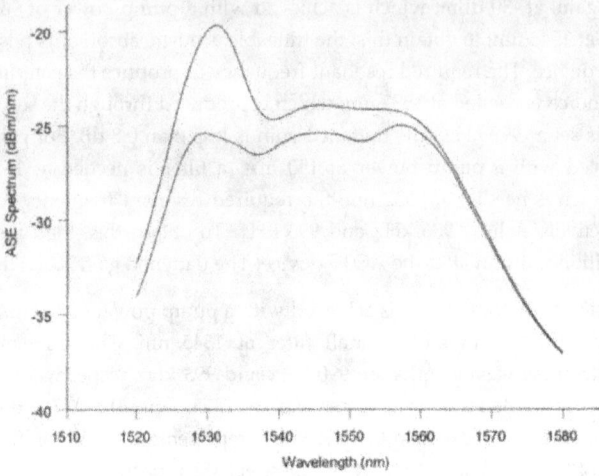

b)

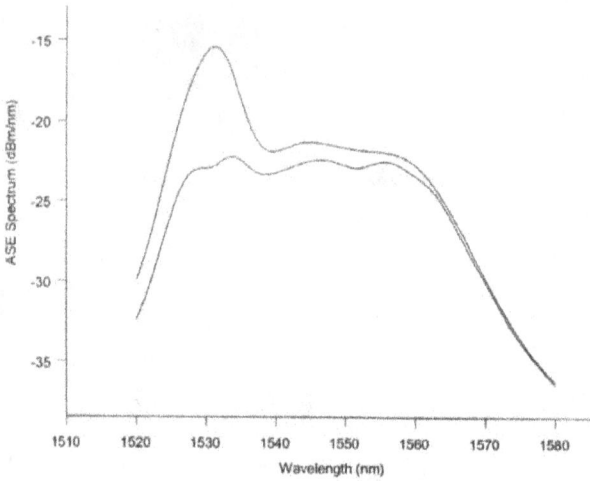

c)

Figure 14. Gain profiles of the ASE spectrum and the flattened gain at various pump powers: a) 96 mA which has a gain of -30 dBm b) 150 mA which has a gain of -25 dBm and c) 220 mA which has a gain of -22 dBm.

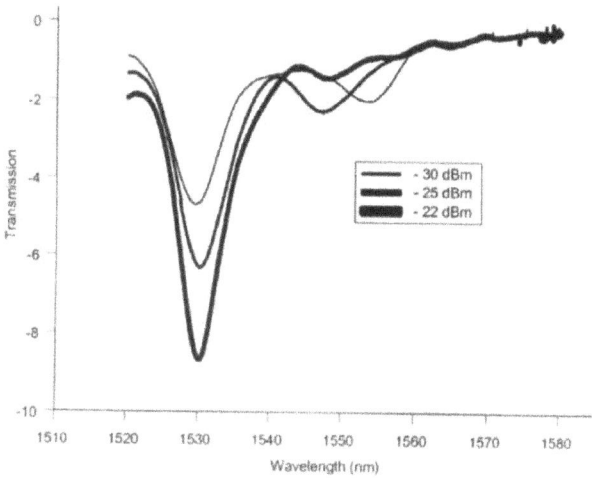

Figure 15. Corresponding filter spectrum to the flattened gain of various gain levels in Fig. 16.

7. Conclusion

The presence of acoustics inside the fiber will create a sequence of bends periodic in nature along the direction of its propagation. Core mode's energy is transferred to a cladding mode's, when it passes through the sequence of bends. The fiber jacket absorbs the coupled energy and this produces a notch filter observed using an optical spectrum analyzer.

Acoustic horn functions to transfer the acoustic wave of the transducer to the fiber. Aluminium horn is preferred over silica horn because it can be easily reproduced. Furthermore, its acoustic impedance almost matches that of silica's. To allow optimum transmission of acoustics to the fiber, the tip of the horn is made small, with its diameter matching that of silica's.

No resonance peaks were observed when the fiber is unetched First peaks are observed when the thickness of fiber is approximately 40 μ m. Overlap integral between the modes is not high in thicker fiber, meaning the transfer of acoustic wave to the fiber is not optimized. Thickness reductions in fibers are observed using a CCD camera. The characteristics of the resonance peaks can be controlled electrically using a RF generator. Voltage of the generator can be used to tune the attenuation depth of the resonance peaks. Frequency of the generator can be used to tune the peak wavelength. Frequency is inversely related to period of bends, thus higher frequencies will shift the peak to lower wavelengths. The 3-dB bandwidth of the resonance peaks can be adjusted by limiting the acoustic bend produced inside the fiber. Introducing a tunable acoustic absorber along the fiber can do this. Frequency used in all experiments was from 800 kHz to 1.1 MHz. All the coupled energy to produce the resonance peaks were to LP11 modes, mode conversion observed using a beam pro filer.

The power fed to cause resonance peaks can be reduced by reducing the thickness of the fiber to a value close to 20 μ m. Allowing light to pass through the acoustic bend region twice, as proposed in the double pass configuration, can increase the attenuation peaks. However, a frequency shift of 13 nm is observed because the light is passing through the bend in opposite directions.

As a spectral shaping tool, the attenuator is efficient as a gain flattening filter for an erbium doped amplifier. The peak of an amplified spontaneous emission at 1531 nm can be reduced to flat levels for various gains of the EDF A pump power. Insertion loss is less than 0.2 dB and polarization dependence loss is less than 0.4 dB.

Author details

Abhilash Mandloi and Vivekanand Mishra
Department of Electronics Engineering, S.V. National Institute of Technology, Surat 395007, Gujarat, India

8. References

[1] Sutharsanan Veeriah, "Design and Characterisation of All Fiber Optical Filters", Master of Science Thesis, Faculty of Engineering, Multimedia University, Malaysia, Feb 2006.

[2] Abdulhalim 1., Pannell C.N., (1993). Acoustooptic in-fiber modulator using acoustic focusing. IEEE Photonics Technology Letters, 9 (5), 999-1002

[3] Au A. A., Liu Q., Lin C.H., Lee H.P., (2004). Effects of Acoustic Reflection on the Performance of a Cladding-Etched All-Fiber Acoustooptic Variable Optical Attenuator. IEEE Photonics Technology Letters, 16 (1), 150-152

[4] Birks T.A., Russell P.SJ., Culverhouse D.O., (1992). The Acousto-Optic Effect in Single-Mode Fiber Tapers and Couplers. Journal of Lightwave Technology, 14 (11), 2519-2529

[5] Culverhouse D.O., Yun S.H., Richardson D.J., Birks T.A, Farwell S.G., Russell P.StJ., (1997). Low-loss all- fiber acousto-optic tunable filter. Optics Letters, 22 (2), 96-98

[6] Engan H.E., (2000). Acousto-Optic Coupling In Optical Fibers. IEEE Ultrasonics Symposium, 625-629

[7] Engan H.E., Kim B.Y., Blake J.N., Shaw H.J., (1988). Propagation and Optical Interaction of Guided Acoustic Waves in Two-Mode Optical Fibers. Journal of Lightwave Technology, 6 (3), 428-436

[8] Engan H.E., Ostling D., Kval Per 0., Askautrud Jan 0 .. Wideband Operation of Horns for excitation of Acoustic Modes in Optical Fibers. 10th Optical Fibre Sensors Conference, 568-571

[9] Feced R., Alegria c., Zervas M.N., (1999). Acoustooptic Attenuation Filters Based on Tapered Optical Fibers. IEEE Journal of Selected Topics in Quantum Electronics, 5 (5), 1278-1288

[10] Jung Y., Lee S.B., Lee J.W., Oh K., (2005). Bandwidth control in a hybrid fiber acoustooptic filter. Optics Letters, 30 (1), 84-86

[11] Keiser G., (1991). Optical Fiber Communications. McGraw-Hill, Inc., (2nd Edition) Kim RY, Blake J.N., Engan H.E., Shaw H.J., (1986). All-fiber acousto-optic frequency shifter. Optics Letters, 11 (6),389-391

[12] Kim H.S., Yun S.H., Kim H.K., Park N., Kim B.Y., (1998). Actively Gain-Flattened Erbium-Doped Fiber Amplifier Over 35 nrn by Using All-Fiber Acoustooptic Tunable Filters. IEEE Photonics Technology Letters, 10 (6), 790-792

[13] Kim H.S., Yun S.H., Kwang LK., Kim B.Y., (1997). All-fiber acousto-optic tunable notch filter with electronically controllable spectral profile. Optics Letters, 22 (19), 1476-1478

[14] Lee S.S., Kim H.S., Hwang LK., Yun S.H., (2003). Highly-efficient broadband acoustic transducer for all-fibre acousto-optic devices. Electronics Letters, 39 (18)

[15] Li Q., Au A.A., Lin C.H., Lyons E.R, Lee H.P., (2002). An Efficient All-Fiber Variable Optical Attenuator via Acoustooptic Mode Coupling. IEEE Photonics Technology Letters, 14 (11),1563-1565

[16] Li Q., Liu X., Lee H.P., (2002), Demonstration of Narrow-Band Acoustooptic Tunable Filters on Dispersion-Enhanced Single Mode Fibers. IEEE Photonics Technology Letters, 14 (11), 1551-1553

[17] Li Q., Liu X., Peng J., Zhou B., Lyons E.R, Lee H.P., (2002). Highly Efficient Acoustooptic Tunable Filter Based on Cladding Etched Single-Mode Fiber. IEEE Photonics Technology Letters, 14 (3), 337-339

[18] Liu Q., Chiang K.S., Rastogi V., (2003). Analysis of Corrugated Long-Period Gratings in Slab Waveguides and Their Polarization Dependence. Journal of Lightwave Technology, 21 (12),3399-3405

[19] Love J.D., Henry W.M., Stewart W.J., Black RJ., Lacroix S., Gonthier F., (1991). Tapered Single-mode fibres and devices Part 1: Adiabaticity criteria. IEE Proceedings, 138, (5), 343-354

[20] Monerie M., (1982). Propagation in Doubly Clad Single-Mode Fibers. IEEE Transactions on Microwave Theory and Techniques, MTT-30 (4), 381-388

[21] Mononobe S., Ohtsu M., (1996). Fabrication of a Pencil-Shaped Fiber Probe for Near-Field Optics by Selective Chemical Etching. Journal of Lightwave Technology, 14 (10), 2231-2235

[22] Ostling D., Engan H.E., (1995). Narrow-band acousto-optic tunable filtering in a twomode fiber. Optics Letters, 20 (11), 1247-1249

[23] Ostling D., Engan H.E., (1995). Spectral Flattening by an All-Fib er Acousto-Optic Tunable Filter. IEEE Ultrasonics Symposium, 837-840

[24] Pannell C.N., Wacogne B.F., Abdulhalim 1., (1995). In-Fib er and Fiber Compatible Acoustooptic Components. Journal of Lightwave Technology, 13 (7),1429-1434

[25] Satorius D.A., Dimmick T.E., Burdge G.L, (2002). Double-Pass Acoustooptic Tunable Bandpass Filter With Zero Frequency Shift and Reduced Polarization Sensitivity. IEEE Photonics Technology Letters, 14 (9), 1324-1326

[26] Yun S.H, Lee B.W., Kim H.K., Kim B.Y., (1999). Dynamic Erbium-Doped Fiber Amplifier Based on Active Gain Flattening with Fiber Acoustooptic Tunable Filters. IEEE Photonics Technology Letters, 11 (10), 1229-1231

[27] Yun S.H., Kim B.Y., Jeong HJ., Kim B.Y., (1996). Suppression of polarization dependence in a two-mode-fiber acousto-optic device. Optics Letters, 21 (12), 908-910

[28] Yun S.H., Kim H.S., (2004). Resonance in Fiber-Based Acoustooptic Devices Via Acoustic Radiation to Air. IEEE Photonics Technology Letters, 16 (1), 147-149

Technological Systems

Chapter 6

Seismic Vibration Sensor with Acoustic Surface Wave

Jerzy Filipiak and Grzegorz Steczko

Additional information is available at the end of the chapter

1. Introduction

Mechanical vibrations are a movement of particles around the state of equilibrium in a solid environment. Vibrations are a common phenomenon in our daily life. These vibrations are often parasite effects threatening our existence. Vibrations of the ground, machines, or a number of technical devices present a process, which require a continuous or a long-term monitoring. In many sectors vibrations are a working factor in a production process.

Mechanical vibrations serve as a source of information in medicine, diagnostics of the structure of many machines and in perimeter protection (monitoring). The knowledge of vibration parameters allows evaluating the technical condition of machines, the quality of their design and manufacture and their reliability. Early detection of ground vibrations serves to predict and warn of earthquakes. Ground vibrations serve to monitor explosions and are used in the reflexive seismology (prospecting for mineral deposits). Detection of ground vibrations in systems of perimeter protection allows detecting an intrusion into an area under surveillance. Mechanical vibrations are characteristic for their differing frequencies and amplitude. The frequency of mechanical vibrations ranges usually from a hundredth of Hz to a dozen or so kHz. Parameters of mechanical vibrations are measured with vibration sensors.

At present practically three types of seismic sensors are used:

- geophones,
- piezoelectric acceleration sensors,
- micro-mechanical silicone acceleration sensors.

Geophones belong to the simplest and most inexpensive vibration sensors. They feature a low mechanical resonance frequency, which ranges usually between 4Hz and 14Hz. They are used in mining, for safety perimeter protection [1] and in the reflexive seismology (Figure 1).

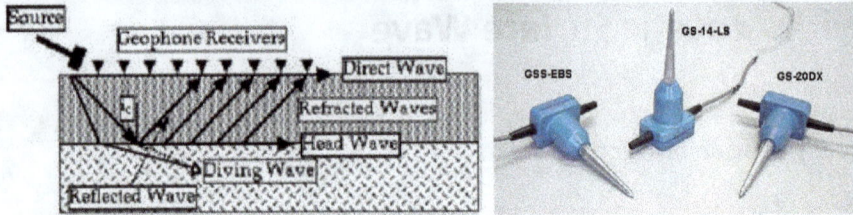

Figure 1. Reflexive seismology and geophones [1]

Micro mechanical silicone acceleration sensors (MEMS) [2-3] are mass-produced and used in many fields (e.g. deploying air bags, laptops). They are characterised by small dimensions. Due to their miniature dimensions their mechanical resonance frequency is very high, usually a dozen or so kHz or a few dozen kHz. With piezoelectric acceleration sensors [4] one can measure variable accelerations. Their mechanical resonance frequency is higher than that of MEMS.

1.   2.   3.

PBB-200S   M2166-VBB   TSA-100S

Figure 2. Seismometers offered by MEROZET 1 - portable broadband seismometer, 2 - very broadband seismometer 3 - triaxal seismic accelerometer.

Figure 2 shows seismometers offered by MEROZET: a portable one, a wide-band and a triaxial one; all these 3 models are based on and include piezoelectric acceleration sensors. Vibration sensors can be built using sensors with the acoustic surface wave (SAW).

These SAW-based sensors are used to measure a number of physical quantities: gas concentration [5-8], temperature, pressure [9-13], and mechanical quantities: torque of a rotating shaft [14], stress [15-17], acceleration [18-19] and vibrations [20-21]. All these SAW–based sensors work on the basis of measuring changes in the delay of a surface wave due to the impact of a physical quantity being measured on its speed and the propagation path. However, depending on the kind of a measured physical quantity, a number of problems occur, which are characteristic for the group of sensors used for measuring that quantity.

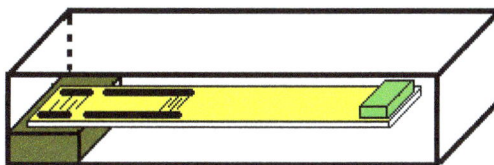

Figure 3. Basic structure of a SAW vibration sensor

Figure 3 presents the idea of the structure of an SAW-based sensor. The main element of this sensor is an anisotropic plate of a piesoelectric material. One end of the piesoelectric plate is made fast to the sensor housing, while on the other free end a seismic mass can be placed. An SAW-delaying line in the form of a four-terminal-network is made on the top surface of the sensor. The movement of the sensor housing causes its plate to vibrate and the SAW-based delaying line delay to change. This is why the phase of a high frequency signal passing thru such a line changes. The magnitude of a signal pahse change will be proportional to the change in the delay of an SAW-based delaying line.

The presented sensor design presents three different issues, which must be solved:

- modeling of the sensor mechanical system, which amounts to the description of plate deformations and stresses occurring in it.
- modeling of the sensor mechanical-electrical converter – delaying line with the SAW, which amounts to the description of the change in parameters of a delaying line with the SAW (first of all of the delay) due to distortions and stresses in the plate
- modeling of the electric circuit cooperating with the sensor, which amounts to the analysis and synthesis of an electric circuit measuring the changes in the delay of the surface wave in a delaying line including the SAW.

The work presents a solution of the mentioned problems, which were further analysed. The work presents executed models of SAW-comprising sensors and the results of a study of their parameters. The use of realised sensors in a system of perimeter protection is described. The structure of an SAW-comprising sensor (Fig. 3) is a combination of a continous system in the form of a piesoelectric, anisotropic support plate and a discrete system in the form of a concentrated mass. In theory such a system can feature an infinite number of free vibration frequencies. Writing a description of the mechanics of the plate of an SAW-based vibration sensor is a complicated proces; what makes it dificult is the tensor description of the plate mechanical properties. The knowledge of the value of an attenuation tensor (viscosity tensor) poses a problem. Therefore an analysis was conducted, which allows to simplify the sensor model presented in Figure 3 and next as a result of this analysis the movement of a piesoelectric, anisotropic plate with a concentrated mass was described with the aid of a discrete system of one degree of freedom. Elastic and viscous properties of the plate material were taken into account. This model was introduced by way of an isotropic descritpion of anisotropic material properties. The model accuracy was evaluated. Explicit realtionships between sensor plate movement parameters and its geometey and parameters describing its elastic an viscous properties were determined, thus a simple analysis and synthesis of the sensor plate movements were possible. The main feature of the sensor mechanical system (a continous system combined with a discrete one) is the occurence of pratically only one resonance frequency. A simple description of this magnitude by design parameters of the system and elastic and viscous plate parameters allow a simple modelling, how these sensors function, and as well to determine elastic and viscous parameters of a plate empirically. These parameters in the form : an equivalent Young's modulus and an equivalent material damping coefficient for a selected direction of a piezoelctric substrate (i.e. the direction of the surface wave propagation) were determined in works [22][23]. In available bibliography they are not knowen or utilised. The above considerations are presented in the work [21]. For a full description of the SAW-comprising vibration sensor designing process in Section 2 we present a modelling process of its mechanical system.

2. Model of mechanical unit for SAW vibration sensor

The object of consideration has been presented in Figure 4. One end of the plate is stiffly attached, and the other is free and without any concentrated mass. The piezoelectric properties of sensor plate will be omitted in the analysis.

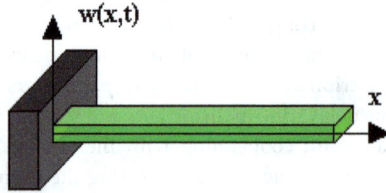

Figure 4. The plate of a vibration sensor.

The equation of a movement of an anisotropic body with the mass density ϱ is:

$$\frac{\partial \sigma_{ij}}{\partial x_j} + \rho \frac{\partial^2 u_i}{\partial t^2} = 0 \tag{1}$$

The stress tensor σ_{ij} depends on the strain tensor ε_{kl} through Hook-Voight equation:

$$\sigma_{ij} = C_{ijkl}\varepsilon_{kl} + \Sigma_{ijkl}\varepsilon_{kl,t} \tag{2}$$

where: C_{ijkl}—is a elasticity tensor, Σ_{ijkl}—is a material damping tensor, u_i—is a displacement vector,

$$\varepsilon_{kl} = \frac{1}{2}\left(\frac{\partial u_k}{\partial x_l} + \frac{\partial u_l}{\partial x_k}\right) \quad \text{– is a strain tensor} \tag{3}$$

The mathematical description of this issue will be closed if the initial and boundary conditions are added to the aforementioned equations. It is complicated to solve the problem. The causes of the complications are huge number of non-vanishing modules of an elasticity and material damping tensor. The material damping tensor Σ_{ijkl} has an identical symmetry as the elasticity tensor C_{ijkl}. For materials used in SAW devices (quartz, lithium niobate), the value of damping constants are difficult to experimental verification. For higher class of symmetry of an anisotropic materials, equations (2, 3) are simple. For the isotropic material the elasticity tensor has only two independent components $C_{1111}=\lambda+2\mu$, $C_{1122}=\lambda$, $C_{2323}=\mu$. Therefore the elastic properties of an isotropic substance describe two quantities (λ, μ). They are often shown in form of a Young's modulus E and a Poisson ratio v. The following relations occur between quantities E, v, and elasticity tensor components [24]:

$$E = c_{2323}\left(2 + \frac{c_{1122}}{c_{1122} + c_{2323}}\right), \quad v = \frac{c_{1122}}{2\left(c_{1122}+c_{2323}\right)} \tag{4}$$

The description of viscous properties of an isotropic body done by material damping tensor is analogical. It is usually described by two quantities [24]:

$$\tau = \frac{\Sigma_{2323}}{E}\left(2 + \frac{\Sigma_{1122}}{\Sigma_{1122} + \Sigma_{2323}}\right), \quad \eta = \frac{\Sigma_{1122}}{2\left(\Sigma_{1122} + \Sigma_{2323}\right)} \tag{5}$$

The Young's modulus E is described as a proportion of a longitudinal stress to longitudinal strain for the direction of the functioning of a stress. To describe the mechanical properties of anisotropic materials taking into account a particular direction of the stress an effective Young's modulus may be used E [24][25](Its magnitude is described by overt dependence). An exemplary expression for a inverse effective Young's modulus for a trigonal unit (lithium niobate, quartz) is:

$$1 / E = (1 - l_3^2)^2 s_{11} + l_3^4 s_{33} + l_3^2(1 - l_3^2)(2s_{13} + s_{44}) + 2l_2 l_3(3l_1^2 - l_2^2)s_{14} \tag{6}$$

where: s_{ij}—is an element of an compliance matrix, l_j—is a cosine of an angle between the chosen direction and the axis − j, in Cartesian coordinates. The compliance matrix s_{ij} is reverse to the stiffness elasticity matrix c_{ij}.

It is possible to calculate the values of material damping coefficients in a chosen crystallographic direction, too. The presented approach allows to model the anisotropic material by the isotropic model. In such model the stresses are the sum of elastic and dissipative components:

$$\sigma = E\varepsilon + E\tau\frac{\partial\varepsilon}{\partial t} \tag{7}$$

We consider equivalent isotropic model of cylindrically bent plate [19]. Equation of free vibrations has the form:

$$\rho \frac{\partial^2 w(x,t)}{\partial t^2} + E_e \frac{h^2}{12}\left(1+\tau\frac{\partial}{\partial t}\right)\frac{\partial^4 w(x,t)}{\partial x^4} = 0 \tag{8}$$

where: ρ—mass density, h—plate thickness, L—plate length, τ—equivalent material damping coefficient, $E_e = \dfrac{E}{1-\nu^2}$

E_e is an equivalent Young's modulus.

At the boundaries we have:

$$w(0,t) = 0, \quad \frac{\partial w(0,t)}{\partial x} = 0, \quad \frac{\partial^2 w(L,t)}{\partial x^2} = 0, \quad \frac{\partial^3 w(L,t)}{\partial x^3} = 0, \tag{9}$$

The solution to the boundary problem (8), (9) has the form:

$$w(x,t) = \sum_{n=1}^{\infty} W_n(x)\cdot A_n\, e^{-\frac{\omega_n^2}{2}t}\cdot\sin(\bar{\omega}_n t + \varphi_n) \tag{10}$$

where: constants A_n and φ_n are determined by initial conditions.

The angular frequency of non-damped vibrations is equal to:

$$\omega_n = k_n^2 \frac{h}{l^2}\sqrt{\frac{E_e}{12\rho}}\,. \tag{11}$$

The angular frequency of damped vibrations is equal to:

$$\bar{\omega}_m = \omega_m\sqrt{1 - \frac{\omega_m^2\,\tau^2}{4}} \tag{12}$$

$$\text{where}: k_1 = 1,875, \ k_2 = 4,694, \ k_3 = 7,855 \tag{13}$$

The orthonormal set of function W_n (eigenfunctions) is taken from [26]. Only some elements in the sum (10) represent vibrations. For $N < n$ where N is the greatest natural number for chich $\omega_n < 2/\tau$, the element of sum represents very strongly dumped movement and there is no resonance at this frequency. Each of the harmonics n=1,2,3... has part of energy. How great is the part depends on the unit (normal vibrations, ω_n, $w(x)$) and depends on activation. In the paper [18][19][27][28] a simplified model with one degree of freedom was presented and it is shown in Figure 5. It has been used to describe the dynamics of sensor plate movement. It was derived according to the Rayleigh method. This method is based on a simplified modeling of a plate with the use of an equivalent circuit with one degree of freedom which is energetically equivalent. The free end of sensor plate has been taken as a point of reduction. The equivalent circuit present Figure 5

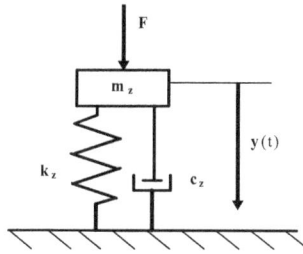

Figure 5. The equivalent circuit of sensor plate with one degree of freedom.

Parameters in the model are as follows [18]:

$$m_z = 0.25196\,\rho b h l, \quad k_z = 3.1169\frac{E_e b\,h^3}{l^3}, \quad c_z = \tau\,3.1169\frac{E_e b h^3}{l^3} \tag{14}$$

where: b — plate width

The model with one degree of freedom has only one resonance frequency. The equation of mass movement is as follows:

$$\frac{d^2 y(t)}{dt^2} + \omega_0^2\tau\frac{dy(t)}{dt} + \omega_0^2 y(l) - \frac{F(t)}{m_z}. \tag{15}$$

The solution of an equation for natural vibrations:

$$y(t) = A e^{-\frac{\omega_0^2\tau}{2}t}\sin(\omega_r t + \phi), \tag{16}$$

where:

$$\omega_0 = 3.5172\left(\frac{h}{l^2}\right)\sqrt{\frac{E_e}{12\rho}}, \tag{17}$$

$$\omega_r = \omega_0\sqrt{1 - \frac{\omega_0^2\tau^2}{4}} \tag{18}$$

It is analogical to the Equation (10) obtained with the use of an isotropic model of sensor plate. Comparison of the first frequency of damped vibrations of the plate obtained in an isotropic model (11) and the frequency of damped vibrations obtained with the use of a model with one degree of freedom (17) are in fulfils relation:

$$\omega_1 = 0.9996\omega_r, \tag{19}$$

First frequency of damped vibrations calculated in an isotropic model is 0.5 per cent lower than frequency calculated with the use of a discrete model. This difference could be smaller in case of a sensor construction with the concentrated mass attached to the movable end of the plate. That is why the model with one degree of freedom may be used to describe the movement of sensor plate. It allows relatively easy simulation of vibrations of the plate with the mass attached to its movable end. Free vibrations of sensor plate are a definite as a sum of damped harmonic frequency vibrations. But, in free vibration damped vibrations with first harmonic frequency will dominate. The amplitudes of the superior harmonic vibrations will be extremely small. As it is shown in [18] their quantity is 40 dB smaller than the first harmonic amplitude. This is the reason why a model with one degree of freedom [18][27][28] has been used to analyze the movement of the plate with concentrated mass. Vibrations of the plate have been activated by the movement of the sensor casing $Y(t)$. The equation of movement is as following:

$$\frac{d^2 y(t)}{dt^2} + \omega_0^2 \tau \frac{dy(t)}{dt} + \omega_0^2 y(t) = \omega_0^2 \tau \frac{dY(t)}{dt} + \omega_0^2 Y(t).$$
(20)

where:

$$\omega_0 = 3.5172 \left(\frac{h}{l^2} \right) \sqrt{\frac{E_e}{12 \rho}} \sqrt{\frac{1}{1 + r\, 3.9689}}$$
(21)

r is a ratio between seismic mass and mass of sensor plate.

The solution of the Equation (20) is a function:

$$y(t) = A \exp\left[-\frac{\omega_0^2 \tau t}{2} \right] \sin\left[\omega_r (t + \varphi) \right] -$$
$$- \frac{4\omega}{4 - \omega_0^2 \tau^2} \int_0^t \left(\tau \frac{dY(\xi)}{dt} + Y(\xi) \right) \cdot \exp\left[-\frac{\omega_0^2 \tau}{2} (t - \xi) \right] \cdot \sin\left[\omega_r (t - \xi) \right] d\xi$$
(22)

where: constants A and φ are determined by initial conditions.

Relations between ω_0 and ω_r are as in identity (18). In both components of the solution (22) appears the following function:

$$\varphi_\delta(t) = A e^{-\frac{\omega_0^2 \tau}{2} t} \sin(\omega_r t + \phi)$$
(23)

It is a product of a harmonic and damping (exponentially decay with time) function. The frequency of a harmonic function is the resonance frequency of the unit. This function describes sensor impulse response and its natural vibrations. It is a sum of:

- convolution of an impulse response of the plate and the component of describing movement of the sensor casing,
- damped vibrations with the resonance frequency of a sensor plate.

It will always have a factor in form of a harmonic function with the frequency equal to the resonance frequency of sensor plate and with variable amplitude. That is why the frequency response of sensor plate may be quantity identifying the sensor. Frequency response of the sensor plate is the ratio of the amplitude of the deflection plate sensor to the harmonic amplitude of its case The frequency response of the sensor plate calculated from the Equation (20) is as follow:

$$H(\omega) = \sqrt{\frac{1+(\omega\tau)^2}{[1-(\frac{\omega}{\omega_r})^2]^2 + (\omega\tau)^2}} \tag{24}$$

Parameters in relation (24) depends on mechanical properties of the sensor plate material. The quantities of elastic and viscous parameters for quartz are shown in Table 1 [22][23].

	ST-cut quartz
Equivalent Young's modulus [GPa]	76
Dynamic critical compressive stress [MPa]	80
Equivalent material damping coefficient [μs]	29.3
Density [kg/m³]	2650

Table 1. Material parameters of quartz.

Theoretical frequency response for plates made of ST-cut quartz with the resonance frequencies of 22 Hz and 100 Hz are shown in Figure 6. The most important is that for low frequency the frequency response has narrower band and higher magnitude so the selectivity of the sensor is high. It decreases with increased resonance frequency.

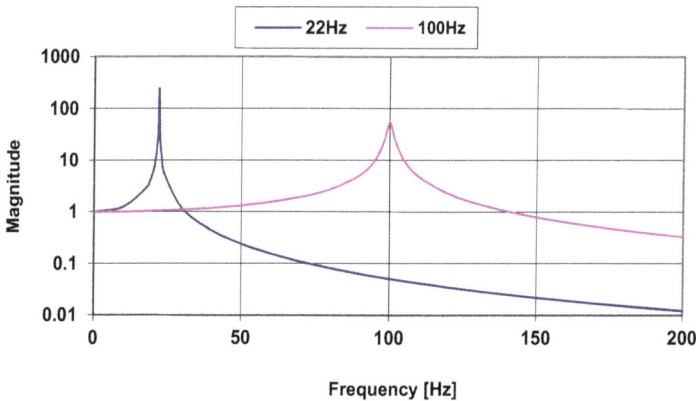

Figure 6. Theoretical resonance characteristics of plates with resonance frequencies of 22 Hz and 100 Hz.

The maximum value of the frequency response of the plate will occur for $\omega = \omega_r$. It has been described with the relationship:

$$H(\omega_r) = \sqrt{\frac{1+(\omega_r \tau)^2}{(0,5\omega_0 \tau)^4 + (\omega_r \tau)^2}} \tag{25}$$

Its value exceeds repeatedly the value of static deflection. (e.g., for the resonance frequency of 22 Hz it is 246 times higher).

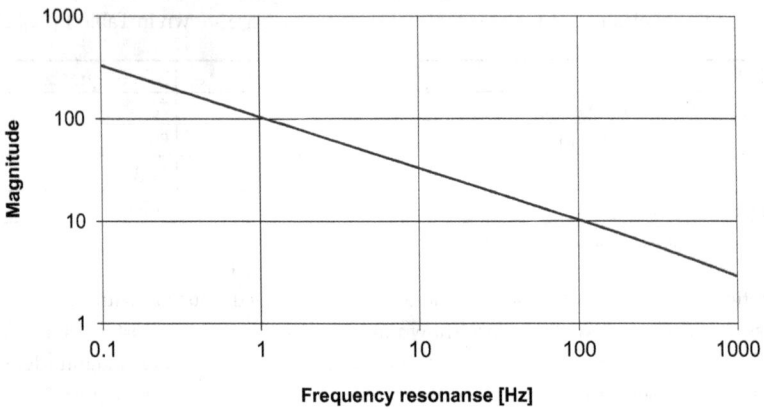

Figure 7. Maximum magnitude of frequency response versus the resonance frequency of the sensor plate.

The change in maximum magnitude of frequency response as a function of resonance frequency of the plate is shown in Figure 7 For the resonance frequency of a plate of 10Hz the vibrations amplitude multiplication is 1,600 higher than static deflection. This property may be used to construct sensors with high sensitivity level. But it is necessary to answer one question beforehand: what is the lowest possible resonance frequency of a plate that we can manufacture? The answer is accessible on the basis of the described model and the length of available plates. The resonance frequency of sensor plate is described by the relation (21). It depends on plate length (l) and on quantity of a concentrated mass (r) attached to the free end of sensor plate. The increase of the concentrated mass lowers resonance frequency of the plate, simultaneously increasing stresses of the plate.

The influence of a change of concentrated mass on resonance frequency of the sensor plate is shown in Figure 8.

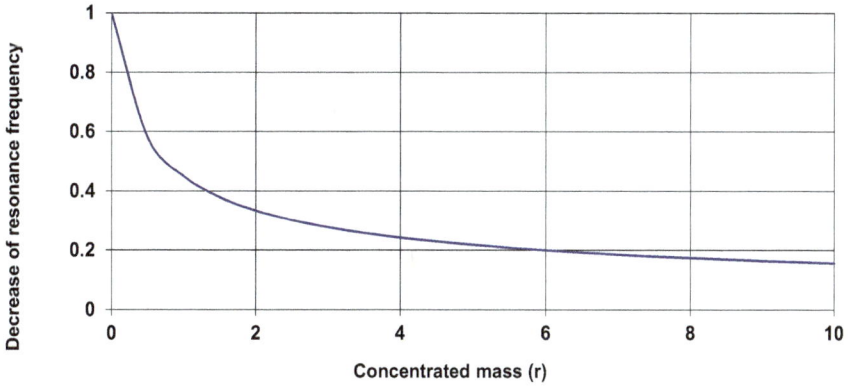

Figure 8. The influence of a change of a concentrated mass on sensor plate resonance frequency.

It is visible that the use of concentrated mass quantities exceeding two times the mass of the plate enables and triple decrease of resonance frequency of a plate. It is the most effective place to decrease the resonance frequency of a plate. Continuous increase of a concentrated mass does not substantially decrease the resonance frequency of sensor plate. The further analysis of sensor parameters will be limited to such range of concentrated mass quantities. The relationship between the value of resonance frequency of a plate made of ST-cut quartz and length of the plate determined by three different concentrated mass values is shown in Figure 9.

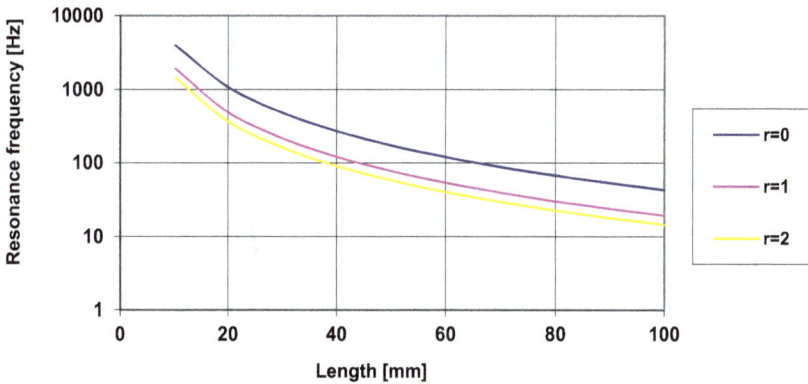

Figure 9. The relationship between sensor plate resonance frequency made of ST-cut quartz 0.5 mm thick, and plate length (l) and the quantity of concentrated mass (r).

From the figure presented above may conclude that it is relatively easy to create plates of resonance frequency form 20 Hz do 4 kHz. For the 0.5 mm thick plates it is necessary to use the concentrated mass up to 1.5 g. The relation between the concentrated mass and the plate length is shown in Figure 10. The sensor impulse response presented by the relation (23) has a damped character. Its fast fading can impose an upper limit of the resonance frequency. The damping value depends on the geometry of the plate and the equivalent damping coefficient. In order to simplify the illustration the impulse response damping measure has been introduced as a relative decrease of its quantity after one period.

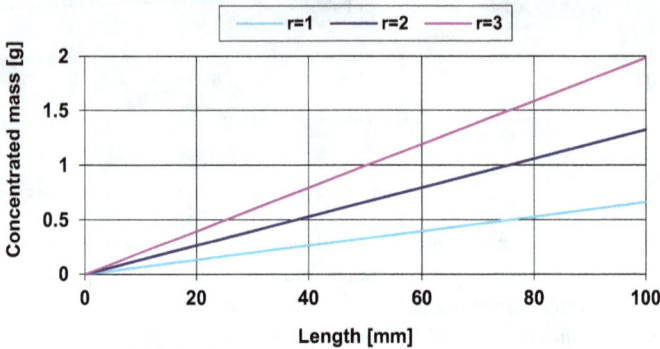

Figure 10. The concentrated mass quantities used in the considered sensor constructions.

The relation of impulse response damping in form of a function of length of ST-cut quartz plate for three different concentrated mass values is presented in Figure 11. For plates longer than 40 mm loaded by the concentrated mass equal to the mass of the plate (r = 1) the

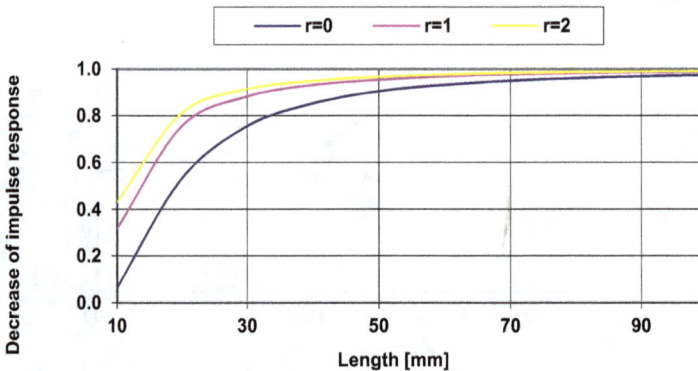

Figure 11. Relative decrease of impulse response amplitude after the time equal to its period in form of a function of plate length for different concentrated mass quantities (r).

damping of free vibrations of the plate is relatively slow. The impulse response of shorter plates is dampened relatively fast. This is why it seems to be beneficial to use possibly long plates loaded by concentrated mass equal to plate mass. The value of resonance frequencies of plates possible to manufacture has changed. It seems that the range of resonance frequency plates available to use is limited to the scope from 20 Hz to 250 Hz. The parameters of resonance frequencies of the plates in the aforementioned range are shown in Figure 12.

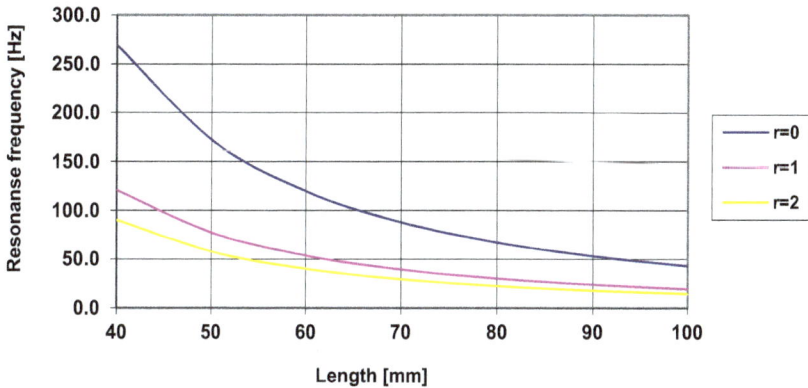

Figure 12. The relation between resonance frequency of a sensor plate and plate length and the value of concentrated mass.

From the above considerations one can draw a conclusion that working with SAW vibration sensors one can utilise their pulse responses or forced vibrations. The sensor resonance characteristics is a basic parameter of the first operation mode. On 0.5 mm thick quartz plates with a concentrated mass equalling the mass of a plate we can achieve plate own vibration frequencies from 20 Hz thru 250Hz. In the second operation mode the vibration sensors with SAW will operate like classical acceleration sensors. The plate resonance frequency should be above the measuring range of a sensor.

The knowledge of the resonance characteristic curve is required for both sensor operation modes. That characteristic can be easily calculated with the aid of the presented model. In the Section 4 we will present experimental examples of the operation of a seismic vibration sensor with SAW, which will enable to evaluate the precision of this model and its usefulness.

3. Vibration sensor electronic components

In order to ensure the transmission of test and supply signals through one coaxial cable there must be a system at the sensor input separating the test signal (74MHz) and supply signal 12VDC (separator). At the output there must be a system summing up the test signal with a constant supply voltage (adder). Test signal (high frequency) after going through SAW delay line must be amplified to input quantity. It will ensure loss compensation caused by SAW delay line. In the entire line of high frequency test signal (74MHz) a characteristic impedance of 50Ω should be retained. Input and output impedance must have the value of 50Ω. Figure 13 shows the basic functional elements of SAW vibration sensor. Depending on the function in the whole system the following components may be distinguished:

- system separating and summing up the test and supply signals;
- systems adjusting the impedance of SAW line to 50Ω;
- SAW delay line;
- amplifier compensating losses caused by SAW delay line.

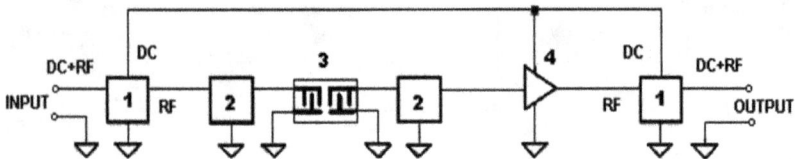

Figure 13. Block diagram of SAW vibration sensor.

The method of making the aforementioned components will be discussed in the next Section.

3.1. System separating or summing up electrical signals (separator/adder)

A system separating test and supply signals is placed at the sensor input. A system summing up these signals is placed at the sensor output. Figure 14 shows a system separating or summing up test and supply signals. The system is in a form of a circulator. It is connected to the line of high frequency signal with a characteristic impedance of 50Ω. A point of separation (or summation) of signals is the place where additional impedance is added to a line of characteristic impedance of 50Ω by inductance L_1. It may change the characteristic impedance of the line and be the reason of signal reflections. In order to avoid this the quantity of added impedance must be much larger than line characteristic impedance (50Ω). In order to fulfill this requirement inductance $L_1=4,7uH$ of its own parallel resonance frequency of 74MHz has been chosen. Figure 15 shows inductance equivalent system diagram.

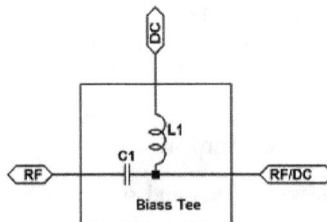

Figure 14. System separating or summing up test and supply signals.

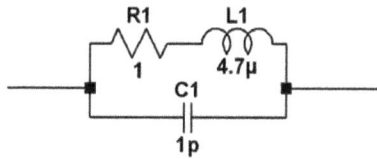

Figure 15. Inductance equivalent system diagram

In reality, the chosen inductance is a parallel resonant circuit. Figure 16 presents change in impedance of such a system in frequency function.

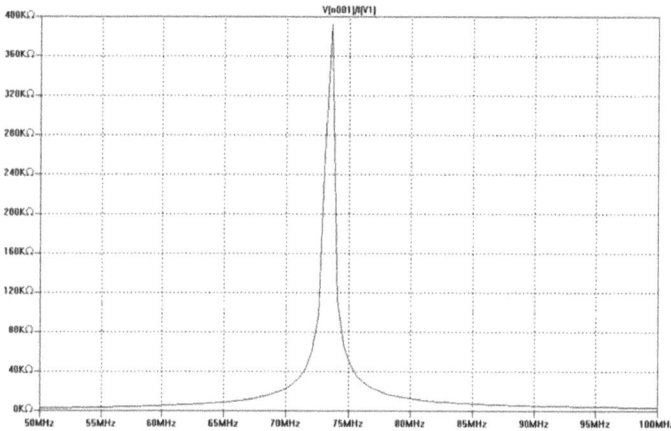

Figure 16. Relationship between impedance and frequency of a system presented in Figure 15

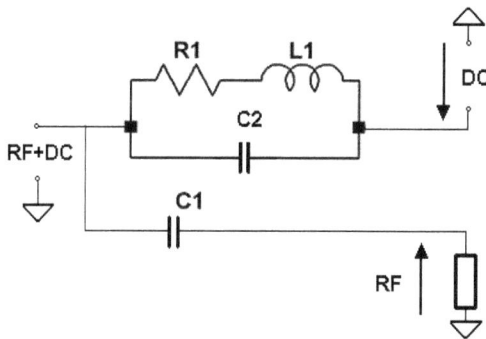

Figure 17. Actual system separating the test and supply signal.

For frequency equal to 74MHz the system impedance value amounts to 400kΩ. It is relatively high in comparison with characteristic impedance of the test signal transmission line (50Ω). It is then possible to obtain considerable attenuation of the test signal entering the supply circuit and it practically eliminates reflections at the point of signal separation or summation. Figure 17 presents an actual system separating the test and supply signals. Impedance of connected in series: C1 capacity and RF output impedance is equal to characteristic impedance of 50Ω. A diagram presented in Figure 17 enables to analyze test signal attenuation in supply circuit. It allows to calculate the change in line characteristic impedance made by the separating system.

Figure 18 shows attenuation of the test signal at DC output and change in line impedance in frequency function for R1=1Ω.

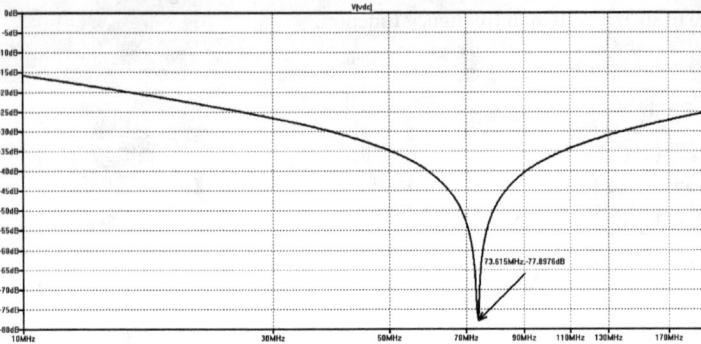

Figure 18. Attenuation of test signal at DC output vs. frequency

Calculations of transmission line impedance change have been done on the assumption that the test signal line has the impedance equaling 50Ω in the entire frequency range. This assumption is correct in the range of line frequency. It substantially simplifies modelling of the system. Figure 19 shows the change line characteristic impedance vs. frequency for R1 = 1Ω.

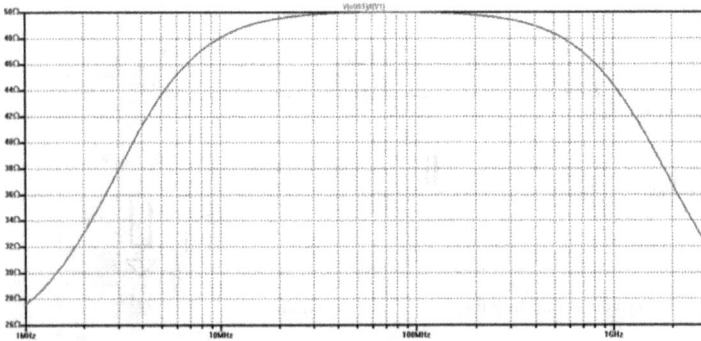

Figure 19. Change in test signal line characteristic impedance in frequency function.

This simplification does not influence the results in the system frequency range. For the sensor frequency equalling 74MHz the test signal attenuation at DC output equals –78dB, and line impedance equals 50Ω. A separating or summing up system prepared in such a manner does not influence the test signal transmission through SAW vibration sensor.

3.2. Delaying line and adapting system

The design of SAW-comprising delaying lines used for vibration sensors (Figure 20A) differ from those used for sensors of other physical quantities (Figure 20B). Since the sensor plate moves, electrodes applying electric signals to converters (bus bar) should be situated on the immobile part of the plate. (Thus a proper strength of electric contacts for these electrodes will be ensured.) Electrodes are long and their resistance is specific. As the plate moves its housing is greater than that of classic filters comprising the SAW. This causes the signal passing directly between the SAW-comprising delaying line inlet and outlet to increase.

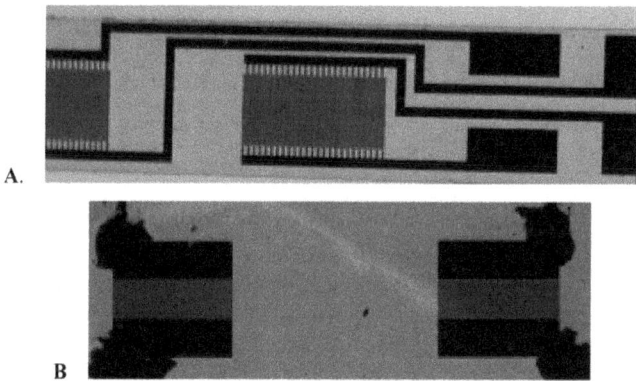

Figure 20. SAW delay lines: for vibration sensor (A), classic (B)

The line was designed in the form of two cooperating, identical, simple, periodical, double-electrode interdigital transducers. Figure 21 shows a system of converter electrodes. Such a structure of converters enables their operation on the third harmonic. The lines were designed to fabricate them with a ST-cut quartz.

Figure 21. Structure of delaying line converters

Due to a low value of the electro-mechanical coupling factor for a ST-cut quartz losses for a mismatch its inlet impedances to 50Ω are significant. In order to cut on these losses operation of interdigital transducers under conditions of matching to impedance of 50Ω at a frequency of 74MHz was selected [29-32]. Figure 22 shows a converter matching system. The element for matching a converter having the conductance of G_p and the capacity C_p to the impedance $R_g = 50Ω$ is the inductance L_1. The matching takes place on the condition that the available power derived from a voltage source E_g of the internal impedance $R_g = 50Ω$ is distributed released/emitted on the converter conductance G_p.

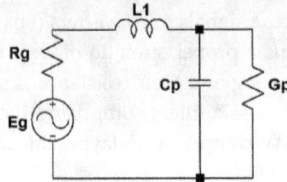

Figure 22. Interdigital transducer matching system to 50Ω impedance..

This condition is met for a conductance resulting from the relationship [29]:

$$L_1 = \frac{C_p}{G_p^2 + \omega^2 C_p^2} \tag{26}$$

With the help of the (26) relationship the converter geometry for a ST-cut quartz was determined. The aperture of the converter A= 2,5 mm was adopted. To fulfill the purpose converters operated on the third harmonic at 74MHz. For such parameters a converter consisting of N=25 pairs of electrodes was received. The electrode width and the gap between electrodes were 16 μm, and the surface length 37 μm. The results of theoretical calculations of conductance and converter capacity versus frequency are presented in Figure 23.

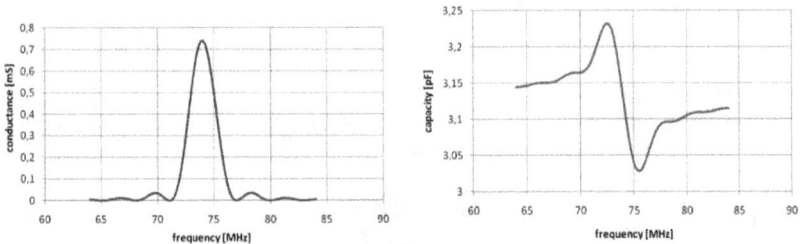

Figure 23. Theoretical dependence of conductance and capacity of simple transducer composed of 25 double electrode pairs on ST-cut quartz.

At 74MHz frequency the converter conductance is 0.74mS and the capacity 3.13pF. For these quantities the inductance L_1= 900nH was calculated, at which value the condition of matching the converter to the impedance 50 Ω is satisfied. Practically, the matching of interdigital transducers to the impedance of 50 Ω was carried out by measurement of the coefficient of reflection Figure 24 shows the change in the coefficient of reflection from a matched converter versus frequency in a system of an impedance of 50 Ω.

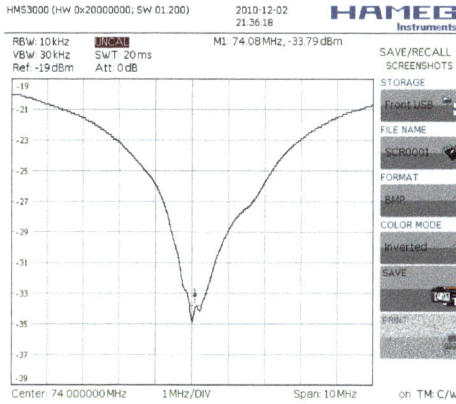

Figure 24. Coefficient of reflection from a matched converter versus frequency.

Figure 25 shows the mounting of the sensor plate.

Figure 25. Mounting of the sensor plate

Figure 26 presents the frequency amplitude curve for the fabricated SAW delaying line. The measurement was conducted after the line was mounted in an SAW-based vibration converter.

Figure 26. Attenuation frequency diagram of SAW vibration sensor.

3.3. Amplifier

The role of an amplifier is to compensate losses caused by SAW delay line. An amplifier has been built on a monolithic system MAR-6 manufactured by Mini-Circuits. Figure 27 shows a diagram of an amplifier being a part of an electronic system of SAW vibration sensor.

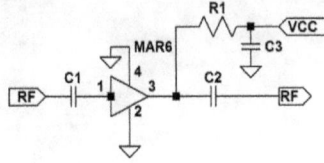

Figure 27. A diagram of an amplifier for SAW vibration sensor.

The amplifier is supplied at the output side by an R1 resistor. The R1 value is selected according to the DC supply voltage. For a supply voltage of 12V the resistance R1 equaled 560Ω. An amplification of 22dB was achieved for test signal frequency of 74MHz. It is the highest amplification value possible to achieve in this system. The value of current input equaled 16mA. Figure 26 shows an experimental frequency characteristics of SAW vibration sensor. Measurement of attenuation frequency diagram of SAW vibration sensor has been conducted on a spectrum analyzer HMS 1010. A supply voltage system has been put at line input. A supply voltage blocking system has been put before the analyzer at line output. Losses of 0.75dB consist of line losses and losses in the connection wiring and discussed separation systems. The value of these losses has been estimated at the level of 1dB. A conclusion may be drawn that an amplifier compensates the losses caused by SAW delay line. A theoretical shape of attenuation frequency diagram of a sensor should be described by the function (sin(x)/x). An experimental characteristic has high-frequency irregularities. Their reasons are the signals going from sensor output to input, omitting sensor electronic components. This signal amplitude is around –36dB lower than useful signal amplitude. The reason of occurrence of signal going from sensor input to output will be discussed in the next Section.

3.4. Parasitic signals

Parasitic signals are the signals going from electronic system input to output, omitting any component which is a part of the test signal transmission line. It is possible due to the occurrence of a parasitic coupling between any place of electronic system. There are two mechanisms leading to the occurrence of couplings [18][33]. The first one is electromagnetic coupling. The second one is ground current coupling. Figure 28 shows the mechanism of electromagnetic coupling. Red lines indicate paths of electromagnetic coupling which may occur in the electronic system of SAW vibration sensor. Electromagnetic couplings occur in all the electronic components constituting a sensor system. Paths of printed circuit are matched to the impedance of 50Ω. They are simultaneously transmitting and receiving aerials. Their efficiency depends on the path length. A similar role is played by inductances occurring in the system and capacities between paths. In order to reduce the electromagnetic coupling the inductances should be placed perpendicularly to each other and placed at a distance. These elements define the manner of making of the printed circuit plates. Any problems are solved individually, in accordance with a chosen construction.

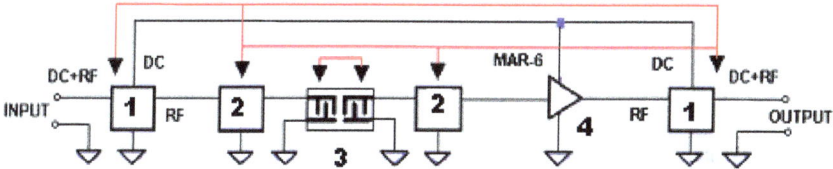

Figure 28. Electromagnetic coupling in electronic system of SAW vibration sensor.

The fundamental problem is an occurrence of electromagnetic coupling between the transducers and SAW delay line. Bus-bars delivering electric signal to transducers are placed on an immobile part of the plate. They are long and they are placed close to each other. It causes an increase in capacity between IDTs. The direct signal going through this way is also strengthened. Because of sensor plate motion its casing is larger than those used in traditional SAW filters. It also causes an increase in direct signal strength in the delay line. This problem and possible solutions are known in literature [29-31]. The most effective solution is symmetrical supply of one of transducers and their functioning in a bridge circuit [34]. Such a solution has been used in the presented SAW vibration sensor. The signal strength at the level of –35dB has been achieved. The second mechanism causing increase in direct signal strength is the ground current coupling [33]. Figure 29 shows the mechanism of this coupling.

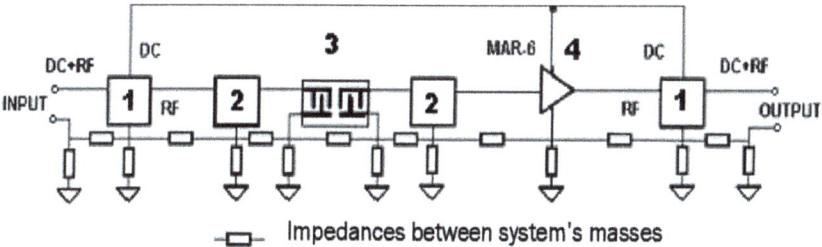

Impedances between system's masses

Figure 29. The mechanism of ground current coupling.

Ground current couplings occur only when the connection between component mass and joint mass is not perfect. A diagram in Figure 29 shows this effect by introduced impedances. An ideal connection is characterised by a null value of all impedances. Introduced impedances change current distribution in the entire system. Values of these impedances are small (fractions of ohm). That is why they are difficult to model. A physical making of component mass connection to the system joint mass must be considered during the design stage of the system. Reduction of this value by careful preparation of the system joint mass is a proper solution. Similarly to the first mechanism, SAW delay line plays an important part. Impedance of bus-bars and impedance of contacts leading the signal to transducer are crucial elements in SAW vibration sensor delay line. The joint mass of the discussed system has been made of 5 mm copper plate to which a printed-circuit board has been soldered. This side of the plate was completely bonded. The electronic system joint mass has been connected to component masses. Figure 30 shows SAW vibration sensor. Only supply voltage of the amplifier system goes through printed circuit paths. Longer segments of test signal line have been made by means of coaxial cables. It allowed to reduce an electromagnetic couplings value in the system. Attenuation frequency diagram of this sensor is presented in Figure 26.

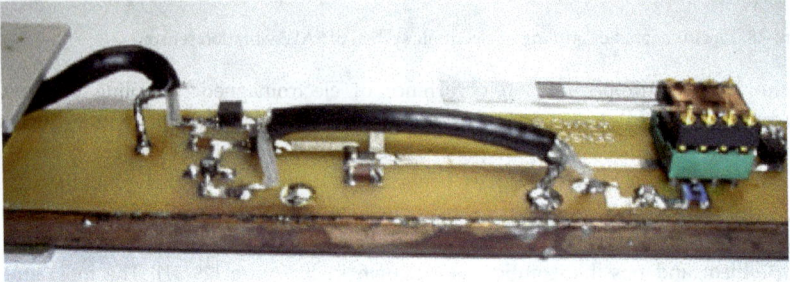

Figure 30. SAW vibration sensor

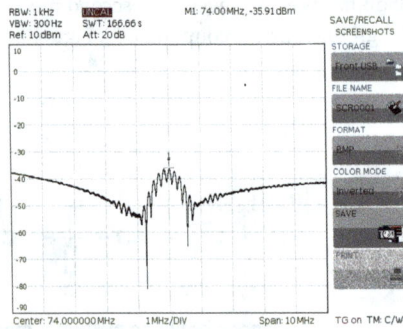

Figure 31. Attenuation frequency diagram of SAW vibration sensor after shut-down of amplifier supply.

Figure 31 presents this characteristic after a shut-down of amplifier supply. Shapes of characteristics in delay line operation band are similar in both figures. It suggests that signal source at the sensor output is situated behind SAW delay line. This signal strength is –35dB below the sensor frequency characteristic signal. Beyond the operation band the signal strength equals –40dB. Conducted measurements make use of the time trace of stationary signals. The lack of information about the delay of these signals does not allow to determine their source. IDTs are selectively matched to the impedance of 50Ω. It can be a signal going directly between interdigital transducers of SAW delay line. In order to explain the origin of the signal, it is important to learn about its delay time. It has been made in a system presented in Figure 32.

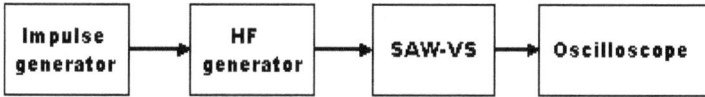

Figure 32. Meter circuit for parasitic signals in SAW delay line.

At sensor input a signal in form of wave packet at frequency of 74MHz has been delivered. The length of the packet is smaller than line delay. In this way a temporary separation of the direct and useful signals has been ensured.

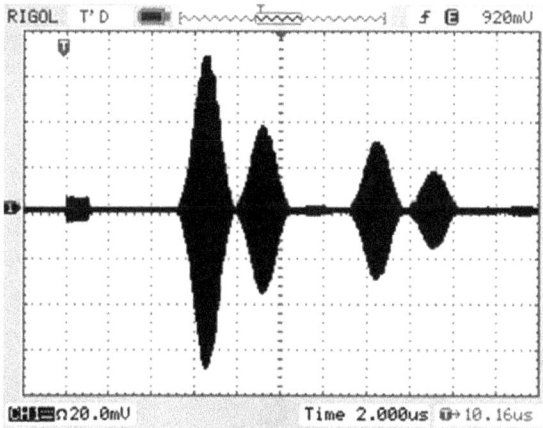

Figure 33. Time run of signals in SAW delay line.

Figure 33 presents timing of signals at the output of SAW vibration sensor. There are five signals at the sensor output. The first signal is going directly from sensor output to its input. The second one is a useful signal. Next three signals are reflected from the plate edge. Reflected signals will be attenuated by a damping paste. The amplitude of direct signal is − 35dB below the useful signal. It leads to conclusion that the signal shown in Figure 31 is a signal going directly between SAW delay line IDTs. When the electronic system is properly made, the elimination of this signal is the most fundamental problem in SAW vibration sensor design.

4. Measurements of vibration sensor parameters

SAW-based vibration sensors with SAW were made as described in Section 3. Delaying lines of a 74MHz middle frequency and various delay quantities of 4.2μs, 6.2μs, 8.2μs were worked out. These delaying lines were made on ST-cut quartz plates of different lengths. These plates were 5.7 mm wide and 0.5 mm thick.

The signal passing thru a sensor is continuous one. In a sensor every parasite signal adds up to a useful signal and causes the amplitude of a useful signal to modulate. This causes the sensor sensitivity to decrease; therefore parasitic signals must be removed. An essential problem is the reduction of a signal passing directly from the delay line inlet to the delay line outlet. The magnitude of this signal for sensors with lines of different delays is shown in Figure 34. The larger the line delay the lower the direct signal level. This level depends on the outlay of electric inlets to the delaying line converters. This is illustrated in Figures 35 (oscilloscope signals A and C). In order to lower the direct signal level additional screening of interdigital transducers or a symmetric supply of one of the converters was used [34]. For executed sensors a direct signal at a level of –36dB was obtained.

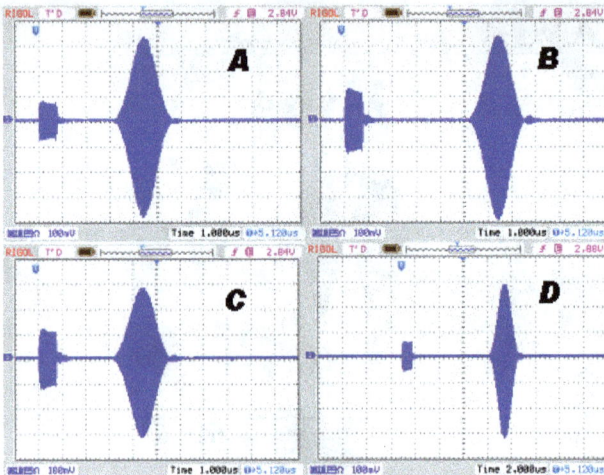

Figure 34. Direct and useful signal in delaying lines of different delays: A – 4,2μs, B – 6,2μs, C – 4,2μs, D – 8,2μs.

To basic parameters of vibration sensors belong their frequency characteristic curves and the static sensitivity. We will present the way they are measured. Determined experimental characteristics will be compared to the theoretical one. This will enable to estimate the model precision and to what extent it can help to model the parameters of SAW-based vibration sensors. The sensor frequency characteristics were determined in two stages. At the first stage the sensor pulse response was registered. At the second stage the spectrum of this pulse response was determined. Its shape corresponds with the sensor frequency characteristics. Pulse responses of sensors were measured and recorded as well as the spectrum of pulse responses from sensors were calculated with the aid of the system shown in Figure 35. Sensors were agitated for vibrations by an impact. Pulse responses were recorded with the help of the Agilent VEE Pro programme.

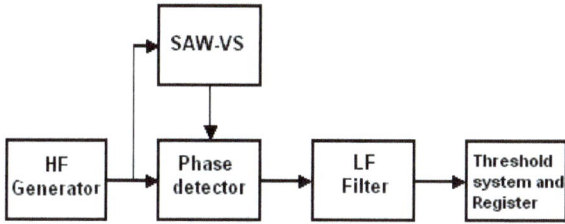

Figure 35. System to measure responses from SAW-based vibration sensor pulse plates.

Figure 36 presents a pulse response of a sensor where a delaying line with a 65 mm long plate was used. The sensor plate was not loaded with a seismic mass. The length of the pulse response was approximately 5s and its frequency was 92Hz.

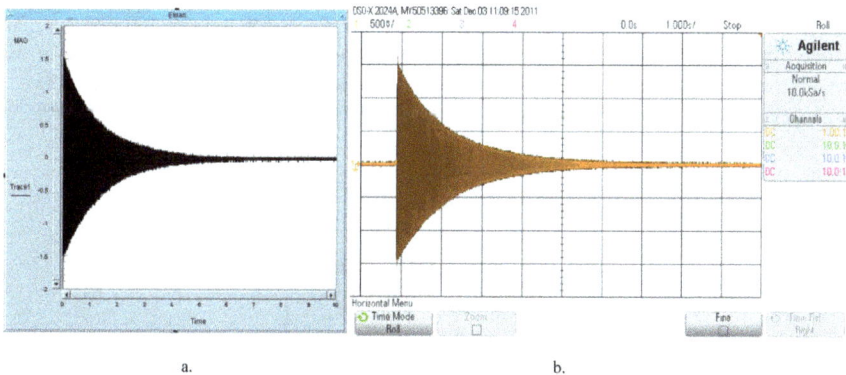

a.   b.

Figure 36. Pulse response of sensor with 65 mm long plate without seismic mass recorded with VEE programme – (a) and oscilloscope – (b)

The spectrum of a pulse response (a Fourier transform) was calculated with the help the Agilent VEE Pro programme, but it can be achieved also directly on an oscilloscope. Figure 37 presents the amplitude of this spectrum. Its shape corresponds to the amplitude of the frequency characteristics of a tested sensor. The resonance frequency equals theoretical values calculated with the relationships presented in Section 2. The frequency characteristics shows a harmonic at a 400Hz frequency. Its level is –26dB below the level of the sensor characteristics for the resonance frequency. This level is higher than its theoretical estimate presented in Section 2. It is difficult to determine the reason for this difference. It can be the inaccuracy of the model. However, it can be also due to a differing effectiveness of incitation of the resonance frequency component and harmonic frequency component.

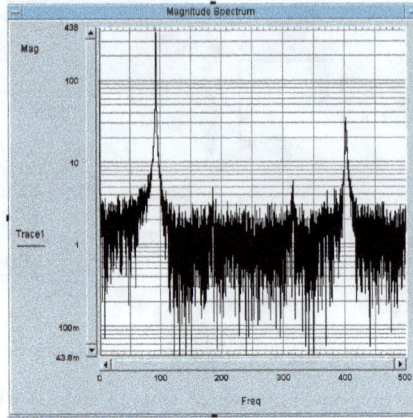

Figure 37. Spectrum of sensor with 65 mm long plate without seismic mass

Figure 38 presents the measurement of the statistical sensitivity of a sensor. It was conducted by recording the sensor output signal during its rotation by 180 degrees. This rotation causes the constant acceleration affecting a sensor to change by a value of two gravitational accelerations – i.e. 2 „g". An estimated static sensitivity of a sensor is 40mV/(„g").

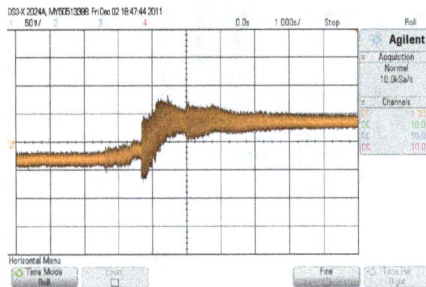

Figure 38. Method of determining static sensitivity of sensor

A 1.07g (r=2,27) seismic mass was placed on a sensor. This caused its resonance frequency to drop to 29Hz, and then for this sensor the above-presented tests were repeated. Fig. 39 shows the pulse response of a sensor with a 65mm long plate with a 1.07 g seismic mass. The pulse response is longer than 10s, with its frequency being 29Hz.

Figure 40 shows the frequency characteristics of the tested sensor. The value of the resonance characteristics equals the theoretical value. The characteristics shows a harmonic at 58Hz. Its level is –4dB below the sensor characteristic value for the resonance frequency, and being low, practically has no impact on the sensor function. Figure 41 shows the measurement result of the sensor static sensitivity as 100mV/(„g"). Against a sensor without a concentrated mass this value rose 2 ½ times. The length of the pulse response increased more than 2 times. These changes are obvious and their quantities determine explicit relationships describing the sensor model. The only difference between the results of experimental tests against the theoretical model is a higher level of the harmonic resonance frequency. To explain this difference classical measurements of the sensor resonance characteristics must be conducted. To conduct these measurements an exciter of stable mechanical vibrations of the sensor housing of an adjusted amplitude and frequency is required. It was not possible for the authors to carry out these tests.

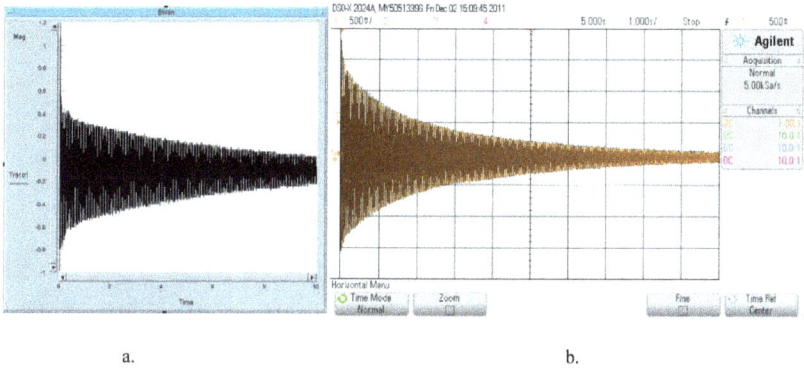

a.   b.

Figure 39. Pulse response of sensor with 65 mm long plate with a 1.07g seismic mass recorded with VEE programme – (a) and oscilloscope – (b)

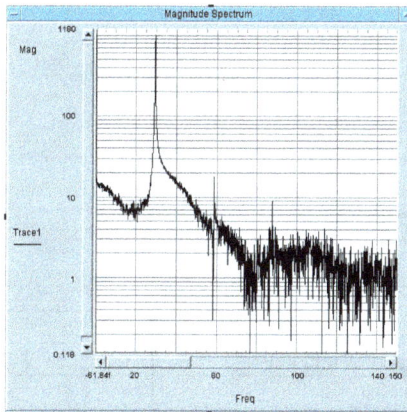

Figure 40. Spectrum of a 65mm long sensor with 1.07 g seismic mass

The test results demonstrated a good compatibility between the theoretical parameters of the sensor pulse response (resonance frequency and the decay time) and their experimental realization. The model presented in Section 2 was used to elaborate SAW-based sensors, which featured required parameters of the pulse response. Figure 42 shows a block diagram of an electronic warning system with SAW-based vibration sensors.

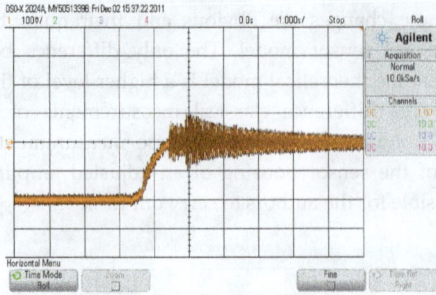

Figure 41. Method of determination the sensor static sensitivity

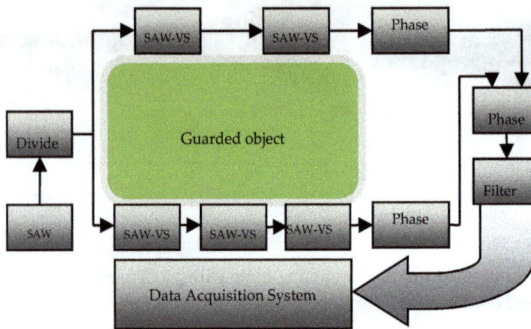

Figure 42. Block diagram of electronic warning system with SAW-comprising vibration sensors.

During operation of the system vibratons of the sensor plate change the phase of the measuring signal. The frequency of these changes equals the plate resonance frequency. Thus the signal from every sensor passing thru a set of filters at the phase detector inlet can be separated. The system operation is discussed further in the following works [35][36]. The said system required SAW-based vibration sensors having parameters given in the Table 2; the authors prepared these sensors.

Sensor No.	1	2	3	4	5
Resonance frequency [Hz]	41	56	73	90	159
Plate length [mm]	86	85	66	68	26
Seismic mass - r	0.22	0	0.14	0	1
Line delay [μs]	6.2	4.2	8.2	4.2	4.2

Table 2. Parameters of SAW-VS sensors made for electronic warning system

By selection of a seismic mass required resonance frequencies of sensors were achieved. In case of sensors No. 1 and 2 as well as No. 3 and 4 plates of similar lengths were used. Figure 43 presents characteristics of four assembled sensors.

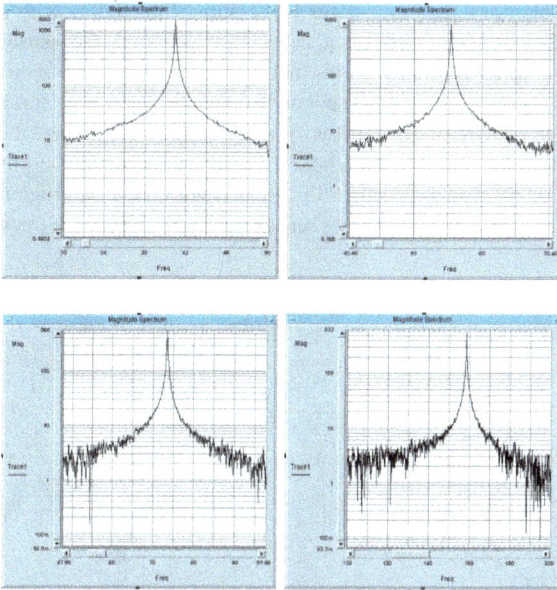

Figure 43. Frequency characteristics of sensors of resonance frequency of 41Hz, 56Hz, 73Hz, 159Hz.

These five SAW-based vibration sensors fabricated and used in an electronic warning system proved the efficiency of the presented modeling method. The system was tested on a stand shown in Figure 44. The sensors were attached to steel ropes tensioned as required. In order to describe the movement of sensors and ropes a model of a string loaded with a sensor mass taking into account its moment of inertia [37] was developed.

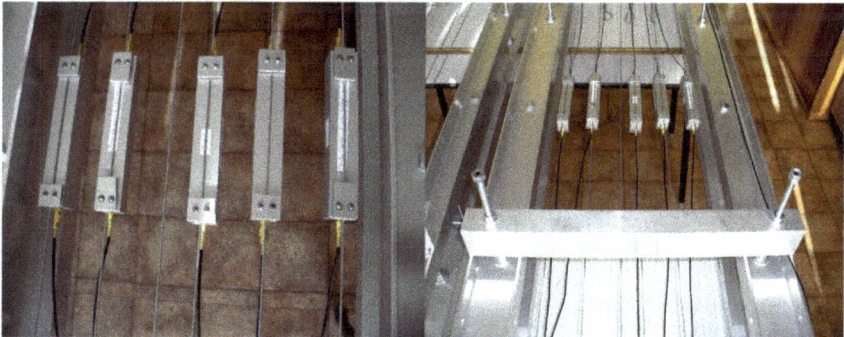

Figure 44. SAW-comprising vibration sensors attached to steel ropes

Vibrations of sensors were stimulated by deflecting them out of the state of equilibrium. The string vibration time was several times as long as that of the decay time of a senor plate pulse response. The movement of a sensor was a sum of the fading with time of the pulse response and vibrations enforced by a cyclic movement of the sensors housing. The frequency of the housing movements was that of the rope vibration frequency, which was selected so that it was lower than the resonance frequency of sensors. Thus it was possible to analyze every component of the sensor movement. For experimental testing a string vibration frequency of some 6 Hz was chosen. For every sensor the output signal from the phase detector (Fig. 42) was recorded and processed with the VEE program. Figure 45 and 46 show the course of signals of vibrating sensor of various resonance frequencies and their spectra.

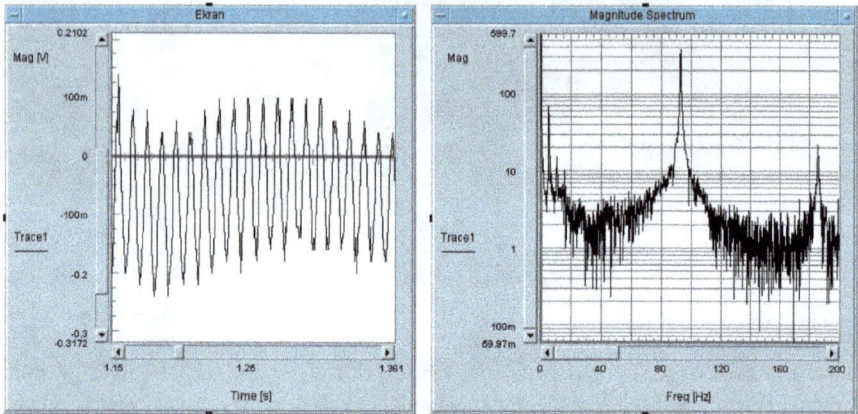

Figure 45. Part of output signal from sensor of plate resonance frequency of 91 Hz and string vibration frequency of 6.7 Hz and its spectrum

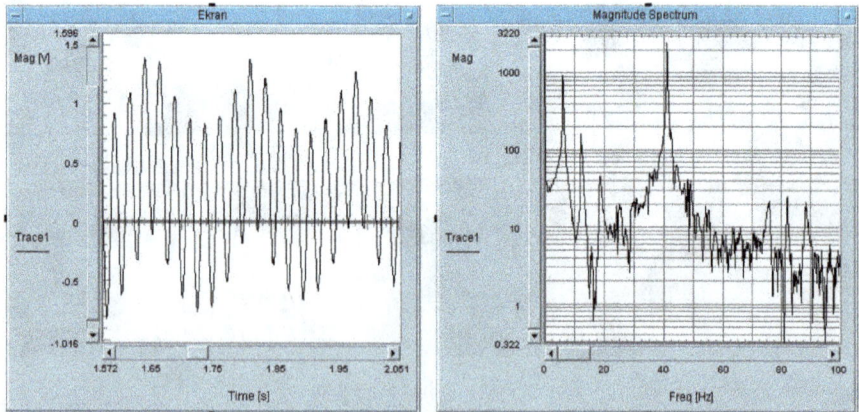

Figure 46. Part of output signal from sensor of plate resonance frequency of 41 Hz and string vibration frequency of 6.7 Hz and its spectrum

Pulse responses of sensors of an 91Hz (Fig. 45) or a 41Hz (Fig. 46) frequency can be easily discerned form signals derived from a string movement of a 6.7 Hz frequency, especially within the signal spectrum. Relations between their amplitudes are visible. After decay of pulse responses sensors measure the rope vibrations; they work then like classical vibration sensors.

In case of a sensor plate without the seismic mass (Fig. 45) the harmonic amplitude occurs at the – 34dB level below the resonance frequency amplitude. With the plate loaded with a small seismic mass (r=0,22, Fig. 46) the level is lower than -40dB. Harmonic vibrations had no impact on the operation of a system presented in Figure 43.

Depending on the use of a sensor by selection of its resonance frequency we can change its sensitivity and linearity. The practically prepared electronic warning system with SAW-based vibration sensors has fully proven the usefulness of the model presented in Section 2.

5. Conclusions

The work presents development, execution and parameters of SAW seismic vibration sensors. A sensor is a two-terminal pair network consisting of a delaying line with SAW and an amplifier compensating losses introduced by a ST-cut quartz. The delay line is fabricated on the CT-cut quartz surface.

A simple vibration model of an anisotropic plate was used to develop sensors. By way of successive simplifications of the description of vibrations of a viscous-elastic sensor plate a model of one degree of freedom was obtained. An explicit description of the movement parameters of the sensor plate was achieved. A material damping of the plate practically causes it to vibrate at only one resonance frequency, thus enabling to design SAW seismic vibration sensors. The results of experiments proved the effectiveness of using this model to design SAW-comprising seismic vibration sensors.

Basic parameters of realised vibration sensors (resonance characteristics, pulse responses, static sensitivity) are presented and analysed.

The range of resonance frequencies of plates made of ST-cut quartz, which were feasible, was determined. The plate length restricts the lower range of resonance frequencies. This range was determined on the basis of an available length of 100mm of a CT-cut quartz. The upper range of resonance frequencies is restricted by the speed of decay of a pulse response of a sensor plate. However, this restriction applies only to a sensor operating on its pulse responses. For the determination of this range the magnitude of stresses occurring in the plate was not taken into account. This magnitude must be smaller than the size of critical stresses presented in Table 1. In the subject work this element was not analysed. The magnitude of dynamic, critical stresses for a ST-cut quartz was determined in works [22] [23]. These values do not conform to standards and they apply to a series of plates cut out with a wire saw. In the course of that determination it turned out that the technology of plates production has a great impact on the value of dynamic critical stresses. The work [18] demonstrated that their value for a given design of a sensor does not restrict the determined range of the plate resonance frequencies.

As the plate resonance decreases the sensor sensitivity increases. Therefore high sensitivity sensors can be designed. Increasing its length can lower the resonance frequency of a plate. This is the most effective way to reduce the plate resonance frequency. It is possible to develop high-sensitivity vibration sensors of resonance frequencies in the order of a few Hz. To design such sensors one can use directly the presented model. Lets compare the SAW (SAW-VS) vibration sensor presented in the work to three kinds of sensors used at present. Resonance frequencies and basic applications of these sensors are presented in the introduction to this work.

The first kind of these sensors is geophones, where the sensor pulse response is utilized, which explains their low resonance frequency (several Hz) and a high sensitivity.

The two remaining kinds of vibration sensors are micro-mechanical silicone acceleration sensors (Micro Electro Mechanical Systems accelerometers - MEMS accelerometers) and piezoelectric acceleration sensors. Lets compare the basic parameters of these sensors: sensitivity, range of measured values and the frequency of acceleration changes, their structure (resonance frequency (natural frequency), weight and the manner of measuring the acceleration. We don't compare acceleration sensors reacting to impacts. They are very light and feature a very broad measuring range. Piezoelectric acceleration sensors work within a range between fractions of Hz thru a dozen or so kHz. The lower and upper range of the measurement dynamics is from 1 "mg" up to 100.000 „g". („g" is a unit of acceleration equal to the gravitational acceleration at sea level, i.e. $9.81 m/s^{-2}$). Depending on the design the sensor sensitivity is 0.2mV/g - 0.7V/g. The natural (resonance) frequency ranges from a few Hz to a few dozen of Hz. These sensors weigh from 3g to 500 g. The structure of these sensors is relatively simple, however, the measuring system is complicated (measuring of charge changes in the order of pC). MEMS accelerometers are characterized by small dimensions (an integrated circuit) and a low price. With these devices constant and variable acceleration up to a frequency of a few hundred Hz can be measured.

The lower and upper range of the measurement dynamics is from fractional „g" to 10 000 „g". Depending on the design the device sensitivity ranges from 0.2mV/g up to 10V/g (this applies to seismology sensors).

The device resonance frequency (natural frequency) is high: several Hz. These devices weigh from a few g up to 2 500 „g" (in case of seismology sensors). The sensor design is relatively simple; as well its measuring system is simple. (Measuring of changes in the charge in the order of aF's.)

The measuring system of a MEMS sensor is similar to a measuring system of an SAW-vibration sensor presented in the work (Figure 35). It consists of a measuring generator of a 1MHz frequency, two measuring paths and a phase detector.

The mechanical frequency of SAW-comprising vibration sensors can be changed within a range from several to a few hundred Hz. This is a significant difference between these both kinds of sensors. The sensitivity of an SAW vibration sensor depends on the resonance characteristics of the sensor plate, the length of the surface wave utilized in a sensor and on the sensor design. One can define here the sensitivity for a constant acceleration and the sensitivity for the resonance frequency of the sensor plate or some other frequency. For comparison's sake we can assume the sensitivity for a constant acceleration.

An example: for a SAW-VS with a plate 57.5 mm long and 0.5 mm thick, loaded with a seismic mass equal to 4 times the plate mass, a static sensitivity of 0.5V/„g" was achieved. The sensitivity of SAW-sensors for changing accelerations can be a couple of times greater than that for constant accelerations. The sensitivity degree is determined by the resonance curve of a sensor plate. The sensitivity of vibration sensors with SAW-VS can be increased by decreasing the length of the surface wave, or changing the sensor design (i.e. reduction of the plate thickness), by increasing the concentrated mass or by increasing the delay of the SAW-delaying line.

SAW vibration sensors can be cascaded which offers many system designs what in turn offers an increase in the sensitivity and a reduction of the cross sensitivity of sensors. This is not possible with a MEMS sensor and a piezoelectric sensor. The high measuring frequency of SAW-comprising sensors allows designing wireless sensor versions.

In terms of the design and the method of measuring MEMS and SAW-VS acceleration sensors have a lot in common. In a MEMS sensor a vibrating plate of silicon changes the capacity of a capacitor. Mechanical properties of silicon (density 2.330kg/m^{-3}, and an equivalent Young's modulus of 106 Gpa) are similar to mechanical properties of quartz. Therefore the parameters of MEMS and SAW-VS must be similar. In our opinion SAW-VS sensors have their place in the measuring technology.

This work presents an example of using SAW-sensors in an electronic warning system. Vibration sensors placed at selected points record vibrations within areas they cover. An alert central station registers signals from a vibration sensor. This is a typical system to be applied for perimeter protection systems.

All sensors included in a system can be placed within one area; then this system will perform the role of an analyzer of vibrations within this area. This is a second prospective application of the system under discussion. The possibility of preparing high-sensitivity vibration sensors of resonance frequencies in the order of a few Hz is a prospective area of application for monitoring vibrations of bridges and buildings, where the frequency of free vibrations is in the order of fractions of Hz up to a dozen or so Hz. The presented application examples apply to working on the pulse response of an SAW-sensor. This kind of a sensor can be used to measure one component of the acceleration vector. Then, the pulse response means for a measurement of acceleration a parasitic signal, and should be eliminated. Therefore thru the use of sensor plates, where the decay of a pulse response is fast, acceleration sensors can be developed. The movement of a plate shows then a character of acceleration changes in time. Works [38] [39] show development of this kind of sensors.

Author details

Jerzy Filipiak and Grzegorz Steczko

Institute of Electronic and Control Systems, Technical University of Czestochowa, Częstochowa, Poland

Acknowledgement

This work was supported by the Polish Ministry of Science and Higher Education as a project "Vibration Warning System with SAW Vibration Sensors" included in the scientific budget of 2009–2011.

6. References

[1] Pakhamov A, Pisano D, Sicignano A and Golburt T (2005) High Performance Seismic Sensor Requirements for Military and Secuirity Applications Proceedings of SPIE, j. 5796: 117-124.
[2] Kaajakari V, (2009) Practical MEMS, Small Gear Publishing
[3] Lee I.G, Yoon G.H, Park J, Seok S, Chun K, Lee K (2005) Development and analysis of the vertical capacitive accelerometer. Sensors and Actuators A j.119: 8-18.
[4] www.metrozet.com
[5] Wohltjen H, Dessy R (1979) Surface Acoustic Waves Probe for Chemical Analysis I. Introductcion and Instrument Design. Analytical Chemistry. j. 9: 1458-1475.
[6] Nakamoto T, Nakamura K, Moriizumi T (1996) Study of Oscillator- Circuit Behavior for QCM Gas Sensor. Proc. Ultrasonics Symposium.j. 1: 351-354.
[7] Gizeli E, Liley M, Love C.R, Vogel H (1997) Antibody Binding to a Functionalized Supported Lipid Layer, A Direct Acoustic Immunosensor, Analytical Chemistry. j. 69: 4808-4813.
[8] Urbańczyk M, Jakubik W, Kochowski S (1994) Investigation of sensor properties of cooper phtalocyanine with the use of surface acoustic waves. Sensors and Actuators B j.22: 133-137.
[9] Cullen C, Reeder T, (1975) Measurement of SAW Velocity Versus Strain for YX and ST Quartz. Proc. Ultrasonics Symposium: 519-522.
[10] Cullen C, Montress T (1980) Progress in the Development of SAW Resonator Pressure Transducers. Proc. Ultrasonics Symposium. j. 2: 696-701.
[11] Pohl A, Ostermayer G, Reindl L, Seifert F (1997) Monitoring the Tire Pressure of Cars Using Passsive SAW Sensors. Proc. Ultrasonics Symposium.j.1: 471-474.
[12] Clayton L.D, EerNisse E.F (1998) Quartz thicknes-shear mode pressure sensor design for enhanced sensitivity. IEEE Transactions on Ultrasonics, Ferroelectrics, and Frequency Control. J.45: 1196-1203.
[13] Jiang Q, Yang X.M, Zhou H.G, Yang J.S (2005) Analysis of surface acoustic wave pressure sensor. Sensors and Actuators A j.118: 1-5.
[14] Drafts B (2000) Acoustic wave technology sensors. Sensors j.10: 1-9.
[15] Seifert F, Bulst W, Ruppel C (1994) Mechanical sensor based on surface acoustic waves. Sensors and Actuator A. j. 44: 231-239.
[16] Hauden D (1991) Elastic waves for miniaturized piezoelectric sensors: applications to physical quantity measurements and chemical detection. Archives of Acoustics. j. 16: 91-106.

[17] Filipiak J, Solarz L, Steczko G (2007) Surface acoustic wave stress sensors. Desinger Analysis. Molecular and Quantum Acoustics. j. 28: 71-80.

[18] Filipiak J (2006) Surface acoustic wave acceleration sensors. Technical University of Czestochowa. Monograph 121, p.198. (in Polish).

[19] Filipiak J, Kopycki C (1999) Surface acoustic waves for the detection of small vibrations. Sensors and Actuators. j. 76: 318-322.

[20] Filipiak J, Solarz L, Steczko G (2009) Surface acoustic wave vibration sensors for linear electronic warning systems. Acta Physica Polonica A j.116: 302-306.

[21] Filipiak J, Solarz L, Steczko G (2011) Surface Acoustic Wave (SAW) Vibration Sensor. Sensors. j.11: 11809-11832. http://www.mdpi.com/journal/sensors

[22] Filipiak J, Zubko K (2005) Determination of damping in piezoelectric crystals. Molecular and Quantum Acoustics j. 26: 75-80.

[23] Kopycki C (1999) Effect of substrates on the parameters of piezoelectric vibration sensor with a surface acoustic wave. Ph.D. Thesis (in Polish), Military University of Technology, Warsaw,

[24] Nye J.F (1957) Physical Properties of Crystals; Clarendon Press: Oxford, GB.

[25] Mason W.P (1958) Physical Acoustics and the Properties of Solids. Van Nostrand,

[26] Bogusz W, Dżygadło Z, Rogula D, Sobczyk K, Solarz L (1992) Vibrations and Waves A. Elsevier: Amsterdam

[27] Zubko K (2006) Applying of the Rayleigh method to determination of elastic and viscoelastic parameters of piezoelectric crystals. Ph.D. Thesis (in Polish), Military University of Technology, Warsaw,

[28] Filipiak J, Solarz L, Zubko K (2004) Analysis of Acceleration Sensor by the discrete model. Molecular and Quantum Acoustics j. 25: 89-99.

[29] Filipiak J (1993) Problems of synthesis of components of a surface acoustic wave signal processing to complex type "chirp". Ph.D. Thesis (in Polish), Military University of Technology, Warsaw,

[30] Matthews H L (1997) Surface Wave Filters. John Wiley and Sons, New York.

[31] Morgan D P (1985) Surface Wave Devices for Signal Processing. Academic

[32] Ruppel C C W, Fieldly T A (2001) Advances in Surface Acoustic Waves Technology, Systems and Applications (vol 2). Word Scientific Pub. Co. Inc,.

[33] www.saw-devices.com

[34] Danicki E, Filipiak J (1982) Bridging method for elimination of the direct signal, Electronics (in Polish), j.10-12: 22-26

[35] Filipiak J, Solarz L, Steczko G(2011) Electronic Warning System Based on SAW Vibration Sensors, Acta Physica Polonica A j.120: 593-597

[36] Filipiak J, Solarz L, Steczko G(2009) Surface acoustic wave vibration sensors for linear electronic warning systems. Acta Physica Polonica A j.116: 302-306

[37] Filipiak J, Solarz L, Steczko G (2010) Analysis of Experimental Stand for SAW Vibrations Sensor, Acta Physica Polonica A j.118: 1118-1123

[38] Filipiak J, Kopycki C, Solarz L, Ostrowski J (1998) The SAW Acceleration Sensor. Proceedings European Frequency and Time Forum j.I :229-232.

[39] Filipiak J, Kopycki C, Solarz L, Ostrowski J (1997) Lithium niobiate as the substratum for the SAW acceleration sensor. Proceedings SPIE, The International Society for Optical Engineering j. 3179: 256-260.

Chapter 7

Techniques for Tuning BAW-SMR Resonators for the 4th Generation of Mobile Communications

M. El Hassan, E. Kerherve, Y. Deval, K. Baraka, J.B. David and D. Belot

Additional information is available at the end of the chapter

1. Introduction

In telecommunication systems, all filters and resonators that constitute the RF part have the tendency to be integrated on the same chip that contains the information treatment.

In order to achieve miniaturization, bulk acoustic wave (BAW) technology is presented. BAW filters are very sensitive to surface contamination, and can exhibit very small sizes. In addition, BAW resonators could be fabricated using compatible material CMOS and BiCMOS [1]. In this context and in order to compensate the variation due to the fabrication process, the work presented in this paper focuses on the tuning of BAW-SMR resonators and filters.

This work is divided in two parts. The first part consists of designing BAW-SMR (Solidly Mounted Resonator) filters. In the second part, we propose the use of two methods to tune this type of filters. Thus, we present the design methodology, the study, and the experimental realization of the BAW-SMR tunable filters.

2. BAW impedance behavior

2.1. SMR impedance behavior

The bulk acoustic wave resonator is basically constituted by a piezoelectric layer sandwiched between two electrodes (Fig.1). The application of an electric field between the two electrodes generates a mechanical stress that is further propagated through the bulk of the structure (acoustic wave). The resonance condition is established when the acoustical path (in thickness direction) corresponds to odd integer multiples of the half acoustic wavelength. The bulk acoustic wave resonator is basically constituted by a piezoelectric layer sandwiched between two electrodes (Fig.1) [2, 3].

The bulk acoustic wave resonator fabrication over silicon substrates imposes its acoustical isolation, confining the acoustic waves into the main resonant structure. Two configurations are proposed: the membrane suspended structure (FBAR – Film Bulk Acoustic wave Resonator) [3], where the resonator is suspended by an air-bridge (Fig.1a); and the solidly-mounted structure (SMR – Solidly Mounted Resonator), where the resonator is mounted over a stack of alternating materials (Fig.1b). This stack is built on a Bragg reflector basis and it has an acoustic mirror behavior [2-3]. Both, air and acoustic mirror, present an optimum discontinuity for reflecting the acoustic waves at the interface with the bottom electrode, confining waves into the main resonant structure.

In the solidly mounted resonator (SMR), the piezoelectric is solidly mounted to the substrate (Fig.1.b). Some means must be used to acoustically isolate the piezoelectric from the substrate if a high quality factor (Q) resonance is to be obtained. In effect, the quarter wavelength layers act as a reflector to keep waves confined near the piezoelectric transducer film [4]. The effect of the reflector on mechanical displacement is to cause the wave amplitude to diminish with depth into the reflector. The number of layers required to obtain a satisfactory reflection coefficient is dependent on the mechanical impedances between layers and, to a lesser extent, the substrate [5]. The number of layers is best determined by an analysis of the resonance response as a function of the number of layers versus resonator 'Q' and coupling coefficient.

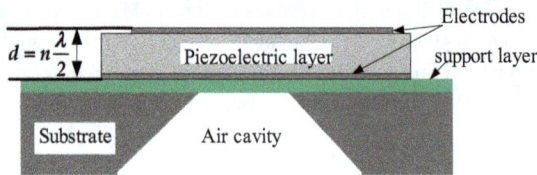

(a)

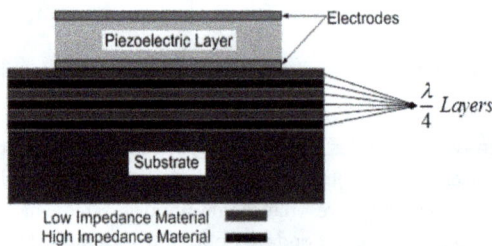

(b)

Figure 1. (a) Film Bulk Acoustic Resonator (FBAR). (b) Solidly Mounted Resonator (SMR).

An important effect of the reflector layers, as demonstrated by Newell [4], is a partial lateral stiffening of the piezoelectric plate that minimizes plate wave generation and spurious resonances normally observed in free plates. However, real resonator structures are inherently 3D and some form of radiation beyond the simple thickness dimension is to be expected. If energy leaves the resonator structure, through radiation, then it counts as a loss mechanism. Reflections of lateral waves at the edge of the resonator can lead to standing waves and spurious responses. The SMR approach requires that the substrate be smooth in order to proceed with the fabrication of reflectors, electrodes, and piezoelectric film [5].

2.2. The Butterworth-Van-Dyke (BVD) model

The Modified Butterworth-Van-Dyke (MBVD) model is an electrical schematic around resonance (Fig.2). The elements R_a, L_a, C_a present the series resonance and the insertion losses. The capacity C_0 represents the piezoelectric material between the two electrodes.

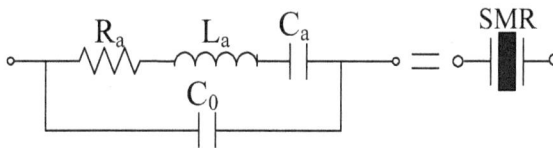

Figure 2. SMR-MBVD model.

The impedance characteristic of a measured BAW-SMR is shown in Fig.3. In this graph, it can be observed that the SMR presents mainly two resonance pulsations: the series resonance (f_s), when the electrical impedance approaches to zero, and the parallel resonance (f_p), when the electrical impedance approaches to infinity. For all other frequencies far from

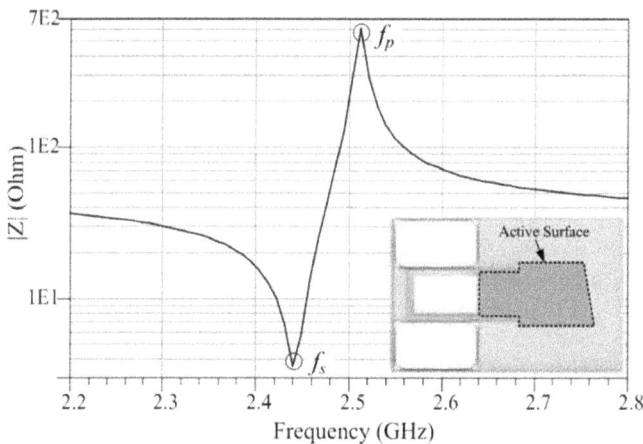

Figure 3. Impedance characteristic of a measured BAW-SMR.

the resonances, the SMR presents static capacitor behavior. f_s is adjusted according to the thickness of the piezoelectric layer and it is spaced by the parallel resonance f_p. The instantaneous frequency deviation between the two resonances is determined by the electromechanical coupling coefficient of the piezoelectric layer. The quality factor of the measured BAW-SMR is 192.5 and its active area is 16800 µm². This relatively small quality factor is due to some technological problems.

The electrical impedance of an SMR is obtained by solving the acoustic boundary problem and applying the transmission line theory [6]. The electrical SMR impedance can be simplified and expressed by the following equation:

$$Z_{SMR} = \frac{1}{sC_0} \frac{(f^2 - f_s^2)}{(f^2 - f_p^2)}$$
(1)

where 's' is the complex variable: $s = jw$.

2.3. Addition of external passive elements

The association of external passive components (L, C) to the resonators was made in two parts. First, the resonators analyses have been carried out using on-wafer measurements. Then, these results have been associated with the capacitors and inductors. The combination of experimental results and modeling of passive elements constitutes the final response of the tuned resonators.

2.3.1. Addition of capacitors

The addition of the capacitors (having an intrinsic quality factor 'Q' > 140) to the SMR circuitry doesn't affect severely the quality factor of the overall design. The resonator's electromechanical coupling coefficient is described indirectly by the capacitor ratio C_0/C_a as determined by the resonator physical configuration and piezoelectric material properties [7]. When changing the electromechanical coefficient of the piezoelectric material, the bandwidth changes. Our goal is to tune the capacitor ratio C_0/C_a by adding series or parallel capacitors to the resonator. Thus, controlling this ratio will enable us to control the electromechanical coefficient, and as a sequence the bandwidth of the resonator.

2.3.1.1. Series capacitor association

The performance analysis of the association of series capacitor to the SMR is based on the BVD model (Fig.4). The analysis of the frequency response of the series capacitor associated to the SMR will lead us to (2) presented below:

$$Z_{in} = \frac{s^2(CC_aL_a + C_0C_aL_a) + C_0 + C_a + C}{sC(s^2C_0C_aL_a + C_0 + C_a)}$$
(2)

Where's' is the complex variable: $s = jw$.

From (2), it is possible to notice the insertion of a pole and zero to the frequency response of the simple resonator. The extraction of the poles and zeroes values of (2), will lead us to (3) & (4) shown below.

$$f_s = \frac{1}{2\pi} \sqrt{\frac{C_0 + C_a + C}{CC_a L_a + C_0 C_a L_a}}$$ (3)

$$f_p = \frac{1}{2\pi} \sqrt{\frac{C_0 + C_a}{C_0 C_a L_a}}$$ (4)

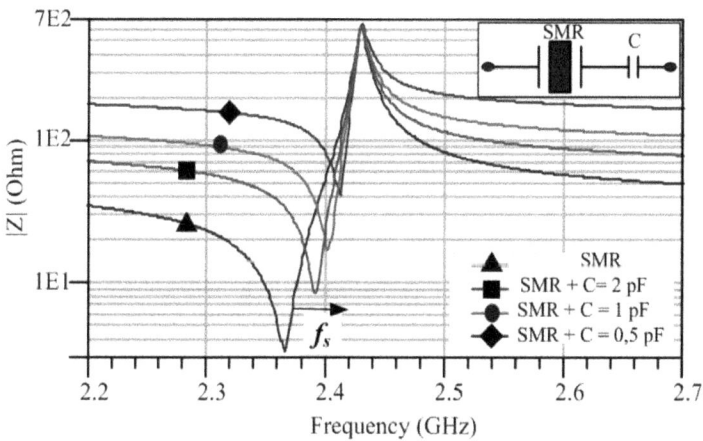

Figure 4. Impedance response of the measured BAW-SMR with series capacitor.

From (3) & (4), it may be noted that the capacitor added in series with the BAW resonator affects only the series resonance frequency. It is inversely proportional to f_s. To illustrate this theoretical study, responses with different values of capacitors added in series with the measured BAW-SMR are shown in Fig.4.

2.3.1.2. Parallel capacitor association

The analysis of the frequency response of the parallel capacitor associated to the SMR will lead us to (5), presented below.

$$Y_{in} = \frac{s(s^2 L_a C_a (C_0 + C) + C_0 + C + C_a)}{s^2 L_a C_a + 1}$$ (5)

$$f_s = \frac{1}{2\pi} \sqrt{\frac{1}{C_a L_a}}$$ (6)

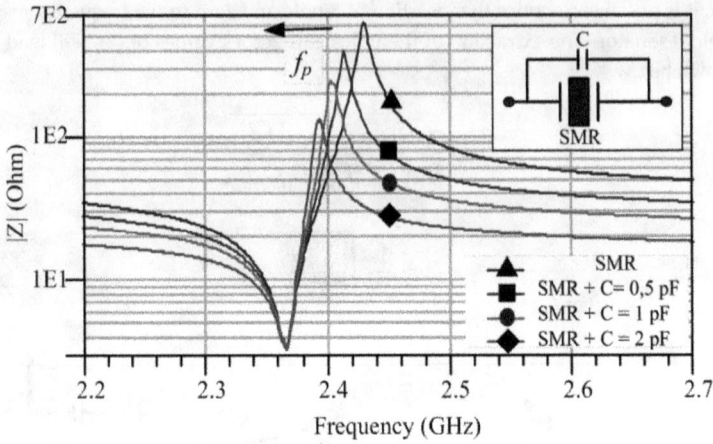

Figure 5. Impedance response of the measured BAW-SMR with parallel capacitors.

$$f_p = \frac{1}{2\pi}\sqrt{\frac{C_0 + C_a + C}{L_a C_a (C_0 + C)}} \tag{7}$$

From (5), it is possible to notice the displacement of the poles of the device's resonance frequency. The extraction of the poles and zeroes values of (5), will guide us to (6) & (7) as shown below.

It is noticed from (6) & (7) that the capacitor added in parallel with the BAW resonator affects only the parallel resonance frequency. It is inversely proportional to f_p. Fig.5 presents the final responses of the measured BAW-SMR with different values of capacitors added in parallel.

2.3.2. Addition of inductors

The association of inductors to the BAW resonators could be realized in series or parallel. We should note that in the case of VLSI-CMOS, these inductors are characterized by a small quality factor with respect to the BAW (400 to 1000) which degrades the quality factor of the overall circuitry, and they occupy a relatively large size.

2.3.2.1. Series inductor association

The performance analysis of the assembly of the inductors in series with the SMRs is based on the BVD model. The analysis of the frequency response of the series inductors associated to the SMR will lead us to (8).

$$Z_{in} = \frac{s^4 C_0 C_a L_a L + s^2 (C_a L_a + C_0 L + C_a L) + 1}{s(s^2 C_a L_a C_0 + C_0 + C_a)} \tag{8}$$

From (8), it is possible to notice the insertion of a zero to the resonator frequency response. The extraction of the poles and zeroes values from (8), will bring (9) & (10). Based on these equations, we can see that the values of the poles are not deteriorated. However, from (9), we notice that the association of series inductors to the SMR will modify the zeroes, and only the series resonance frequency is affected. It is inversely proportional to f_s.

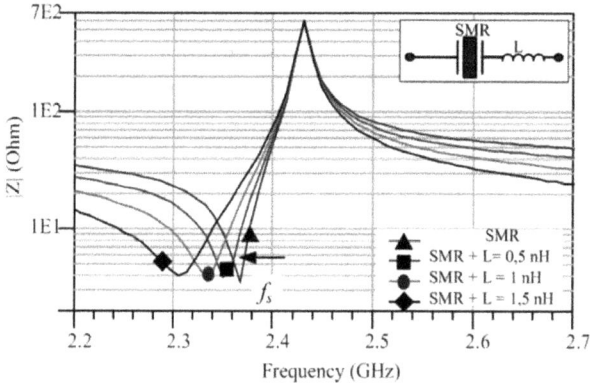

Figure 6. Impedance response of the measured BAW-SMR with series inductors.

To illustrate this theoretical study, impedance behavior of the measured BAW-SMR with different values of inductors associated in series is shown in Fig.6.

$$f_s = \frac{1}{2\pi}\sqrt{\frac{(C_a L_a + C_0 L + C_a L) \pm \sqrt{(C_a L_a + C_0 L + C_a L)^2 - 4C_0 C_a L_a L}}{2C_0 C_a L_a L}} \qquad (9)$$

$$f_p = \frac{1}{2\pi}\sqrt{\frac{C_0 + C_a}{C_a L_a C_0}} \qquad (10)$$

2.3.2.2. Parallel inductor association

Based on the same procedure used above, the analysis of the frequency response of the parallel inductors associated to the SMR will lead us to (11), presented below.

$$Y_{in} = \frac{s^4 C_0 C_a L_a L + s^2 (C_0 L + C_a L + C_a L_a) + 1}{sL(1 + s^2 C_a L_a)} \qquad (11)$$

From (11), it is possible to notice the insertion of a pole to the resonator frequency response and the displacement of another. Also, we can notice the composition of a double pole near the frequency response of the device. The extraction of the poles and zeroes values from (11), will lead us to (12) & (13).

From (12) & (13), we can see that that the inductor added in parallel with the BAW resonator affects only the parallel resonance frequency. It is inversely proportional to f_p. Fig.7 presents the final responses of the measured BAW-SMR with different values of inductors associated in parallel.

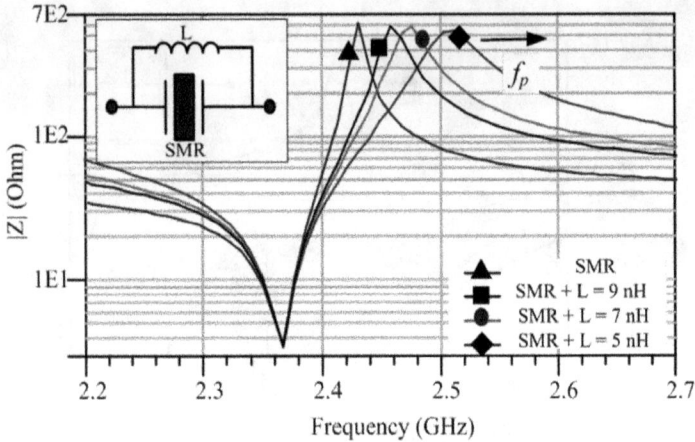

Figure 7. Impedance response of the measured BAW-SMR with parallel inductors.

$$f_p = \frac{1}{2\pi} \sqrt{\frac{\left(C_0 L + C_a L + C_a L_a\right) \pm \sqrt{\left(C_0 L + C_a L + C_a L_a\right)^2 - 4C_0 C_a L_a L}}{2C_0 C_a L_a L}} \qquad (12)$$

$$f_s = \frac{1}{2\pi} \sqrt{\frac{1}{C_a L_a}} \qquad (13)$$

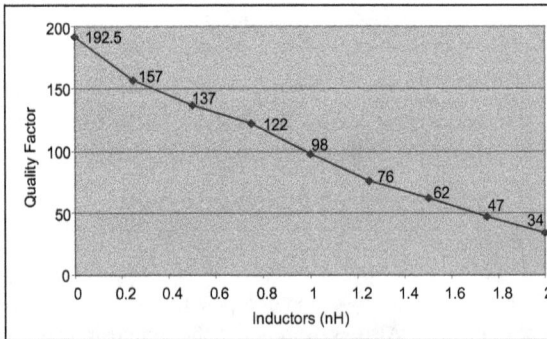

Figure 8. Impact of the quality factor of the inductors added in series with the BAW resonator on the series quality factor 'Qs' of the overall resonator circuitry (SMR+L).

The addition of inductors to the BAW-SMR increases the insertion losses and degrades the quality factor of the overall circuitry. Fig.8 illustrates the influence of inductors added in series with the SMR on the overall quality factor of the assembly.

3. BAW filter design

3.1. Bulk Acoustic Wave (BAW) filters topologies

Bulk acoustic wave filters are basically divided in three topologies: ladder, lattice and ladder-lattice (Fig.9) [3].

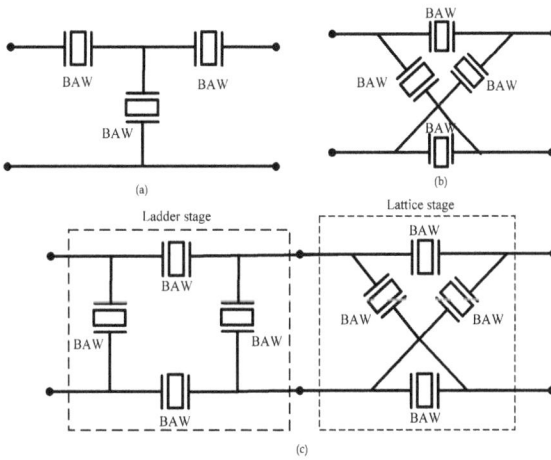

Figure 9. Bulk acoustic wave resonator filter topologies: (a) ladder stage, (b) lattice stage and (c) ladder-lattice stage.

Ladder filter are characterized by an unbalanced operation mode and very small size able to deliver very high selectivity filtering responses, however presenting low rejection or isolation out-of-band. Typically, ladder bulk acoustic wave filters are quite effective for blocking signals close to the passband, but poor at rejecting undesired bands [3-4].

On the other hand, lattice bulk acoustic wave filters are characterized by a balanced operation mode. In contrast to the first one, this topology presents typically a low selectivity close to the passband, followed by a high rejection out-of-band. They present slower roll off coefficient and higher rejection. Thus, lattice networks are not interesting to block signals close to the passband, but they are more effective for rejecting undesired bands [3].

Ladder-lattice bulk acoustic wave filters are also characterized by a balanced operation mode. Ladder-lattice filters ally advantages of both network types, enabling high isolation at undesired bands and steep responses close to the passband [3]. This topology is able to strongly reduce the linearity constraints of the receiver RF chain. Fig.10 shows a comparison between the typical theoretical transmission responses (S_{21}) of these three network types.

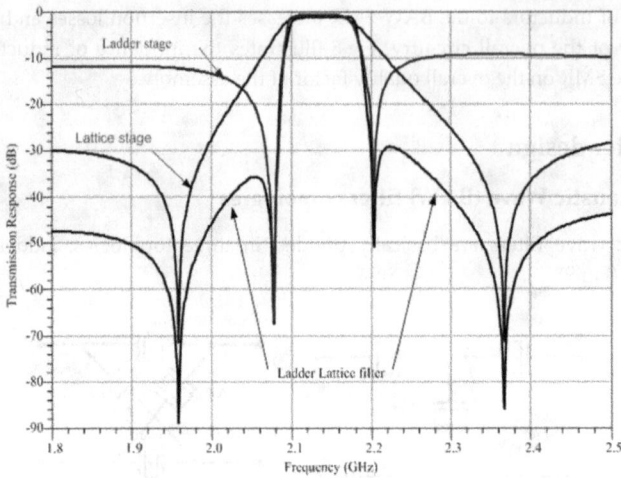

Figure 10. Theoretical transmission responses for main bulk acoustic wave (BAW) filter topologies.

3.2. Ladder BAW-SMR filter design

The ladder filter is an association of resonators in series and in parallel. The shunt resonators are loaded and their resonance frequencies are smaller than the series resonators. Ladder BAW topology presents a good selectivity which enables to block undesired signals near the pass band. In this context, and in order to support the theoretical study, a tunable BAW-SMR ladder filter was designed for the 802.11b/g standard (2.40 – 2.48 GHz). The resonators and filters were fabricated at the CEA-Leti in the framework of the project 'EPADIM'. The filter is composed of five SMRs, associated in ladder topology. The filter stack can be divided into resonators' layers and Bragg reflector's layers. The resonator's layers are composed by the classical couple AlN-Mo [8]. However, in contrast to [8], the Bragg reflector was implemented using an exclusive dielectric stack composed by SiOC:H and Si_xN_y [9].

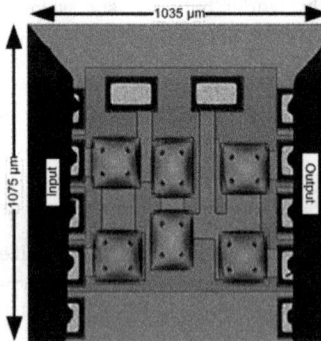

Figure 11. Microphotography of the Ladder BAW-SMR for the 802.11b/g standard.

The acoustical performance of the fully dielectric stack is comparable to the traditional SiO2-W reflectors; however, it strongly reduces the coupling between resonators through the Bragg reflector. Furthermore, the filter stack was realized on a high resistive silicon substrate in order to reduce losses due to the capacitive coupling [10].

In order to optimize the filter performance, a double resonator and apodized geometries have been used. Indeed, double resonators present large electrodes' areas, which results in lower resistive losses. Also, the filter resonators present apodized geometries in order to avoid spurious resonances caused by the parasitic lateral acoustic modes [11]. Fig.11 and 12 show the microphotography and a comparison between the measured and simulated results of the tunable ladder BAW-SMR filter, respectively. The filter occupies a small area and has reduced dimensions (1035*1075 μm^2). Electromagnetic simulation of the overall filter structure has been performed using the ADS-Momentum software. Next, the acoustical effects have been considered using the Mason Model [12] and included in the simulations.

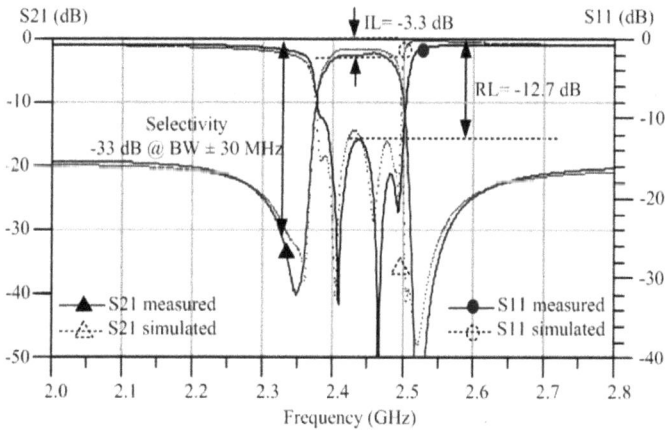

Figure 12. Comparison between measured and simulated results of the 802.11b/g ladder BAW-SMR filter.

The filter design was realized for implementation in SiP context. The performances of the tunable BAW-SMR filter are in concordance with the simulation results. Mainly, the filter fulfills the requirements for the WLAN 802.11 b/g standard, presenting -3.3 dB of insertion loss, -12.7 dB of return loss and a selectivity higher than 33 dB at ± 30 MHz of the bandwidth. The filter high insertion losses are mainly due to the low resonators quality factor obtained in the fabrication (Q = 200). Therefore, these losses can be strongly reduced using mechanical energy concentration techniques in the resonator acoustical cavity [13].

4. Tunability of the ladder filter

The shunt resonators of the ladder filter determine the position of the zeroes at the left of the center frequency and the series resonators determine the position of the zeroes at the right of the center frequency. Thus, changing the impedance of the parallel and series resonators leads to a change in the zeroes' positions.

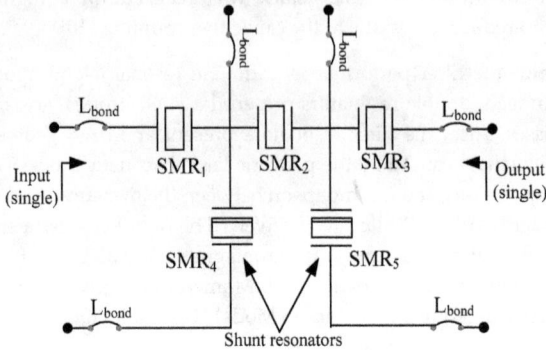

Figure 13. Ladder BAW-SMR filter.

Based on this theory and in order to tune the BAW-SMR filter (Fig.13), we propose to add passive elements to the shunt and series resonators that constitute the filter.

4.1. Shift towards higher frequencies

To shift the center frequency of the filter towards higher frequencies we have to move all the zeroes towards these frequencies. Thus, we have added inductors in parallel with the series resonators and capacitors in series with the shunt resonators that constitute the ladder BAW-SMR filter (Fig.14).

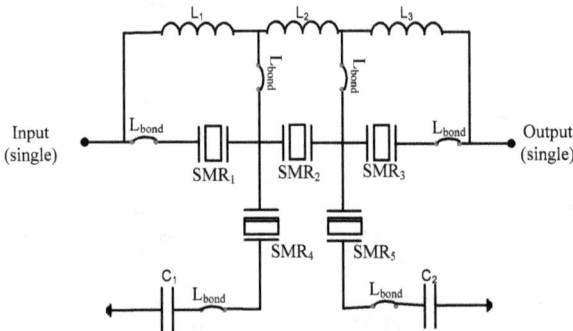

Figure 14. Tunable BAW-SMR filter with inductors added in parallel with the series resonators and capacitors added in series with the shunt resonators.

Table 1 presents the values and the quality factor of the external passive elements used in the circuitry designed to move the center frequency of the tunable BAW-SMR filter towards higher frequencies.

	Value	Quality Factor 'Q'
Capacitors	$C_1 = C_2 = 2.2$ pF	140
Inductors	$L_1 = 3.9$ nH	47.7
	$L_2 = 5.6$ nH	40
	$L_3 = 4.7$ nH	45

Table 1. Values of the passive elements used to shift the center frequency of the tunable BAW-SMR filter towards higher frequencies and their quality factors.

The tunable BAW-SMR filter and the passive components are mounted on a FR4 substrate as shown in Fig.15.

Figure 15. Tunable BAW-SMR filter and passive elements mounted on FR4 PCB.

Fig.16 shows the simulation results of the tunable filter. The insertion loss (IL = -2.4 dB) obtained in the simulation is due to quality factors of resonators (Q = 500). The return loss (RL) is -10 dB, and the out of band rejection is 26 dB at 2.0 GHz. The simulation results shows that a shift of +1% of the initial central frequency (2.44 GHz) is obtained.

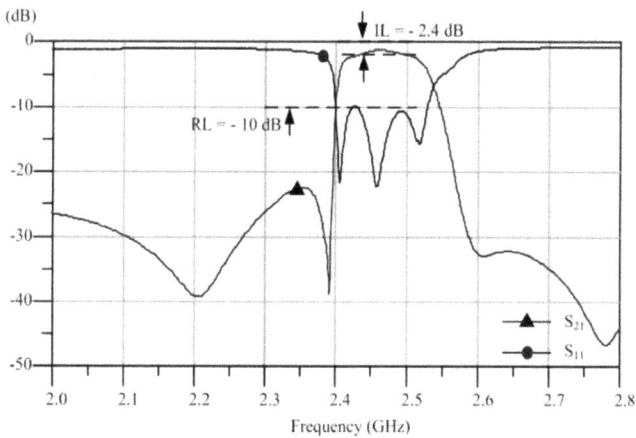

Figure 16. Simulation of tunable BAW filter with inductors added in parallel with the series resonators and capacitors added in series with the shunt resonators.

Comparisons between the measurements of the tunable filter with the original one are shown in Fig.17 and Fig.18.

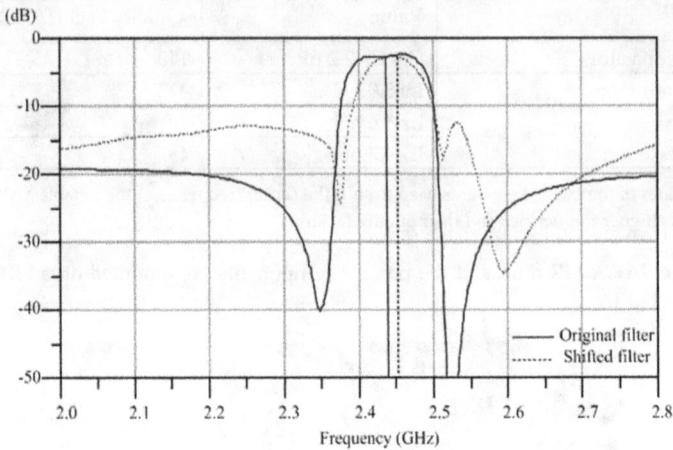

Figure 17. Shift towards higher frequencies: Comparison of transmission characteristic (S₂₁)between the tunable BAW-SMR filter (passive elements added) and the original one.

Based on the measurements of the tunable filter, we can note -4.5 dB of insertion losses and a shift of +0.6% of the center frequency (2.44 GHz) towards the higher frequencies (Fig.17). As well, a return loss of -7 dB is obtained (Fig.18). The filter high insertion losses are mainly due to the low resonators quality factor obtained in the fabrication (Q = 200) and to the low quality factor of the passive element used. Moreover, the parasitic capacitors generated by the FR4 PCB and the bonding wires used to connect the tunable filter with the passive elements caused a reduction of 15 MHz to the bandwidth of the tunable filter.

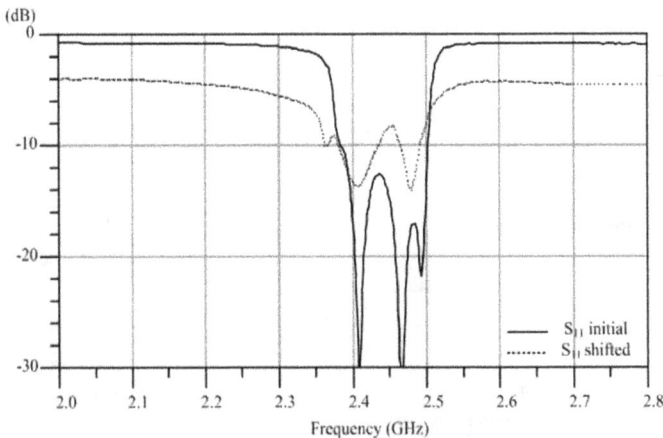

Figure 18. Shift towards higher frequencies: Comparison of reflexion characteristic (S₁₁) between the tunable BAW-SMR filter (passive elements added) and the original one.

4.2. Shift towards lower frequencies

This time and in order to shift the center frequency of the filter towards lower frequencies, we have to move all the zeroes towards these frequencies.

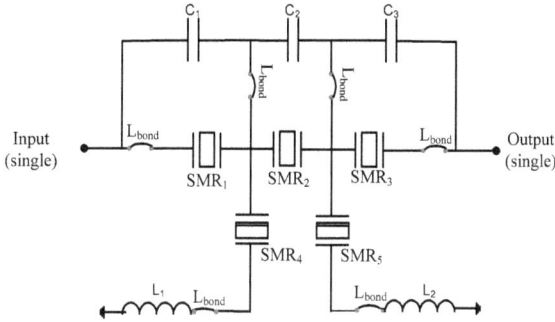

Figure 19. Tunable BAW-SMR filter with capacitors added in parallel with the series resonators and inductors added in series with the shunt resonators.

Thus, we have added capacitors in parallel with the series resonators and inductors in series with the shunt resonators that constitute the ladder BAW-SMR filter (Fig.19). Table 2 presents the values and the quality factor of the external passive elements used.

	Value	Quality Factor 'Q'
Capacitors	$C_1=2.2$ pF	140
	$C_2=1$ pF	200
	$C_3=2$ pf	150
Inductors	$L_1=L_2=2.2$ nH	57

Table 2. Values of the passive elements used to shift the center frequency of the tunable BAW-SMR filter towards lower frequencies and their quality factors.

As same as before, the tunable BAW-SMR filter and the passive components are mounted on a FR4 substrate. Fig.20 shows the simulation results of the tunable filter.

The insertion loss (IL = -2.4 dB) obtained by simulation is due to quality factors of resonators (Q = 500). In addition the return loss (RL) is -9.5 dB, and the out of band rejection is 16 dB at 2.0 GHz (Fig.20). The simulation results shows that a shift of -1% of the initial central frequency (2.44 GHz) is obtained. A comparison between the measurements of the tunable filter with the original one is shown in Fig.21.

Based on the measurements of the tunable filter, we can note -4.5 dB of insertion losses and a shift of -1.3% of the center frequency (2.44 GHz) towards the lower frequencies (Fig.21). In addition, the parasitic capacitors generated by the FR4 PCB and the bonding wires used to connect the tunable filter with the passive elements caused a reduction of 13 MHz to the bandwidth of the tunable filter.

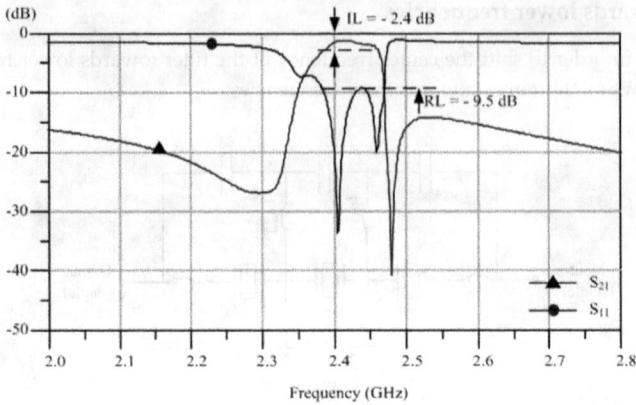

Figure 20. Simulation of tunable BAW filter with capacitors added in parallel with the series resonators and inductors added in series with the shunt resonators.

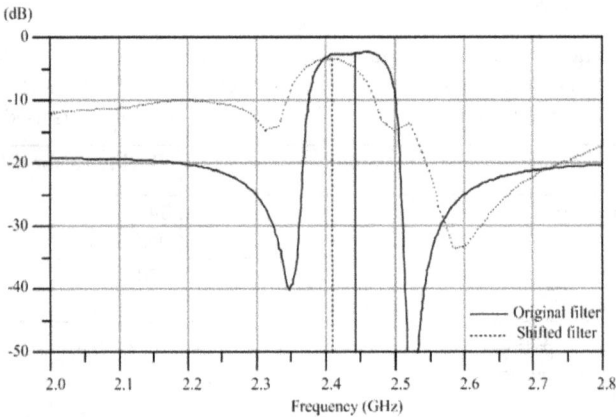

Figure 21. Shift towards lower frequencies: Comparison between the tunable BAW-SMR filter (passive elements added) and the original one.

As a conclusion, one should note that in contrast to [13], where lumped elements (inductors or capacitors) were proposed to be added at a time, in this paper the use of capacitors and inductors together have shown how to shift the center frequencies towards higher or lower frequencies.

5. Digitally tuning BAW filters

To validate the concept of digitally tuning BAW filters using passive elements controlled by CMOS transistors, we present in this part the use of CMOS switches at the terminals of capacitors (Fig.22) [6].

When a transistor is ON, the capacitor is short circuited, and when a transistor is OFF, the capacitor will be considered in series with the shunt resonator. Thus, the bandwidth and the characteristics of the filter will be modified. The circuitry of the filter, capacitors, transistors and the associated "bonding wires" are shown in Fig.22.

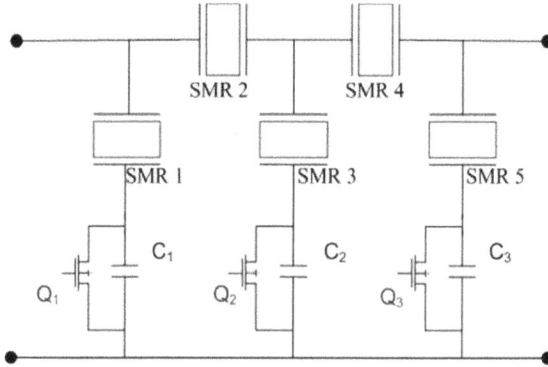

Figure 22. Tunable BAW-SMR using CMOS transistors.

The BAW-SMR filter used in this study is a fifth order filter designed for the W-CDMA standard in the ladder topology. This topology is composed by the resonator in series and parallel, the parallel resonators are loaded and their resonance frequencies are smaller than the series resonators. The die photography of the tunable BAW-SMR filter is shown in Fig.23. The filter has reduced size, and the die area is 1450*985 µm². Moreover, many passive pads connecting the filter with the active chip were taken in consideration.

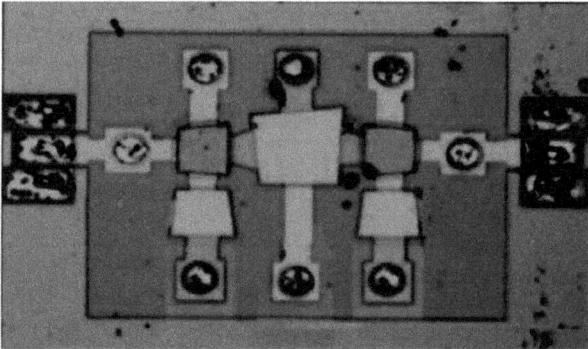

Figure 23. Microphotography of ladder BAW-SMR for the W-CDMA standard.

Fig.24 presents the comparison between the measurement and the simulation of the BAW-SMR filter. Electromagnetic simulation of the overall filter structure has been accomplished by using the ADS-Momentum software, where the acoustical effects have been included using the Mason Model. The filter is designed for implementation in SiP context. As shown in Fig.24, the measurements are in concordance with the simulations. However, the filter fulfills the requirements for the W-CDMA standard, exhibiting -2.77 dB of insertion loss, -8.75 dB of return loss and selectivity higher than 38 dB at 40MHz offset from the operating frequency.

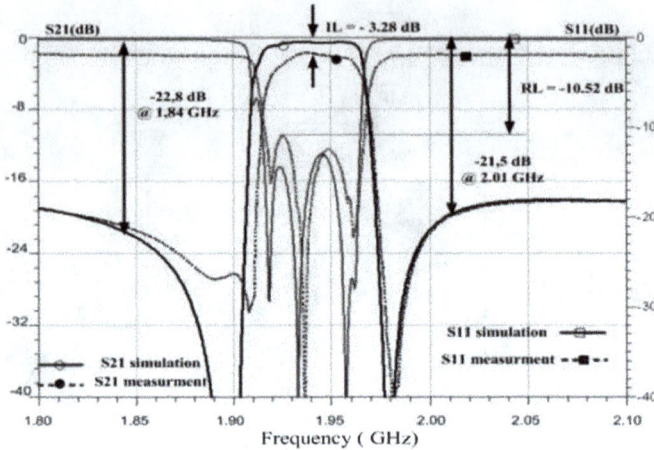

Figure 24. Simulation and measurement results of the W-CDMA ladder BAW-SMR filter.

5.1. Switches design

To adjust the bandwidth of the BAW filter, a chip is realized in 65nm CMOS technology. This chip is composed by the capacitors mounted in series with the MOS transistor, and these transistors are controlled by a 2 to 4 decoder (Fig.25).

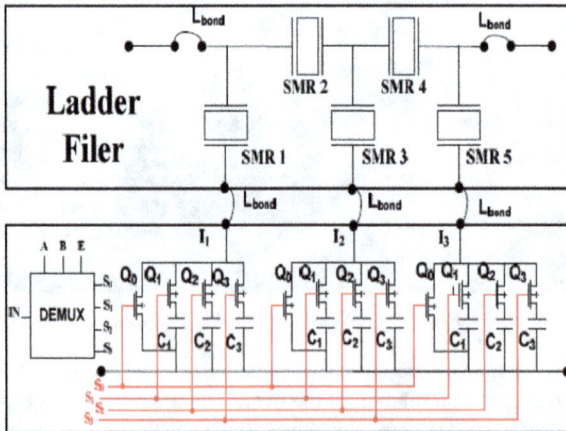

Figure 25. Circuitry of the tuning mechanism.

Fig.26 shows the layout of the tuning circuit. I_n symbolize the pad for connection with ladder filter. V_G, V_{dd1}, V_{dd2} and GND correspond to the gate voltage, the command of decoder and the ground respectively. The size of the Silicon area is $335*330\mu m^2$.

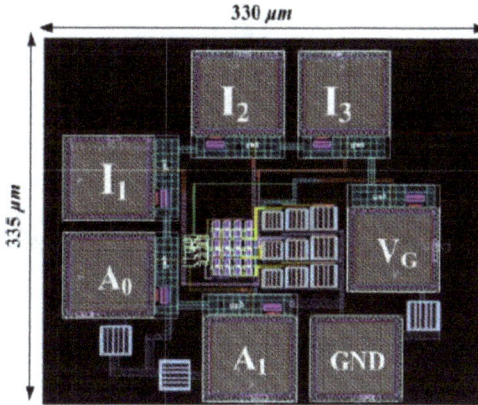

Figure 26. Layout of the tuning mechanism.

The tuning is attained by controlling the MOS transistors and capacitors in series with shunt resonators. Each transistor is open or short circuited by obtaining different outputs of the 2 to 4 decoder. Table 3 shows the truth table of the realized decoder. A, B, E symbolize the input of a decoder commanded by V_{dd1}, V_{dd2} and V_G respectively, S_n (n = 0, 1, 2 or 3) represent the output of this decoder used to control Q_n. All transistors used in the tuning mechanism are provided by STMicroelectronics (CMOS 65 nm). The width and length of the gate are: W = 50 μm and L = 0.06 μm. The main parasitic elements are taken into account in the simulation (C_{gs}, C_{gd}, C_{ds} and R_{on}). The length of bonding wire is 2 mm. It represents an inductive effect of approximately 2 nH at 2 GHz. R_{on} of the MOS transistor is function of its dimensions and of the gate voltage (V_G). Thus, with an external adjustment of V_G, R_{on} value is regulated.

A	B	E	S_0	S_1	S_2	S_3
	0	1	0	0	0	1
0	1	1	0	0	1	0
1	0	1	0	1	0	0
1	1	1	1	0	0	0
X	X	0	0	0	0	0

Table 3. Truth table of the 2 to 4 decoder.

5.2. Co-design: BAW filter- 65nm CMOS chip

Fig.27 shows the microphotography of the association of the ladder filter with the active chip. The devices are connected with bonding wires. The capacitors C_1, C_2 and C_3 values are fixed to achieve 12, 9 and 6 MHz tuning range, respectively. When the output S_0 of the decoder is ON, the transistor Q_0 is ON and all of the capacitors are short circuited. When the output S_n (n = 1, 2 or 3) is ON, the transistor Q_n is ON and the capacitor C_n will be considered in series with the shunt resonators.

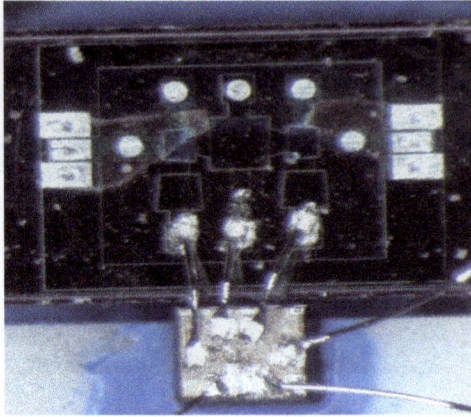

Figure 27. Microphotography of the digitally tunable ladder BAW-SMR for the W-CDMA standard.

The comparison between the simulation and the measurement results is shown in Fig.28. of the tunable BAW-SMR filter combined with the active chip presents a tuning range of 12 MHz, when the output S_3 of the decoder is ON. It show also -1.52 dB of insertion loss and 12 dB of the out-band rejection at 1.85 GHz. This out-band rejection is improved by 3 dB. This degradation is due to the length of bonding wire associating the active chip and PCB [10].

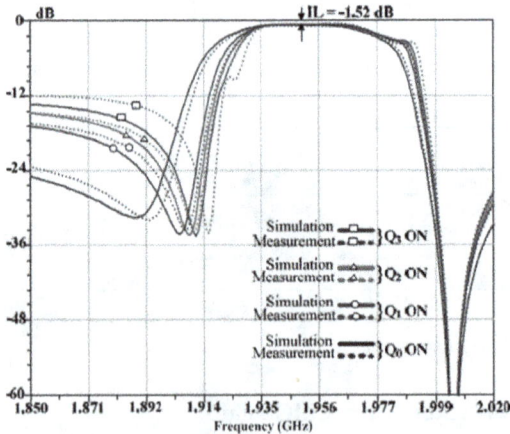

Figure 28. Comparison between the simulation and the measurement of the digitally tunable BAW filter.

6. Conclusion

In this paper, the impedance behavior of the BAW-SMR has been shown. Also, the effects of the addition of passive elements (L, C) to this type of resonators have been illustrated. In addition, a tunable BAW-SMR filter realized in a ladder topology used for the 802.11b/g standard (2.40 - 2.48 GHz) was shown. Mainly, the filter fulfilled the requirements for the WLAN 802.11 b/g standard, presenting a measured -3.3 dB of insertion loss, -12.7 dB of return loss and selectivity higher than 33 dB @ ± 30 MHz of the bandwidth. This tunable BAW-SMR filter has reduced dimensions (1035*1075 μm2). Moreover, the center frequency of this tunable filter was shifted towards higher and lower frequencies by adding passive elements. Measured shifts of -1.3% of the center frequency (2.44 GHz) towards lower frequency and +0.6% of the center frequency towards higher frequencies were obtained. Furthermore, digitally tunable BAW-SMR filter implementation was shown. The tunable filter was designed for the W-CDMA standard. The filter fulfilled the requirements for the WCDMA standard, presenting a measured -2.77 dB of insertion loss, -8.75 dB of return loss and selectivity higher than 38 dB @ ±40 MHz of the bandwidth. Moreover, the center frequency of this tunable filter is digitally shifted towards higher frequencies by adding capacitors in series with transistors that act as switches. These switches are controlled by a 2to4 decoder, and they are added to the shunt resonators.

Author details

M. El Hassan
University of Balamand –Al Kura, Lebanon

E. Kerherve, Y. Deval and K. Baraka
IMS Laboratory – UMR 5218 CNRS – University of Bordeaux, France

J.B. David
CEA-Leti – Minatec – Grenoble, France

D. Belot
ST Microectronics – Crolles, France

Acknowledgement

IMS laboratory is acknowledged for all facilities offered and the access to obtain the filter measurements. Also, CEA-LETI (Grenoble, France) and STMicroelectronics (Crolles, France) are acknowledged for the technology access and filter fabrication.

7. References

[1] Bradley, P. et al. "A Film acoustic bulk resonator (SMR) duplexer for USPCS Handset Applications", *IEEE MTT-S*, pp.367–370, 2001.

[2] Carpentier, J. F. et al. "A SiGe:C BICMOS WCDMA zero-IF RF front-end using an above-IC BAW filter", *IEEE ISSCC*, 2005, pp. 394-395.

[3] Shirakawa, A. A. Jarry, P. Pham, J-M. Kerherve, E. Dumont, F. David, J-B. Cathelin, A. "Ladder-Lattice Bulk Acoustic Wave Filters: Concepts, Design and Implementation", *RF and Microwave Computer Aided Engineering*, 2007.Newell, "Face-mounted piezoelectric resonators", Proc. IEEE, Vol 53, June 1965, pp.575-581.

[4] Newell, "Face-mounted piezoelectric resonators", Proc. IEEE, Vol 53, June 1965, pp.575-581.

[5] Lakin, K.M. Kline, G. McCarron, K.T. "High-Q microwave acoustic resonators and filter", *IEEE Trans. On Microwave Theory and Techniques*, vol. 41, pp. 2139-2146, Dec. 1993.

[6] Lakin, K.M. Belsick, J. McDonald, J.F. McCarron, K.T. "Improved bulk wave resonator coupling coefficient for wide bandwidth filters", *IEEE Ultrasonics Symposium*, Atlanta, GA, USA, pp.827-831, October 9, 2001.

[7] Fattinger G. G. et al. "Thin Film Bulk Wave Devices for Applications at 5.2 GHz", *IEEE UFFC Symposium*, Honolulu, Hawaii, pp. 174-177, 2003.

[8] Ancey, P. "Above IC RF MEMS and BAW filters: fact or fiction", *IEEE BCTM Proceedings*, Maastricht, Netherlands, pp. 186-190, 2006.

[9] Shirakawa, A. A. Pham, J-M. Jarry, P. Kerherve, E. Dumont, F. David, J-B. Cathelin, A."A High Isolation and High Selectivity Ladder-Lattice BAW-SMR Filter", *36th European Microwave Conference*, 2006, Manchester, UK, pp. 905 – 908, 10-15 September.

[10] Tsutsumi, J. Iwaki, M. Iwamoto, Y. Yokoyama, T. Sakashita, T. Nishihara, T. Ueda, M. Satoh, Y. "A Miniaturized FBAR Duplexer with Reduced Acoustic Loss for the W-CDMA Application", *IEEE Ultrasonics Symposium Proceedings*, Rotterdam, Netherlands, pp. 93-96, 2005.

[11] Lakin, K. M. Lakin, K. G. "Numerical Analysis of Thin Film BAW Resonators", *Proceedings of Ultrasonics Symposium*, Honolulu, Hawaii, Vol. 1, pp. 74-79, 2003.

[12] Mason, W. P. "Electromechanical Transducers and Wave Filters", Princeton, New Jersey, Van Nostrand, 1948.

[13] El Hassan, M. Kerherve, E. Deval, Y. David, J.B. Belot, D. "Reconfiguration of Bulk Acoustic Wave Filters Using CMOS Transistors: Concept, Design and Implementation", 2010 IEEE RFIC Symposium, Anaheim, CA, USA, 23 - 25 May, 2010. pp. 241 – 244.

Chapter 8

Acoustics and Vibro-Acoustics Applied in Space Industry

Rogério Pirk, Carlos d'Andrade Souto,
Gustavo Paulinelli Guimarães and Luiz Carlos Sandoval Góes

Additional information is available at the end of the chapter

1. Introduction

During flight, Expandable Launch Vehicles (ELV) are excited by severe acoustic loads in three phases of flight: lift off, transonic flight and maximum dynamic pressure instant [1]. As such, principles to make onboard equipment compatible with the mission environments must be adopted. At lift off, the highly intense acoustic loads occur; and these levels are usually adopted to qualify payloads and equipments. However, during the transonic flight and maximum dynamic pressure phase, acoustic excitation is also present and such characteristics are also significant for performance evaluation as well as for specific system dynamic qualification/acceptance programs. In this way, noise control treatments (NCT) shall be adopted to alleviate internal vibro-acoustic environments, in view of decreasing costs and developments.

The hostile in-flight environments can damage sensors/conditioners as well as make measurements unreliable. In this way, installation adapters must be designed to protect the sensors. The acoustics of such protective cavities influence the measured sound pressure level (SPL) As such, the cavities must be analyzed and their amplitude-frequency characteristics evaluated. Finally, the measurement corrections, necessary to obtain the actual external SPL, are determined.

Concerning the internal environment found during flights, important launcher subsystems as payload fairing (PLF) and equipment bays shall be investigated and vibro-acoustic analysis can be done, as pointed by [2], [3] and [4]. The PLF is the structural compartment of a launcher where the payload is placed during the flight mission. PLF inner acoustics and its attenuation designs, using virtual prototypes are analyzed using deterministic and statistic techniques. However, when in-flight loading are not characterized, the accounted external air-borne excitation can be those described in [5]. In a similar way, SPL along the launcher structure at lift-off can also be estimated [6]. Furthermore, an alternative procedure to characterize external SPL during flight can also be adopted as described by [7]. Passive vibration control techniques can be used to attenuate structure-borne vibration and the use of viscoelastic materials adding structural damping to reduce the magnitude of vibrations is a well-known solution, usually applied in space and aeronautical industries. On the other hand, the use of active vibration control (AVC) is still considered difficult to be implemented in space industry.

For acoustic noise attenuation, the standard practice is to use passive techniques like blankets ([8] and [9]), which attenuate sound by trapping the energy in the blanket material and dissipating it as heat [10] and Helmholtz resonators tuned to absorb acoustic energy at one or some specific frequencies, typically the cavity frequencies as done by [11].

Another acoustic crucial subject in space industry is combustion instability, since it can severely impair the operation of Liquid Propelled Rocket Engine (LPRE) [12]. In this way, solutions for instability problems in combustion chambers of LPRE as well as solid rocket motors (SRM) are of large interest. In [13], it is described that combustion instability can be verified when the power spectrum of the acoustic pressure measured during tests is analyzed. When an oscillation is observed, i.e., combustion instability, well-defined sound pressure peaks, summed to the background noise are present. Such peaks are correlated to the resonance frequencies of the combustion chamber. In this way, the coupling of acoustic natural frequencies and burning oscillations of the combustion chamber occurs, which can cause instabilities and consequent unexpected behavior as efficiency loss or even the explosion of the engine. In the early developing phases of liquid rocket engines, it is usually proposed the investigation of different combustion chamber configurations [14]. This is usually done in two steps as follows: using theoretical calculation and through experimental measurements. In this way, theoretical and experimental natural frequencies of the acoustic cavity are obtained. Further studies must be performed, applying devices and techniques to attenuate pressure oscillations inside combustion chambers and devices as Helmholtz Resonators, baffles and ¼ wave filters are largely used ([15] and [16]).

This chapter describes three case studies applied on space industry. Firstly, analytical and numerical modelings of in-flight external microphone protection devices are described. Testing procedures and the SPL measurement correction factors determination are also presented. As a second case study, deterministic and statistical coupled vibro-acoustic analysis techniques are used to estimate PLF internal SPL at lift-off as well as to assess the effect of including NCT (blanket materials) on its skin. The modeling procedures and experimental ground test are described. Finally, in the third case, the acoustic characterization of combustion chambers is presented. Cold tests are described as well as the theoretical modeling procedures. The pressure attenuation control technique using Helmholtz Resonators are also presented. In all three case studies, theoretical x experimental results are depicted.

2. External on board microphones installation devices

At lift off, the source of the acoustic noise is the gas stream ejected by the motors (Fig. 1). Such acoustic pressure lies in the range of 140 to 180 dB near the rocket and is very close to an acoustic diffuse field (ADF) noise. At transonic flight the launcher is excited by the turbulent boundary layer (TBL) in the neighborhood of the shock waves. According to [17], when the maximum dynamic pressure occurs, the unsteady pressure field applied to the launcher is due to aerodynamic noise. The characteristics of such noise are very different from those at lift off. Non-attached flows increase the pressure in low-frequencies, which excite the launcher first structural modes. A simulation of the VLS-1 flight aerodynamics was done by Academician V. P. Makeyev State Rocket Centre (SRC- Makeyev), as shown in Fig. 2. Notice that the upper nose and 1st stage noses are the most exposed regions to aero-acoustic noise.

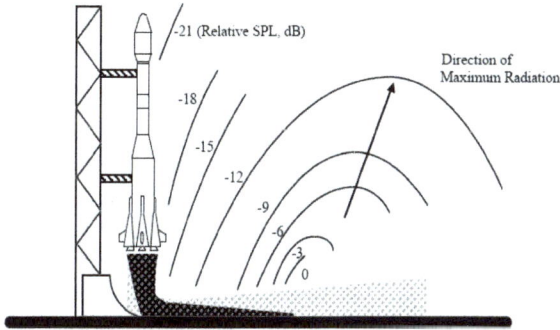

Figure 1. Acoustic noise at lift off

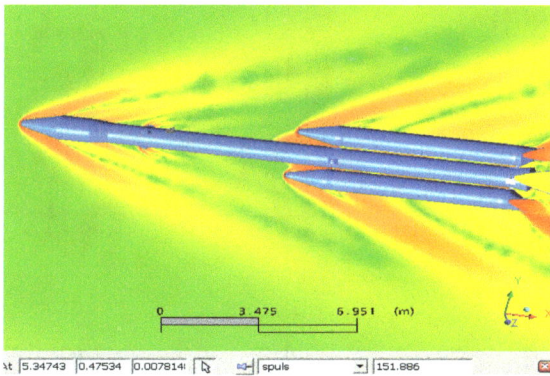

Figure 2. In-flight aero-acoustic noise

In view of having a good characterization of the in-flight acoustic loads acting upon the launcher structure, external acoustic measurements are required. Due to the high SPL and hostile environments found during flight, special microphones and adapters are specified. Such adapters must be designed in order to provide appropriate microphone/pre-amplifier installation and protection. Besides, when necessary, measurement correction procedures must be adopted. In this way, measurement programs for in-flight external acoustic characterization shall be developed, which may take into account three main phases as: preparation for experimental studies and acoustic testing sensors, ground development testing of acoustic sensors and methodology for reading acoustic pressures during flight.

Three different adapters for ¼" microphones were conceived, as described by [18]. On the upper parts of the launcher, two different configurations can be adopted. Firstly, at the PLF, heating and propellant dust effects are not significant; therefore, a structure flush installation (Fig. 3a) can be used. In this case, the measured SPL can be read directly. The second configuration, straight adapter (Fig. 3b), is applied for microphones installed near the equipment bays, where one has temperature and dust influences and, therefore, the sensor/conditioner must be protected. For such an assemblage, the protection channel dynamics directly affect the sensor response and, as a result, a measurement correction must be done.

On the bottom, the intense SPL at lift off generate a severe acoustic excitation of the first stage back modules region. Highly hostile dust, hot gas flow, heat flux and temperature environments are present during the motors operations. Nevertheless, the angular adapter must be used to install acoustic microphone/pre-amplifier, as shown in Fig. 3c. Notice in Figs. 3b and 3c, that the adapters were designed with small acoustic straight and angular cavities, respectively. When acoustically excited, the acoustical responses of such cavities directly influence the measured SPL, since the external pressure excitation profile and the measured signal are related by the cavity transfer function.

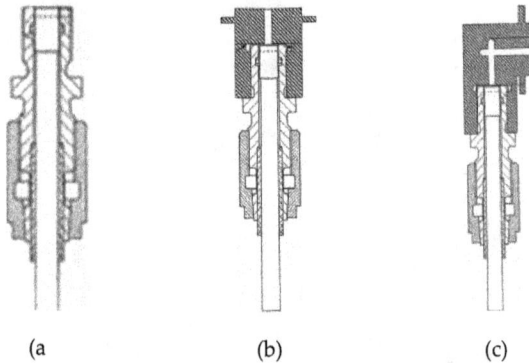

(a)   (b)   (c)

Figure 3. (a) Flush adapter; (b) Straight adapter; (c) Angular adapter

In order to determine straight and angular adapters' dynamics, analytical and numerical calculations are done. The transfer functions of these channels are evaluated during ground acoustic tests, on which acoustic excitation with SPL close to those expected during flight is used to excite the cavities. Consequently, the measured SPL as well as the channels transfer functions are determined. Finally, the measurement corrections are determined, which may be applied when these adapters are used.

2.1. Mathematical models

2.1.1. Analytical model

In view of describing the dynamical behavior of the protective channels, one can assume the straight and angular channels as Helmholtz Resonators, which the channels and the space for microphone installation are accounted as the resonator throat and volume, as described by Eq. (1) [19].

$$f_0 = \frac{c_0}{2\pi}\sqrt{\frac{S}{V_0(l+l_c)}} \tag{1}$$

where: f_0 : natural frequency, c_0 : sound speed, S : cross section of the resonator throat, V_0 : volume of the resonator cavity, l : length of the resonator throat, $l_c = 0.8r$: end correction and r : radius of the resonator throat cross section.

By considering the dimensions of the adapters into Eq. (1), one can calculate the natural frequencies of the straight and angular channels shown in Figs. 3b and 3c. These calculations are the starting point to assess the accuracy of the numerical models, built by using the Finite Element Method, once analytical x numerical natural frequencies can be compared.

2.1.2. Numerical model by Finite Element Method (FEM)

In a similar way as in structural dynamics, an acoustic cavity FEM model will have an acoustic stiffness matrix [K_a], an acoustic mass matrix [M_a], acoustic excitation vectors {F_{ai}} and an acoustic damping matrix [C_a]. The combination of these components yields the acoustic finite element model, which can be solved for the unknown nodal pressure values p_i [20].

$$\left(\left[K_a\right] + j\omega\left[C_a\right] - \omega^2\left[M_a\right]\right)\cdot\{p_i\} = \{F_{ai}\} \tag{2}$$

Acoustic finite element models of the three adapters cavities are built. All cavities' surfaces were considered as rigid walls but the openings that are in direct contact with the external acoustic environment. In such cases, opened surfaces approximated using prescribed nodal pressures (equal to 0 for the eigenanalysis) were considered. The fluid inside the cavities is assumed as air at 15° C (c=340 m/s, ρ=1.225 Kg/m³, values used in this entire chapter).

Linear tetrahedral fluid elements are used in all three meshes. In order to have good prediction accuracy in the frequency range of interest, the general rule of thumb that requires at least 6 elements per wavelength is adopted. The main meshing characteristics are described in Table 1.

The acoustic load generated at lift off is simulated as an acoustic diffuse field (ADF). According to [21], an ADF is defined as an acoustic field in which the SPL is equal at any location and have an identical energy distribution in all directions. Such ADF can be obtained in an acoustic reverberant chamber, where the reflections along the rigid walls lead to this field. A formal way to describe an ADF consists on superimposing an infinite number of uncorrelated plane waves through different directions. In a FEM model, a finite number of uncorrelated plane waves can be generated and the pressure due the superposition of all the uncorrelated plane waves can then be applied as prescribed nodal pressures on the cavity's open surface (see [22]).

Parameters	Mesh		
	Flush adapter	Straight adapter	Angular adapter
Number of elements	4,419	3,986	10,266
Number of nodes	1,009	980	2,367
Maximum frequency (6 elements/wavelength)	45,374 Hz	44,593 Hz	44,199 Hz

Table 1. Meshes data

The FEM models of the three adapters are shown in Figs. 4a, 4b and 4c. The cavity's transfer functions are calculated by imposing prescribed nodal pressures in the nodes marked with small green arrows. In order to save time and computational efforts, a modal solution method is adopted using the first 14 modes. A modal damping of 5% is considered in these calculations.

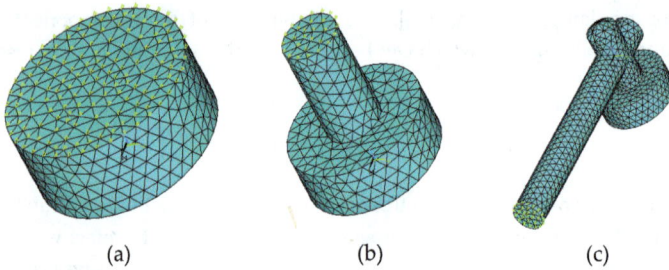

(a)   (b)   (c)

Figure 4. a) Flush adapter mesh; b) Straight adapter mesh; c) Angular adapter mesh

2.2. Experimental set up

An experimental unit is conceived to characterize all three adapters, as shown in Fig. 5. The experimental unit is placed into an acoustic reverberant chamber and submitted to an ADF, with a frequency profile shown in Fig. 6, which impinges the plate where the adapters and microphones/conditioners are installed. Care is taken to assure that the plate has structural response similar to that found along the launcher skin. Accelerometers are installed on the plate to measure the acoustically induced structural vibration.

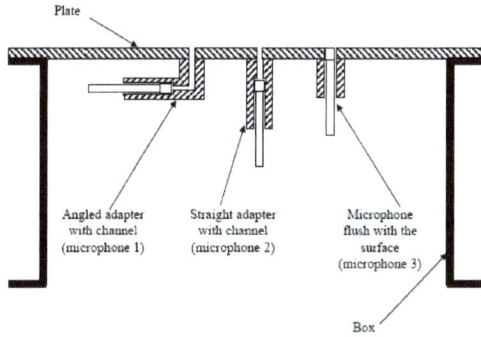

Figure 5. Experimental unit

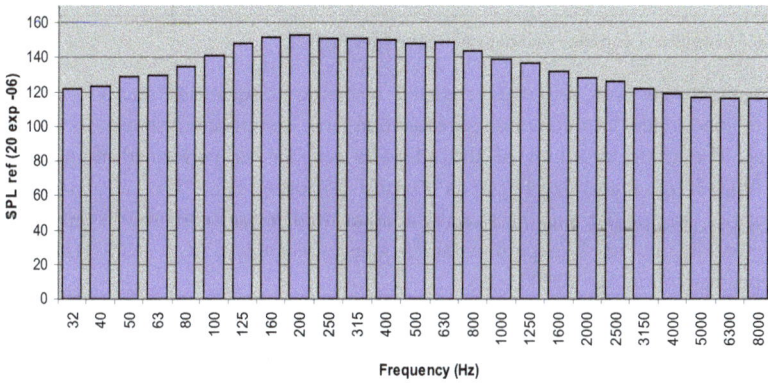

Figure 6. Excitation profile Overall Sound Pressure Level (OSPL) – 160 dB

2.3. Methodology for the external measured SPL correction

2.3.1. Characterization of the adapters´ cavities

The amplitude-frequency characteristics of such adapters must be accounted to determine the acoustic levels measured during the flight mission, since the device channels operate like filters. The SPL measured by the microphone installed with the flush adapter (Fig. 4a) is considered as the reference. Then the adapters transfer functions relating the input signal given by the microphone with the flush adapter and the output signal given by the microphones with the protective adapters (angular or straight) can be obtained by Eqs. (3) and (4)

$$M1(\omega) = H_{M1}(\omega)M3(\omega) \tag{3}$$

$$M2(\omega) = H_{M2}(\omega)M3(\omega) \tag{4}$$

where: $H_{M1}(\omega)$ and $H_{M2}(\omega)$: angular and straight adapter transfer functions, respectively. M1, M2, M3: measured SPL into angular, straight and flush adapters.

2.3.2. *Results*

Analytical and numerical natural frequencies results are compared with those obtained by acoustic testing in Table 2. As pointed out before, the measured SPL with the flush adapter can be read directly. In this way, only the adapters shown in Figures 4b and 4c are considered.

	Straight Adapter				Angular Adapter			
Mode	Analytical	FEM	Testing	FEM-Testing Difference (%)	Analytical	FEM	Testing	FEM-Testing Difference (%)
1	5,000	5,396.6	5,202	3.74	2,300	2,204.7	1,953	12.88
2	-	27,032	-	-	-	8,660.3	7,831	10.58

Table 2. Theoretical x Experimental natural frequencies (in Hz)

Disagreements between test and predicted resonance frequencies can be explained by possible inaccuracy in microphone installation and complicated shape of the angular channel. The characterization of the adapters' cavities is performed numerically, by calculating $H_{M1}(\omega)$ and $H_{M2}(\omega)$ (Eqs. (3) and (4)) using FEM. The calculated response functions of the angular and the straight adapter models are compared to those obtained experimentally in Fig. 7 (experimental frequency resolution Δf = 18.78 Hz; numerical frequency resolution Δf = 20.00 Hz).

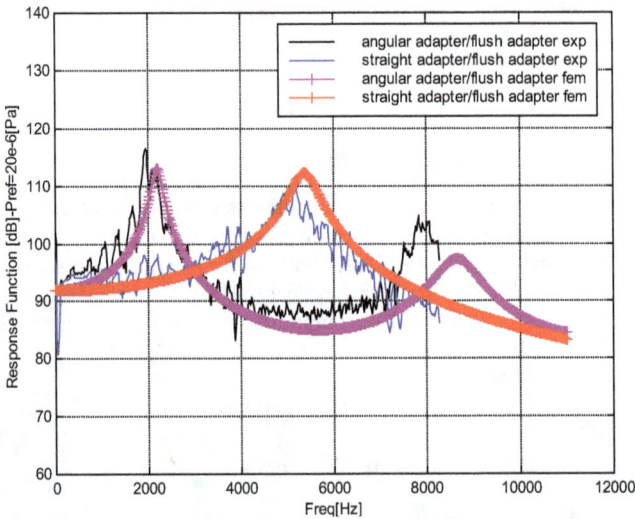

Figure 7. Microphones with adapters transfer functions (numerical and experimental)

The theoretical and experimental response functions show good agreement, with a shift in the second resonance peak for the microphone with angular adapter. The errors can be caused by: bad characterization of the ADF spectral distribution into the FEM model; adapter's geometry complexity; microphone installation inaccuracy or a combination of some of these factors.

Equations 3 and 4 show that the external noise (given by $M3(\omega)$) can be identified by knowing the internal noise and the inverses of the transfer functions for the angular and straight adapters.

3. Vibro-acoustic modelling of payload fairings (PLF)

A complete survey of PLF vibro-acoustic environment must be carried out, in order to determine its inner SPL. In this respect, it is important to have reliable numerical tools that can predict the responses of ELV systems, subjected to in-flight acoustic loads and that enable NCT design.

Low-frequency coupling techniques are used to estimate a PLF dynamic behavior. The fairing body and its inner acoustic domain are analyzed by using Finite Element Method (FEM) and Boundary Element Method (BEM). Structural FEM/fluid FEM and structural FEM/fluid BEM modeling techniques are then applied. In order to simulate the lift off acoustic excitation, an ADF of 145 dB OSPL is applied on the fairing body and coupled calculations are done from 5 to 150 Hz, which yielded the acoustic and skin responses for both models. Modal expansion and semi-modal expansion model techniques are applied, respectively.

For the high-frequency analysis, it is applied the Statistical Energy Analysis (SEA) technique, for a frequency range from 5 to 8,000 Hz. The 145 dB OSPL excitation is applied to the structural panels of the fairing and the acoustic and structural mean responses are calculated.

In view of validating the numerical predictions for the fairing, acoustic test is done to measure the acoustics inside the PLF. The PLF is submitted to 145 dB OSPL in a 1,200 m^3 acoustic reverberant chamber and microphones are positioned in its inner domain.

The implementation of sound absorption blankets is applied as a control technique to attenuate acoustic noise from medium- to high-frequency bands. SEA is a technique for high-frequency analysis; therefore, adequate to assess the influence of blankets on space systems. The generated SEA fluid-structure model is used to calculate internal SPL, with single-, double-, and multi-layered noise control treatments (NCT). Two NCT modeling approaches are used to simulate the effect of blanketing the fairing cavity:

i.   acoustic materials Biot's parameters, given by the manufacturer;
ii. material samples absorption coefficient, measured in a Kundt Tube.

3.1. Model description

The analyzed fairing is hammerhead type geometry and is composed of the body structure, functional components as electric and pyrotechnic components of the ejection system, mechanisms as well as the exterior cork liner. Figures 8a and 8b show the Brazilian VLS fairing structure.

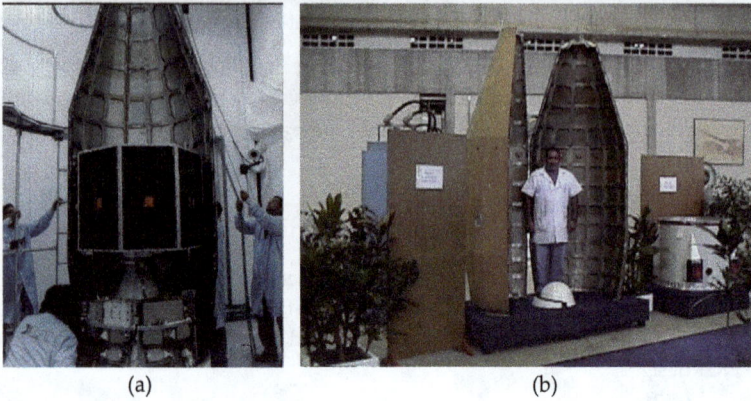

(a)   (b)

Figure 8. a) PLF structure; b) PLF structure

3.2. Modelling methodology

3.2.1. Low-frequency modeling (deterministic) techniques

In view of predicting the operational fairing cavity SPL, both the dynamic displacements of the fairing structure as well as the acoustic pressure fields at the interior and the exterior side of the fairing should be considered. In this study, however, the fluid-structure coupling interaction between the structural displacements and the exterior acoustic pressure field is neglected. The exterior acoustic pressure is assumed to be a known external excitation for the vibro-acoustic system, consisting of the fairing body and the internal acoustic cavity. The FEM and BEM are the most appropriate numerical techniques for the (low-frequency) dynamic analysis of this type of vibro-acoustic system.

FEM based models for coupled vibro-acoustic problems are most commonly described in an Eulerian formulation, in which the fluid is described by a single scalar function, usually the acoustic pressure, while the structural components are described by a displacement vector. The resulting combined FEM/FEM model in the unknown structural displacements and acoustic pressures at the nodes of, respectively, the structural and the acoustic FEM meshes are [20],

$$\left(\begin{bmatrix} K_S & K_C \\ 0 & K_A \end{bmatrix} + j\omega \begin{bmatrix} C_S & 0 \\ 0 & C_A \end{bmatrix} - \omega^2 \begin{bmatrix} M_S & 0 \\ -\rho K_C^T & M_A \end{bmatrix} \right) \begin{Bmatrix} w_i \\ p_i \end{Bmatrix} = \begin{Bmatrix} F_{Si} \\ F_{Ai} \end{Bmatrix} \tag{5}$$

Where:

K_S – Structural stiffness matrix M_S – Structural mass matrix C_S – Structural damping matrix K_C – Fluid – structure coupling matrix

M_A – Fluid mass matrix K_A – Fluid stiffness matrix C_A – Fluid damping matrix
w_i – Structural displacements nodal vector p_i – Acoustic pressures nodal vector
F_{Si} – Force nodal vector F_{Ai} – Acoustic sources nodal vector

In comparison with a purely structural or purely acoustic FEM model, the coupled stiffness and mass matrices (Eq. (5)) are no longer symmetrical due to the fact that the force loading of the fluid on the structure is proportional to the pressure, resulting in a cross-coupling term K_C in the coupled stiffness matrix, while the force loading of the structure on the fluid is proportional to the acceleration, resulting in a cross-coupling term $M_C = -\rho K_C^T$ in the coupled mass matrix.

Low-frequency vibro-acoustic problems can also be modeled by describing the structural behavior in a FEM model and the fluid behavior in a BEM model. In the same way as in the FEM/FEM technique, deterministic FEM/BEM models are usually described by acoustic double layer potential and structural displacement, which are the field variables. Equation (6) presents the resulting combined FEM structural displacements and BEM acoustic pressure differences at the nodes, for a coupled FEM/BEM mesh [23].

$$\begin{pmatrix} K_S + j\omega C_S - \omega^2 M_S & L_C \\ L_C & \dfrac{D}{\rho_0 \omega^2} \end{pmatrix} \begin{Bmatrix} w_i \\ \mu_i \end{Bmatrix} = \begin{Bmatrix} F_{Si} \\ F_{Ai} \end{Bmatrix} \tag{6}$$

Where L_C is the fluid-structure coupling matrix, D is the BEM acoustic matrix of coefficients and μ_i is the nodal vector of double layer potentials.

In deterministic models, the dynamic variables within each element are expressed in terms of nodal shape functions, usually based on low-order (polynomial) functions. Since these low-order shape functions can only represent a restricted spatial variation, a large number of elements is needed to accurately represent the oscillatory wave nature of the dynamic response. A general rule of thumb states that for fluid-structure interactions, at least 6 (linear) elements per wavelength are required to get reasonable accuracy. Since wavelengths decrease for increasing frequency, the FEM model sizes, computational efforts and memory requirements increase also with frequency. As a result, the use of FEM and BEM models is practically restricted to low-frequency applications. In comparison with uncoupled structural or acoustic problems, this practical frequency threshold becomes significantly smaller for coupled vibro-acoustic problems, since a structural and an acoustic problem must be solved simultaneously. Moreover, the matrices in a coupled model are no longer symmetrical, so that less efficient non-symmetrical solvers must be used. As a consequence, the computational effort, involved with the use of coupled FEM/FEM and FEM/BEM models for real-life vibro-acoustic engineering problems, becomes large at very low frequencies.

In order to obtain coupled vibro-acoustic response predictions within reasonable computational efforts, the dimensions of the FEM/FEM problem (Eq. (5)) have to be reduced. The most applied technique for model reduction is the modal superposition technique, which expresses the unknowns of the system in terms of a modal basis, resulting in a set of unknown modal participation factors, whose size is much smaller than the size of the original set of unknowns. A modal expansion in terms of uncoupled structural and acoustic modes is performed by using computationally efficient symmetric eigenvalue algorithms and requires much less computational effort than the use of vibro-acoustic (coupled) modes. However, a large number of high-order uncoupled acoustic modes is required to accurately represent the normal displacement continuity along the fluid-structure interface.

In a FEM/FEM virtual prototype, a modal expansion in terms of uncoupled structural and uncoupled acoustic modal bases is used, in order to keep the computational efforts within reasonable limits. On one hand, structural wavelengths are much smaller than acoustic wavelengths, so that the structural FEM mesh of the fairing must be finer than the acoustic FEM mesh of the inner cavity. On the other hand, due to the continuity of the normal structural and fluid displacements along the fluid-structure coupling interface, both meshes must be compatible in this region. In this framework, the following modeling methodology is adopted: 1) A fine FEM mesh of the fairing is used for the construction of the uncoupled structural modal data basis. 2) The resulting modes are then projected onto a FEM coarse mesh of the fairing structure. 3) For the acoustic cavity FEM mesh, the same mesh density is used along the fluid-structure coupling interface as the PLF structural coarse mesh, while the mesh density is slightly decreased towards the central axis of the cavity. 4) The uncoupled modes, resulting from this acoustic FEM mesh, together with the projected structural modal basis are used in a coupled FEM/FEM model. It is important to highlight that the coarse structural mesh has only the shells of the fairing structure, while all reinforcing beams are omitted, since it is assumed that these stiffeners have no significant effect on the fluid-structure coupling interaction, while their presence would increase the computational load of the modeling process.

For the case of the FEM/BEM problem (Eq. (6)), the modal expansion cannot be used, since the frequency dependency of the matrix coefficients in the acoustic part prohibits a standard eigenvalue calculation. Such as, the semi-modal approach, which uses only the expansion of the structural modal data basis, is applied. As mentioned above, BEM drawbacks as fully population of the matrices, complex and frequency dependent models result in a coupled FEM/BEM model less efficient than coupled FEM/FEM model. Therefore, the rule of thumb of 6 (linear) elements per wavelength becomes prohibitive for the actual fairing fluid-structure study. Such a way, a coarsest structural mesh may be generated and the same adopted frequency range for FEM/FEM model is kept for this FEM/BEM model, even considering that the structure has not enough discrete density. However, the modal data basis calculated using the structural fine mesh can assure good results for the structural displacements, in this fluid-structure model, since the expansion in terms of such a data basis is used on these Frequency Response Analysis (FRA) computations. The displacement continuity of the structural and acoustic meshes (same density in the fluid-structure interface) is considered to perform the link and calculate the coupled dynamic skin displacements and acoustic cavity pressure responses, for both FEM/FEM and FEM/BEM models.

3.2.2. High-frequency modeling (statistical) technique

A characteristic of high-frequency analysis is the uncertainty in modal parameters. The resonances and mode shapes show great sensitivity to small variations of geometry, construction and material properties. In addition, programs used to evaluate mode shapes and frequencies are known to be inaccurate for higher modes. In light of these uncertainties, a statistical model of the dynamic parameters seems natural and appropriate. As an alternative method for higher frequency analysis of the inner cavity of fairings, Statistical Energy Analysis (SEA) approach is proposed. This approach is the description of the dynamic system as a member of a statistical population or ensemble, whether or not the temporal behavior is random. SEA emphasizes the aspects of this field dynamical study.

The SEA equations express the energy balance of different subsystems in a model [24]. Some subsystems have direct power input of an independent source, e. g. an excitation force on a structural component, a sound power source in an acoustic medium etc. In general, subsystems can receive power (input power from external sources), dissipate power (internal losses due to damping) and exchange power with other subsystems to which they are coupled (losses due to coupling). SEA fundamental hypothesis as dissipation losses in relation to the energy variable and modal energy proportionality from connected subsystems are used to yield the SEA matrix equation of complex structures. The distribution of the dynamical response in the system due to some excitation is obtained from the distribution of the energy among the mode groups, based on a set of power balance equations for the mode groups.

3.3. FEM structural meshes

The fairing body is divided in five surfaces. The surfaces are discretized by using 4-noded quadrilateral shell elements, while 2-noded beam elements are used for the circumferential and the axial stiffeners. To account for the mass of the cork blanket on the exterior fairing surface, a distribution of concentrated mass elements are attached to the fairing nodes.

A total of 174 structural modes in a frequency range up to 220 Hz have been identified. Table 3 describes the main structural modes calculated by using FEM in the range up to 150 Hz.

mode	frequency (Hz)
first bending	38.637
first breathing	77.821
second breathing	92.694
first longitudinal	108.749
first torsion	125.772
second bending	150.735

Table 3. Fairing body structural modes

Table 3 shows that the first structural bending mode of the fairing is identified at 38.6 Hz, while the second mode is at 150.7 Hz. Figures 9a and 9b show the referred structural modes.

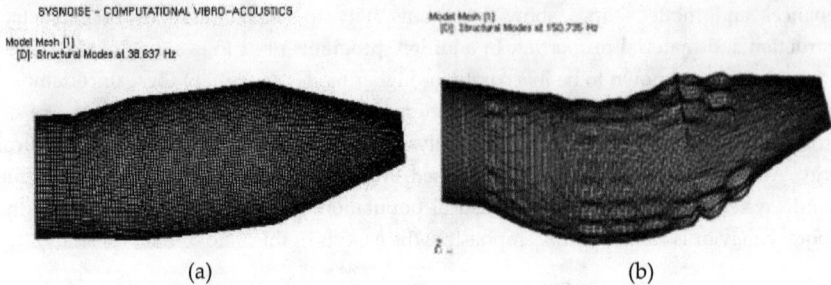

(a)   (b)

Figure 9. a) First structural bending mode; b) Second structural bending mode

3.4. FEM and BEM acoustic meshes

The acoustic FEM mesh consists of 119,577 nodes and 110,238 elements (106,050 8-noded hexahedral elements and 4,188 6-noded pentahedral elements). The cavity is considered filled with air at 15° C. The cavity's bottom and top faces are assumed to be acoustically closed (rigid walls). The acoustic mesh generation takes into account the meshes compatibility on the fluid-structure interface.

A total of 80 acoustic modes in a frequency range up to 566 Hz were identified. Acoustic wavelengths are bigger than structural wavelengths. Such that, a large number of high-order uncoupled acoustic modes is required to accurately represent the normal displacement continuity along the fluid-structure interface. That is why higher frequency range is used to describe the acoustic modal behavior of the fairing. Table 4 describes the acoustic modes in the frequency range up to 150 Hz.

Mode	Frequency (Hz)
rigid body	0.000
first longitudinal	63.491
second longitudinal	112.129

Table 4. Fairing cavity acoustic modes

It can be noticed in table 4 that the first and second acoustic modes of the fairing cavity are identified at 63.5 Hz and 112.1 Hz, respectively. Figures 10a and 10b show the referred acoustic modes.

The BEM acoustic mesh is a 2-D coarsest mesh. Therefore, as the coupled FEM/BEM equation is frequency dependent (Eq. (6)), the acoustic modes are not considered in the acoustic pressure calculations (semi-modal reduction model).

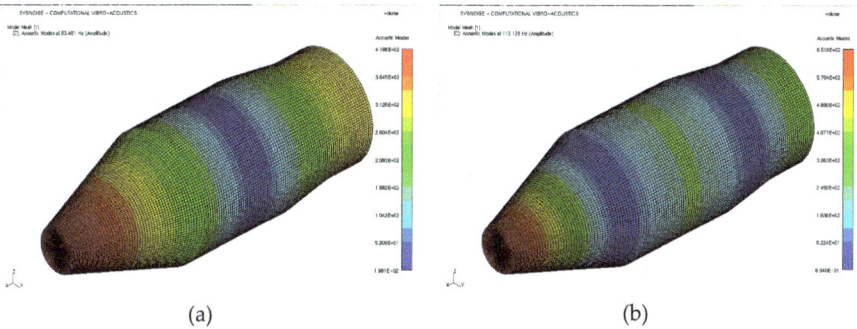

(a)   (b)

Figure 10. a) First acoustic longitudinal mode; b) Second acoustic longitudinal mode

3.5. Model excitation

A uniform pressure loading is simulated by applying a normal point force varying harmonically on all nodes of the fairing shell elements. The force value is defined such that the total load is equivalent to a uniform pressure of 145 dB OSPL. Link of the acoustic and structural parts is done as well as the structural modal data basis is projected to the coarse and coarsest meshes.

In this way, all the meshes, modal data bases and excitation, needed to perform low-frequency calculations, using coupling fluid-structure techniques are ready. Next step is to perform FRA calculations for both models.

3.6. SEA fairing vibro-acoustic model

The fairing body is divided in four surfaces, as shown in Fig. 11a. To account for the rib-stiffened plates of the surfaces 2, 3 and 4, the SEA structural fairing model considers connected plates and beams (longitudinal and circular). The plate structural subsystems are generated as singly curved shells and uniform plates. Shell surface 1 has a thickness of 3mm and is modeled as a simple plate of aluminum (E=72 GPa, ν=0.29, ϱ=2750 kg/m^3), while the other three surfaces are 0.8 mm thick and made of an aluminum alloy (E=72 GPa, ν=0.29, ϱ=7000 kg/m^3). The circular and longitudinal beams are modeled by assigning the same material as the shells of the surfaces 2, 3 and 4 (Figure 11a). Damping loss factors of 1% (for flexure, extension and shear propagating waves) are assigned to the plates and beams subsystems, in order to account for the internal loss factors.

A total of 72 beams (44 longitudinal and 28 circular) and 8 shells (02 singly curved shells of the adaptor, 02 singly curved shells of the lower cone, 02 singly curved shells of the main cylinder and 02 singly curved shells of the upper cone) compose the structural SEA model. Figure 11a shows the SEA plates and beams generated to model the VLS-1 fairing structure. The external blanketed treatment of cork on the surfaces 2, 3 and 4, was simulated in this model as material addition. The layered area and the density of the cork were considered to assign this mass.

The acoustic environment inside the fairing was generated by starting from the structural model. This acoustic cavity was created considering air at 15° C as the fluid as well as the dimensional parameters of the fairing. The top and bottom face of the cavity were assumed to be acoustically closed. Figure 11b presents the 3D acoustic cavity of the fairing.

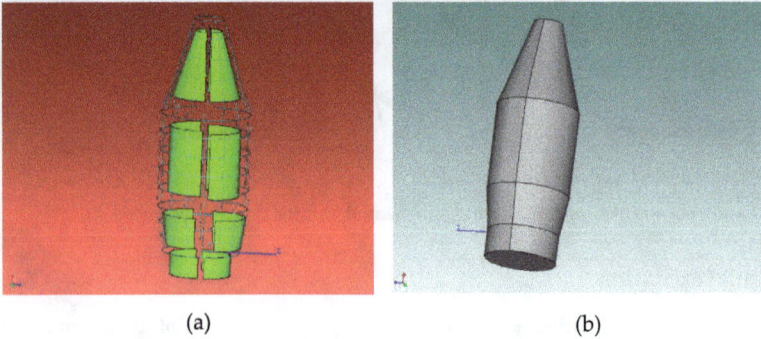

(a)   (b)

Figure 11. a) Structural subsystems (shells, circular beams and longitudinal beams) b) Acoustic cavity of the fairing

The coupling boundary between all the structural and acoustic subsystems is modeled to consider the transmission of power across the junctions. A junction is comprised of connections to any number of coincident subsystems. As a result, all the subsystems that share common nodes are connected by point, line and area junctions and all the appropriate wavefields are connected as well as the corresponding coupling loss factors (CLF) between subsystems are created.

The estimated sound pressure levels at the lift off are assigned to the SEA model. Only elements with large surface areas, as plates and panels, are considered to be susceptible to acoustic excitation ([11], [24] and [25]). An ADF of 145 dB OSPL (Fig. 12) is applied to the plates of the SEA fairing model, which simulates the power input into a structural plate or shell element.

3.7. Analysis results

In view of having a complete knowledge of the fairing dynamic vibro-acoustic behavior, the fairing structural skin as well as its inner acoustic domain responses should be presented. However, since this chapter concerns acoustics, the body structural displacements are not presented here. Below, the obtained results of the acoustic behavior applying vibro-acoustic low-frequency and high-frequency analysis techniques predictions are presented.

3.7.1. Low-frequency techniques

3.7.1.1. FEM/FEM response calculations

A modal expansion in terms of 174 uncoupled structural and 80 uncoupled acoustic modes is used for the coupled calculations. A modal damping of 1% is assigned to all structural modes. All calculations are performed with a frequency resolution of 1 Hz. Figure 13 shows the low-frequency acoustic pressure spectra of the PLF for the case of a uniform exterior pressure loading, using FEM/FEM coupling analysis. It can be seen that the low-frequency pressure is dominated by the first longitudinal mode around 63.5 Hz and the second longitudinal mode around 112.1 Hz.

3.7.1.2. FEM/BEM response calculations

The same structural modal expansion as used for FEM/FEM is used for this FEM/BEM response calculations. Due to the frequency dependency of the boundary integral equation, the acoustic modal basis can not be used. A damping of 1% is assigned to all structural modes. All calculations are performed with a frequency resolution of 2 Hz. Figure 13 presents a comparison of the computed inner cavity space averaged acoustic pressure using FEM/FEM and FEM/BEM techniques.

3.7.2. High-frequency technique

The energy balance (levels and interactions) between different subsystems of the SEA model is calculated. The interest frequency range is 5 to 8,000 Hz, by third octave bandwidth. As mentioned before, SEA technique is more effective in higher frequencies, where dynamic systems present higher modal density. The vibro-acoustic responses of the fairing, using SEA technique, are shown in Fig 13.

It is important to highlight that for the low-frequency range, SEA analysis results are not reliable, since the accuracy of the SEA technique is proportional to the modal density [24].

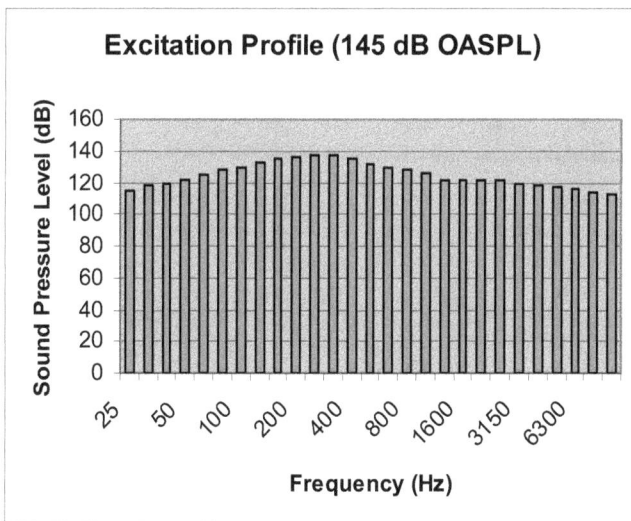

Figure 12. Acoustic diffuse field at lift off

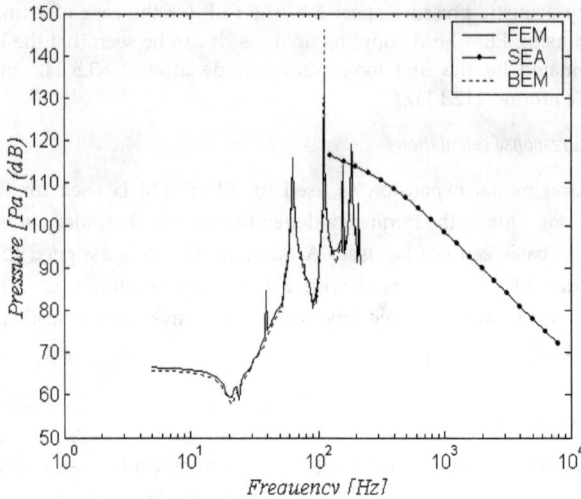

Figure 13. Acoustic Response Inside PLF

Considering the accuracy, advantages and drawbacks of the deterministic and statistical techniques, each of them is successfully applied in different frequency ranges. Such that, for the analyzed PLF, valid response results using deterministic techniques are assumed up to 150 Hz, while valid SEA results are assumed from 300 up 8,000 Hz. It is important to mention that in the "twilight zone" or medium frequency bandwidth (from 150 to 300 Hz), where deterministic models are inaccurate and present prohibitive computation time for the calculations and where the high modal density requirement is not yet accomplished for SEA, both results may be considered, as shows figure 13.

3.8. Model validation

The fairing structure was positioned inside an acoustic chamber and excited with an ADF of 145 dB OSPL. Eight control microphones were positioned inside the reverberant chamber, which feedback the control system. Four measurement microphones were located in the acoustic cavity of the PLF. The measured space averaged SPL is compared with the theoretical acoustic responses, computed using the virtual prototypes (FEM and SEA models) (Fig. 14).

The calculated internal acoustic frequency response function shown in figure 13 may be transformed into 1/3 octave band responses to be compared with the experimental (measured) results. Figure 14 presents the 1/3 octave comparisons for the frequency bandwidths ranging from 31.5 up to 8,000 Hz. It can be noticed, that experimental and calculated low-frequency responses have good agreement, presenting more significant differences only on the 1/3 octave bands 31.5, 40 and 50 Hz. This is because the low-frequency modes of the acoustic chamber are not well excited. However, in the regions where the cavity response is dominant (63 Hz and 112 Hz), differences are pretty small.

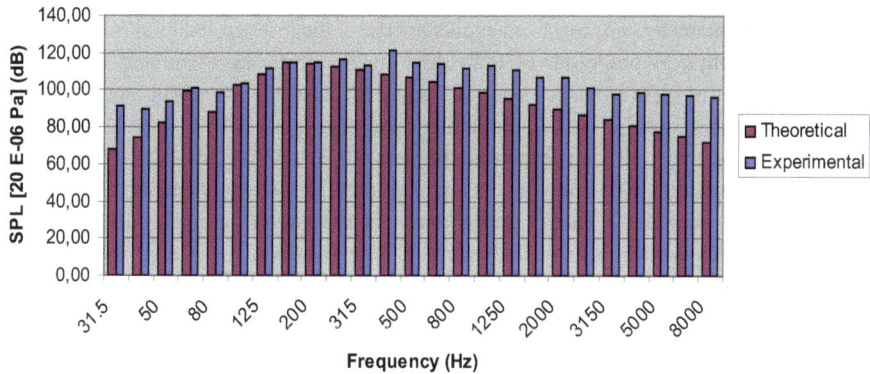

Figure 14. High-frequency theoretical X experimental comparison

For the higher frequencies, a more reasonable comparison should be done using Power Spectral Density (PSD) [28], since SEA calculations may be interpreted as mean values of energetic response functions when averaged at a given frequency over an ensemble of similar systems, differently of peak values resulting from deterministic approaches. However, a qualitative comparison can be presented for 1/3 octave bands from 160 up to 8,000 Hz, since one keeps in mind that mean values and the predicted magnitudes yielded by SEA should under estimate the dynamic response with a certain (acceptable) variance. Valid SEA results are assumed from 300 up 8,000 Hz, since the minimum five modes by bandwidth (modal density) requirement becomes true starting from 300 Hz.

At the beginning phases of space projects, the assessment of the effect of using different passive techniques for acoustic environment alleviation to be applied to PLF is an important issue. One of the main applications of numerical control prediction is the decision, still in the early product development phase, which design version is the most appropriate from the noise control point of view. By introducing the concept of sensitivity analysis, product development can be performed in a more systematic way. In order to predict the efficiency of a NCT, one compares the effects of design modifications.

In this framework, different blanket layers are implemented on the PLF elasto-acoustic virtual prototype and the effects of these NCT implementations are assessed. Since blankets acoustic absorption depend on certain material parameters, two blanket modeling approaches are assessed as follows: material physical Biot's parameters as density, porosity, resistivity, tortuosity, viscous and thermal characteristic lengths, given by the blanket manufacturer and measured normal incidence absorption coefficients of material samples.

Furthermore, the influence of the NCT thicknesses and the presence of air-gaps between blankets are analyzed. Biot's parameters and absorption coefficient approaches are implemented in the coupled elasto-acoustic SEA model of the PLF.

For the Biot's parameters approach, an explicit model of the inserted material is considered, based on the physical properties of individual layers, which are accounted in the SEA model. Six types of glass wools are analyzed and the SPL inside the fairing are calculated. The wools' densities are given in pounds per cubic feet (pcf - 1 lb/ft^3 = 16.02 Kg/m^3). A thickness of 7,62 cm is adopted for almost all glass wools, but the two 1.2pcf glass wools, that presents particular behavior, which adopted thicknesses were 0.19 cm and 0.38 cm. The best performing material is chosen and a comparison between different thicknesses and percentages of layered surfaces of the fairing is done, considering the final weight of the applied NCT. The materials used were glass wools described in Table 5. The wools' Biot's parameters can be found in [4].

Material characteristcs						
Density (pcf)	0.34	0.42	0.60	1.20	1.20	1.50
Thickness (inches)	3.00	3.00	3.00	0.75	1.50	3.00

Table 5. Glass wools used

On the other side, the measured absorption coefficient of multi- and single-layered samples of glass wools of 0.42 and 1.0 pcf were considered. According to [26], air gaps between materials increase the acoustic absorption at low-frequencies. For this case, samples with two different air gaps are positioned into a Kundt tube. The single-layered samples are 3.50 cm thick, while combinations are done with samples of 1.75 cm thick. Other configurations were assembled with air gaps of 1.0 and 3.0 cm between samples. Figure 15 shows the sample combinations. All the measured absorption coefficients are shown in Fig. 16. These absorption coefficients are assigned on the fairing vibro-acoustic model and SPL are calculated.

The PLF acoustic responses for different NCT configurations are shown in Fig. 17. Notice that the insertion of the 0.34pcf glass wool – 7.62 cm yields almost 20 dB of attenuation (chosen as the best performing material). The assessment of the thickness influence is done by assigning 0.34pcf glass wools of 7.62, 10.16 and 12.7 cm thicknesses, with total NCT weights of 3.90, 5.30 and 6.60 Kg, respectively. Figure 18 shows the internal SPL one-third octave distribution, as well as the OSPL.

Figure 19 shows SPL and OSPL from 50 to 8,000 Hz, for the NCT described in Fig. 16, without air gaps. A 3.50 cm double-layered blanket (0.42pcf/1.0pcf) is compared with two single-layered NCT. Notice in this figure that NCT decrease the internal OSPL from 132 dB to 128 dB. Figure 20 shows that single-layered treatment with 1.0 pcf and air gaps presented better results. One can see the air gap effect, since the SPL close to 100, 315 and

500 Hz are higher mainly when the 1.0 pcf material with 3.0 cm air gap is applied. The calculations yielded 127.5 dB OSPL inside the fairing cavity. This means that a gain of approximately 3.0 dB at 100 Hz bandwidth can be obtained, yielding an overall gain of 1.0 dB, approximately. However, air gap installation can be limited, due to fairing internal space. In this case, it is preferable to install the blanketed treatment distant from the panels by small air gaps, instead of bonded, once this installation configuration presents higher transmission loss [10].

Figure 15. Double-layered 0.42pcf (yellow)/1.0pcf (orange), 1.75 cm thick each.

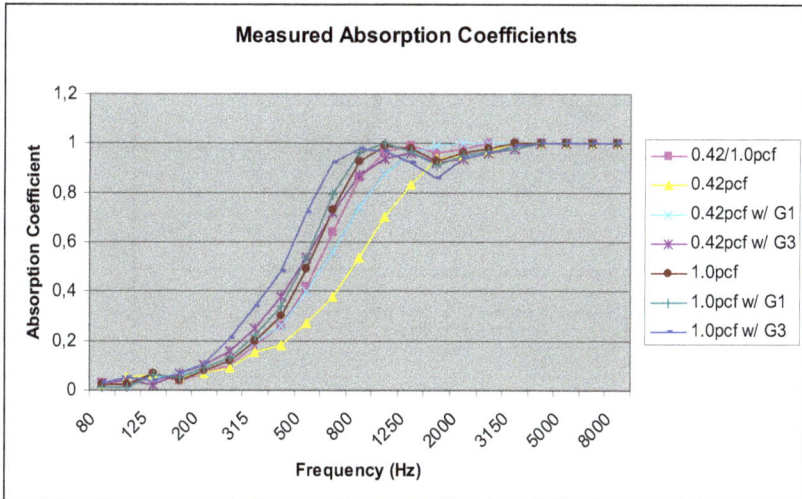

Figure 16. Measured absorption coefficients*G1 and G3: air gaps of 1.0 cm and 3.0 cm

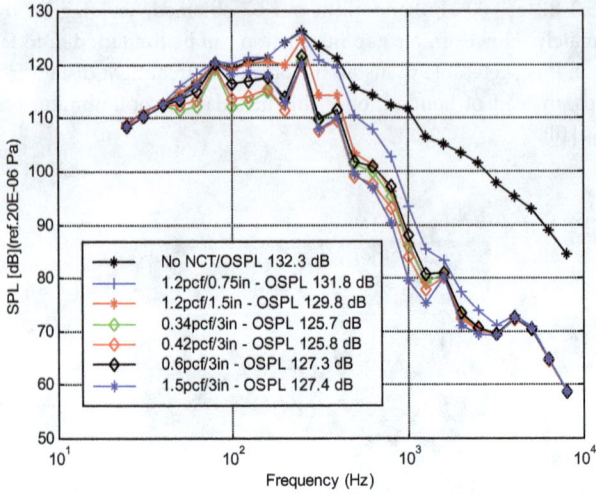

Figure 17. SPL for different blankets

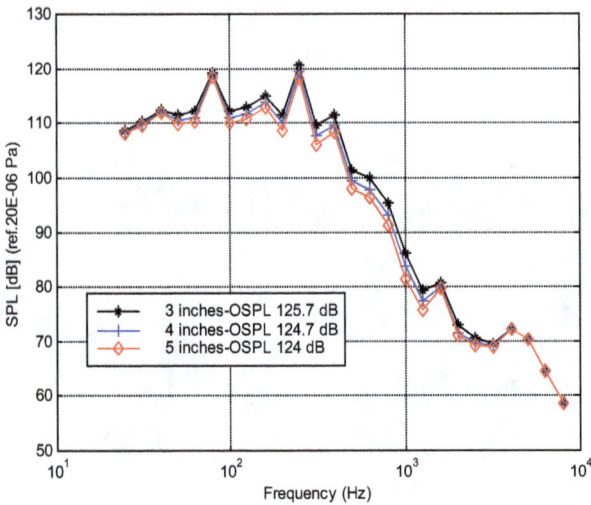

Figure 18. SPL for different thicknesses

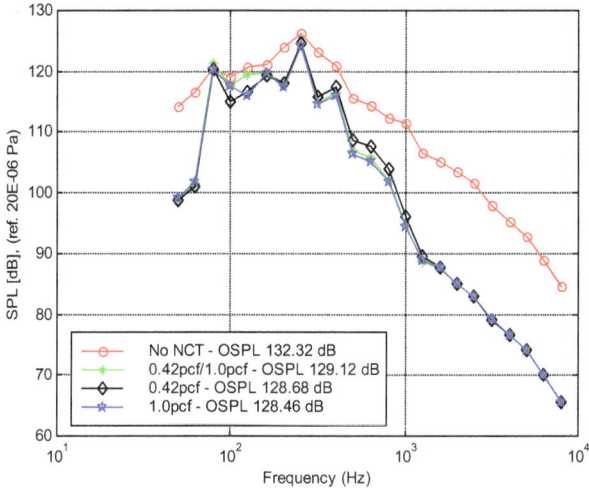

Figure 19. SPL for NCT without air gap.

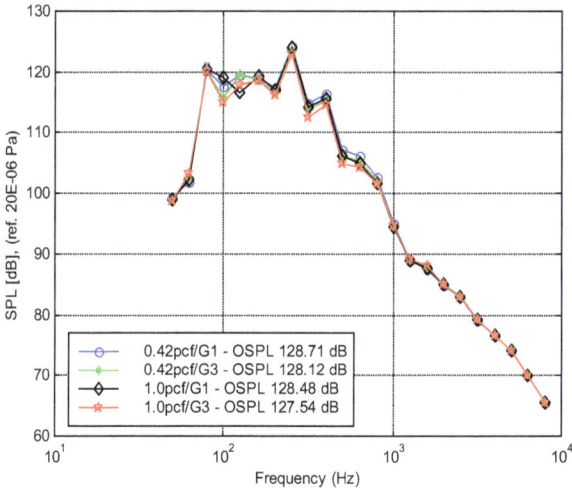

Figure 20. SPL for NCT with air gaps (*G1 and G3: air gaps of 1.0 cm and 3.0 cm)

4. Combustion instabilities of liquid propelled rocket engines due to chamber combustion acoustics

Combustion instabilities have been present in the development of LPRE over the last decades. There are basically three types of combustion instabilities: low-frequency (LF), medium-frequency (MF) and high-frequency (HF). LF instabilities, also called chugging, are caused by pressure interactions between the propellant feed system and the combustion chamber. MF instabilities, also called buzzing, are due to coupling between the combustion process and the propellant feed system flow. The HF instabilities are the most potentially dangerous and not well-understood ones. It occurs due to coupling of the combustion process and the chamber acoustics [27].

The presence of acoustic combustion instabilities must be considered still in development phase, although combustion instabilities can be clearly identified only during firing tests. In [13], it was described that instability can be verified when the power spectrum of the acoustic pressure levels, measured during burning tests, is analyzed. When an oscillation is observed, i.e., combustion instability, sound pressure peaks with well-defined magnitudes summed to the background noise are present. These peaks are correlated to the resonance frequencies of the combustion chamber. This phenomenon can cause instabilities and consequent unexpected behavior such as efficiency loss or even explosion of the engine. In this framework, the engine acoustic cavity characterization becomes an important issue to be investigated.

Acoustic behaviour of chambers is usually determined by doing cold tests measurements (without combustion). Acoustic dynamics in combustion environments are obtained by shifting the cold test resonant frequencies by a scalar factor defined by the ratio of sound velocity at the cold test temperature and at real operation temperature [15].

In view of attenuating acoustic pressure oscillations inside combustion chambers, reactive techniques as Helmholtz Resonators (HR), among others, are widely used ([13] and [16]). These devices are specially designed to attenuate oscillations at discrete resonance frequencies (pure tones). HR have been applied as combustion stabilization devices for solid motors and liquid rocket engines, with success. It could be noted in literature that they are used in a set of dozens or even hundreds in each chamber cavity, distributed along the walls or in a single row along the injector periphery [28].

This item describes a procedure for cold test acoustic characterization of LPRE combustion chambers. Firstly, the acoustical dynamic characterization of a combustion chamber is done and a typical longitudinal resonant frequency is chosen to be attenuated. A HR is designed (tuned at the chosen frequency) and applied to the mock up face plate. A LPRE mock-up [14] was used as experimental model. This test rig faithfully represents the internal acoustic cavity of the original engine. This procedure is followed by doing virtual prototypes of the combustion chamber. The acoustic natural frequencies and mode shapes are numerically calculated by a FEM model and validated through acoustic experimental modal analysis [29].

4.1. Experimental Acoustic Modal Analysis (EAMA)

Experimental Modal Analysis (EMA) is a well-applied technique in structure dynamics. However, due to the development of commercial acoustic sources, EAMA can be a suitable choice in view of extracting the acoustical Frequency Response Functions (FRF). In addition, the mathematical approach of structures modal parameters extraction can be applied to acoustic systems. [29].

In order to check the mutual orthogonality among modes from a modal model and to compare modes between different modal models (i.e., experimental and numerical solutions), the Modal Assurance Criterion (MAC) was used. This criterion indicates the degree of linear dependence between two eigenvectors and can be described as Eq. (7) [30].

$$MAC_{ijm} = 100 \cdot \frac{\left| \{\phi_{im}\}^T \{\phi_{jm}^*\} \right|^2}{\{\phi_{irm}\}^T \{\phi_{im}^*\} \{\phi_{jm}\}^T \{\phi_{jm}^*\}}$$

(7)

where: indexes i and j denotes modes obtained by different methods.

4.2. Helmholtz resonator

Helmholtz Resonators are widely applied in order to suppress or attenuate the acoustic pressure inside cavities, rooms and other volumes. A HR consists of a small volume connected to a bigger cavity (the combustion chamber, in this case) through an orifice by a flanged neck. The dimensions of the HR must be much smaller than the acoustic wavelength of interest, in order to consider the resonator as lumped elements coupled to a geometric discontinuity. The coupling condition is that the oscillatory volume flow in the neck is equal to that imposed on the fluid inside the cavity, neglecting the elastic property of the air in the neck [30].

A typical HR is shown in Fig.21 (left), being d the neck diameter, D the cavity diameter, V_c the volume cavity, l the neck length and L the cavity length. P_1 is the incident acoustic pressure and P_2 is the cavity pressure. The gas motion in the HR coupled in an acoustic cavity can behave equivalently to a mass-spring-dashpot system (Fig.21, centre). The system can be divided into three distinct elements. The fluid enclosed in the neck behaves as an uncompressible gas, and its mass correspond to the m element of the mechanical system. The air inside the cavity is compressible and stores potential energy, representing the mechanical stiffness k. The mechanical damping element (c) is represented by two factors: (i) the open-end of the neck radiates sound, introducing a radiation resistance and (ii) the gas movement in the neck introduces a viscous resistance. Considering the electrical analogue (Fig.21 right), the acoustic compliance C (analogous to electrical capacitance) is related to the stiffness of the air in the cavity, the acoustic inertance M (analogous to electrical inductance) is associated to the inertia element (mass) and the acoustic resistance R (analogous to electrical resistance) is related to the dissipative components stated above.

Considering that the gas beyond the end of the neck moves as a whole with the gas inside the neck, it is necessary to use an effective length l_{eff} which is bigger than the true length l of the neck [19]. The effective length l_{eff} is obtained by adding a mass end correction δ, which is empirically determined. In [28], it was presented a complete set of recommended equations for mass end correction, depending on the adopted considerations. For the purpose of this work, the appropriated equation is defined by:

$$\delta = 0.85d\left(1 - 0.7\sqrt{AR}\right) \text{ for } AR < 0.16 \tag{8}$$

where AR is the Area Ratio (A_n / A_c), being A_n and A_c the neck cross-sectional area and the cavity cross-sectional area, respectively. The effective length is calculated as $l_{eff} = l + \delta$.

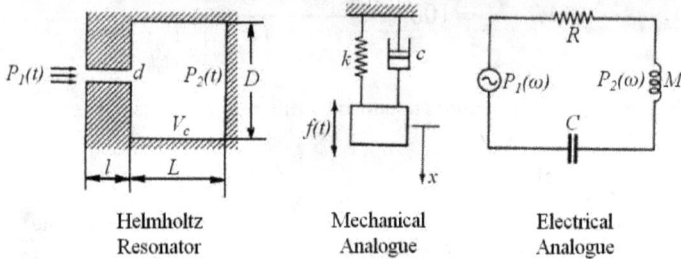

Helmholtz   Mechanical   Electrical
Resonator   Analogue   Analogue

Figure 21. Helmholtz Resonators scheme and its analogues

The definition of acoustic inertance (M) applied to the Helmholtz resonator gives:

$$M = \frac{m}{A_n^2} = \frac{\rho l_{eff} A_n}{A_n^2} = \frac{\rho l_{eff}}{A_n} \tag{9}$$

where ρ is the air density and m is the effective mass.

The acoustic compliance C is defined as the volume displacement that is produced by the application of unit pressure [19]. By applying this definition to HR, one obtains:

$$C = \frac{V_c}{\rho c^2} \tag{10}$$

where c is the velocity of sound.

The acoustic resistance in the neck (R) was approximated as the dissipation associated with viscous forces, considering the dynamic viscosity μ [28]:

$$R = \frac{8\pi \mu l}{A_n} \tag{11}$$

The acoustic impedance Z of the HR is:

$$Z = R + j\left(\omega M - \frac{1}{\omega C}\right) \tag{12}$$

As can be seen, the acoustic impedance is determined by the geometric and mechanical properties of the resonator. The resonance will occur when the acoustic reactance equals zero:

$$\omega M - \frac{1}{\omega C} = 0 \tag{13}$$

The resonance frequency can be determined by considering that the dimensions of the resonator are much smaller than the wavelength of interest:

$$\omega_0 = \sqrt{\frac{1}{MC}} = c\sqrt{\frac{A_n}{l_{eff}\, V_c}} \tag{14}$$

The resonance's sharpness of a HR can be quantified by its quality factor Q, given by:

$$Q = \frac{\omega_0 M}{R} \tag{15}$$

4.3. Finite Element Model

The cavity of a LPRE combustion chamber was analyzed using FEM in configurations without and with resonators. The first was modeled using 11,136 linear solid hexahedral elements, 12,510 nodes (12,093 degrees of freedom) and the second was modeled using 38,052 linear solid tetrahedral elements, 7,493 nodes (7,399 degrees of freedom). Both meshes are shown in Fig. 22. The fluid is air at 15° C. The eigenfrequencies were calculated from 0 to 2,400 Hz. Nodal pressures on the openings were assigned to zero.

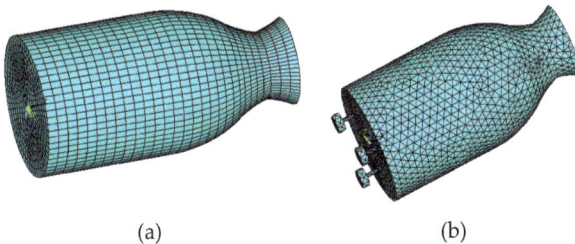

(a)   (b)

Figure 22. a) Cavity without resonators; b) Cavity with resonators

4.4. Experimental setup

Figure 23 shows the experimental setup. The mid-high frequency volume acceleration source is composed by a driver, a tube and a nozzle, where it is installed a volume acceleration sensor. This source produces a voltage signal proportional to the volume acceleration [m³/s²] variation, with a nominal frequency range of 200 up to 8,000 Hz. This source nozzle was installed in the mock-up plane surface as shown in Fig. 23.

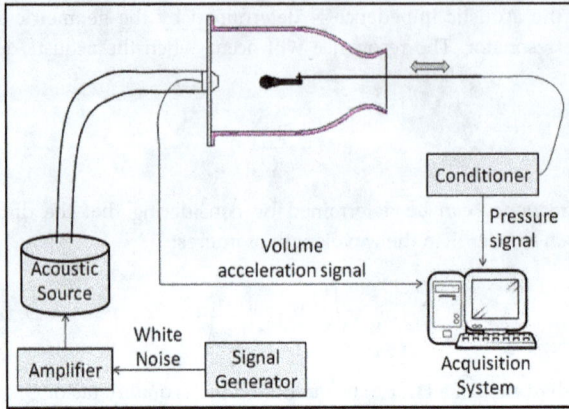

Figure 23. Measurement setup

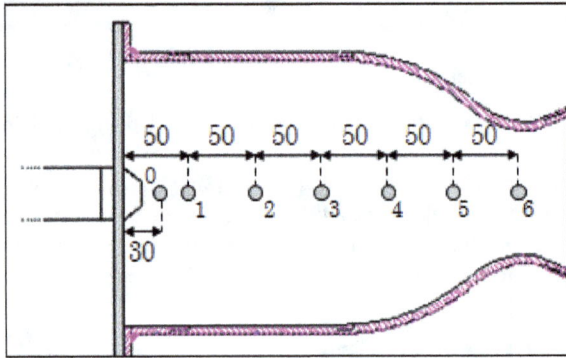

Figure 24. Source and microphone positions (distances in mm)

The chamber excitation was provided by a signal generator, a power amplifier and the source. The microphone was supported by a thin rod, placed in each measurement point inside the cavity. The pressure oscillations inside the cavity were captured by the microphone and registered by the data acquisition system. The volume acceleration source channel was settled as the reference channel. A white noise was used as excitation signal and the FRF were acquired at 7 points along the longitudinal axis (Fig. 24), being the point 0 the FRF driving point. The FRF were obtained by considering the volume acceleration as the excitation and the sound pressures as the responses. In order to make compatible theoretical x experimental comparisons, the volume velocity was assessed (instead of volume acceleration).

4.5. HR design

The objective is to tune the HR resonance as the same frequency that must be attenuated. Due to construction facility; it was chosen a cylindrical shape to develop the HR. Not only the resonance frequency must be observed during the design process, but also several factors that influence directly the behavior of the HR:

- Resonance frequency of interest = 730 Hz (second longitudinal mode to be attenuated);
- Relation HR dimensions and wavelength λ (must be at least 10 times smaller than λ);
- Area Ratio (AR) must be smaller than 0.16, in order to assure an end mass correction
- Quality Factor (Q): monitored to be used in future designs (compare with other HR shapes);
- Constructive factors. Define such dimensions that can be feasible constructively.

The parameters used in the design were updated considering the room temperature (28 °C) observed during the experiment:: sound speed: $c = 348.3$ m/s; air density: $\rho = 1.1839$ kg/m^3; air dynamic viscosity: $\mu = 1.983x10^{-5}$ kg/ms.

Considering the geometric dimensions in Eq. (14), one obtains the tuned resonance frequency as 726 Hz, with a quality factor of $Q = 372.8$. The acoustic parameters were calculated $M = 1019$ kg/m^4, $C = 4.72x10^{-11}$ s^2m^4/kg, $R = 12468$ kg/sm^4, using the Eq. (9), (10) and (11), respectively.

Three HR were manufactured in nylon. The presented measurement methodology was repeated in order to acquire the same FRF, considering the new configuration, with the resonators.

4.6. Results

The identified natural frequencies are summarized in Table 6. The transversal modes frequencies were identified only numerically. Considering the first four longitudinal modes, the maximum error comparing the numerical and experimental estimation of natural frequencies was 6.6%.

Figures 25a, 25c and 25e present the three first longitudinal numerical modes.

Mode	Experimental	Numerical	Error (%)
Longitudinal 1st	164.50	170.28	3.51
Longitudinal 2nd	730.85	721.90	1.23
Transversal 1st	-	1126.00	-
Transversal 2nd	-	1126.00	-
Longitudinal 3rd	1272.64	1296.00	1.84
Transversal 3rd	-	1382.00	-
Transversal 4th	-	1382.00	-
Longitudinal 4th	1689.52	1801.00	6.60

Table 6. Experimental and Numerical Natural Frequencies (Hz)

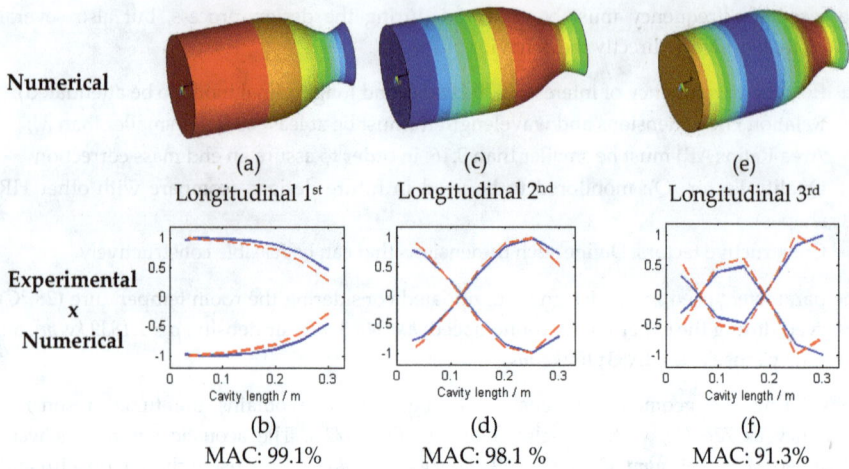

Numerical

(a)   (c)   (e)
Longitudinal 1st   Longitudinal 2nd   Longitudinal 3rd

Experimental
x
Numerical

(b)   (d)   (f)
MAC: 99.1%   MAC: 98.1 %   MAC: 91.3%

Figure 25. a) to f) Numerical modes and comparison between experimental (blue line) and numerical (red line) longitudinal mode shapes

Experimental versus numerical modes comparison, considering the normalized amplitude is also shown (Figs. 25b, 25d and 25f). In the numerical modes represented by the collor maps pictures (Figures 25a, 25c and 25e) the nodal regions are in green. The MAC (Eq. (7)) is also presented.

Notice in Fig. 25 that MAC values are bigger than 91% for the three first modes. For the first and second modes MAC values reach about 99%. Figures 25d and 25f show that the nodes in these modes are almost at the same point. After the introduction of the HR, the attenuation of the second mode is clearly noted in Fig. 26a, when comparing measurement results of the original cavity. At least 9 dB of attenuation can be observed in the new configuration. The FRF with and without HR are almost the same, but the second mode region (about 730 Hz), where the HR is tuned.

Figure 26 depicts the behavior of the chamber with HR. Numerical mode shapes of the configuration with HR were plotted and correlated to each part of the experimental FRF. The Fig. 26b was zoomed from the squared highlight in Fig.26a (point 1). This allows visualize the entire system behavior.

In Fig. 26 it can be noticed that mode shapes (c) and (d) have similar behavior, but different natural frequencies: it can be realized that the pressures inside all resonators vary in phase. In this case, the pressure inside the chamber remains almost unchanged. In the mode shapes (e) and (f), the resonators act close its tuned frequency. In these modes, the whole chamber behaves as a nodal region and the pressure inside the resonators varies out of phase.

As a result of the movement of the air mass inside the neck in the four modes represented in Figs. 26, the acoustic energy on the resonators behaves as expected, reducing the energy inside the chamber.

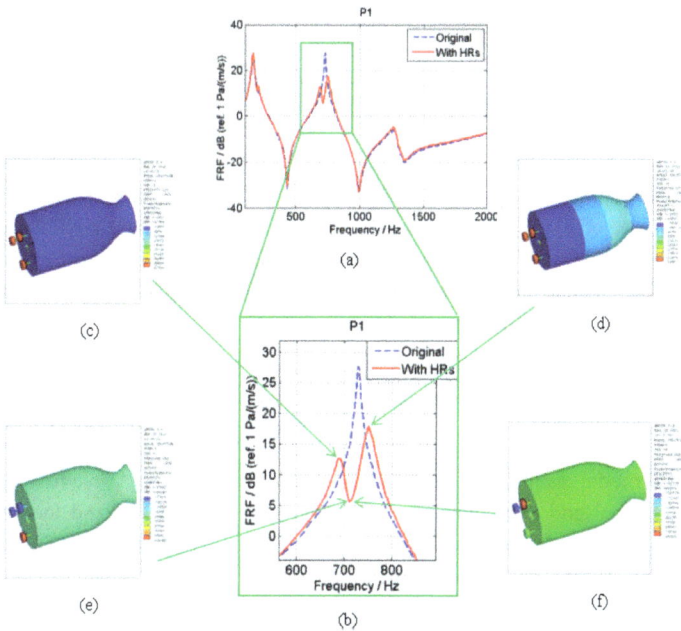

Figure 26. Numerical mode shapes near the resonances of the HR

5. Conclusions

This chapter presents three acoustic case studies applied on rocketry design.

Firstly, microphone protection devices design procedures for in-flight measurements are described. The modeling techniques using analytical and FEM numerical tools are presented as well as the validation acoustic testing procedures are presented. Good agreement among numerical and experimental results was obtained. A procedure to asses the SPL outer the launcher's structure by using the adapters' acoustic transfer functions and internal SPL measurements was also described.

Vibro-acoustic virtual prototypes were used to predict the acoustic response of a PLF cavity when excited by an ADF of 145 dB OSPL ranging from 5 to 8,000 Hz, generated at lift off.

Coupled deterministic techniques, using FEM/FEM and FEM/BEM, were applied to the fairing problem in a low-frequency band, considering accurate and efficient modeling techniques. The modal and semi-modal superposition techniques were applied to perform a FRA. In the higher frequencies, SEA coupling technique was applied to obtain the fairing acoustic responses in 1/3 octave bands.

The fairing was submitted to the lift off excitation in acoustic reverberant test and the internal acoustic pressure levels were measured. Experimental and numerical results show good agreement, except for the frequencies below 50 Hz and above 4,000 Hz.

The sensitivity analysis of acoustic blankets showed to be an effective tool for the development of the fairing NCT design. The effectiveness of a NCT considering its weight and performance can easily be evaluated using SEA, still in the development phase, when detailed subsystems are not required.

By analyzing many NCT configurations one can provide a library of performances and weights, important parameters that describe the ELV performance book. As in space industry the cost of a mission is a major issue, a trade-off between NCT weight and efficiency must be accounted. Acoustic testing in reverberant chamber may be conducted to validate the presented results and other porous-elastic materials may be investigated to complement the fairing NCT design library.

Finally, for the third case study, the use of a volumetric source in Experimental Acoustic Modal Analysis has important role in the process, once allows the accurate measurement of acoustic FRF.

The numerical model results were used as the basis for the HR design, in a first moment. In addition, numerical and experimental models were used to identify and localize, with a level of security, the node and maximum amplitude regions of each mode. The HR design seemed to be adequate, once it was verified an attenuation of 9 dB or bigger, depending of the location inside the chamber.

Author details

Rogério Pirk*, Carlos d´Andrade Souto and Gustavo Paulinelli Guimarães
Institute of Aeronautics and Space (IAE) and
Technological Institute of Aeronautics (ITA), São José dos Campos, Brazil

Luiz Carlos Sandoval Góes
Technological Institute of Aeronautics (ITA), São José dos Campos, Brazil

6. References

[1] Jorge P. Arenas, Ravi N. Margasahayam, 2006, Noise and Vibration of Spacecraft Structures, Ingeniare. Revista Chilena de Ingeniería, vol. 14 Nº 3, pp. 251-264
[2] Pirk, R, Sas, P., Desmet, W. and Góes, L. C. S., "Vibro-Acoustic Analysis of the Vehicle Sattelite Launcher´s (VLS) Fairing", Ph. D. Thesis, 2003, Technological Institute of Aeronautics (ITA), São José dos Campos, Brazil.
[3] Pirk, R. and Góes, L. C. S., 2006, "Acoustic Theoretical x Experimental Comparison of the Brazilian Satellite Launcher Vehicle (VLS) Fairing", Proceedings of the ISMA 2006, Leuven, Belgium.

* Corresponding Author

[4] Pirk, R. and Souto, Carlos, d´A, 2009, Implementation of Acoustic Blankets to the VLS Fairing – A Sensitivity Analysis Using SEA, Proceedings of the ISMA 2008, Leuven, Belgium.

[5] Scott, J. M., Jay, B. G., Robert, C. B., 1996, General Environmental Specification for STS and ELV – Payloads, Subsystems and Components, GEVS-SE, Rev. June 1996, NASA Goddard Space Flight Center, Maryland, 20771.

[6] Souto and Pirk, 2009, Estimation of the external sound pressure levels generated during VLS – 1 lift off, Institute of Aeronautics and Space, Technical Report RT 001/AIE/R/2009).

[7] Pirk et al. 2010, Alternative External Acoustic Loads Estimation of the Brazilian Satellite Launcher (VLS-1), Using In-Flight Experimental Data, International Congress on Sound and Vibration, Rio de Janeiro, Brazil

[8] Glaese, R. M. and Anderson, E. H., 2003, "Initial Structural-Acoustic Modeling and Control Results for a Full-Scale Composite Payload Fairing for Acoustic Launch Load Alleviation".
http://www.csaengineering.com/techpapers/technicalpaperspdfs/CSA1999_Initial Structural

[9] Weissman, K., McNelis, M. E. and Pordan, W. D., 1994, "Implementation of Acoustic Blankets in Energy Analysis Methods with Application to the Atlas Payload Fairing, Journal of the Institute of Environmental Sciences – IES, July/August 1994.

[10] Bolton, J. S., 2005, "Porous Materials for Sound Absorption and Transmission Control", Proceedings of the 2005 Congress and Exposition on Noise Control Enginnering, INTERNOISE 2005, Rio de Janeiro, Brazil.

[11] H. Defosse, M.A. Hamdi, "Vibro-acoustic study of Ariane V launcher during lift-off", Proc. of Inter-Noise 2000, vol 1, 9-14 (Nice, 2000).

[12] Culick, F.E.C. Combustion Instabilities in Liquid Rocket Engines: Fundamentals and Control. California Institute of Technology, 2002]

[13] Burnley, V.S.; Culick, F.E.C., The Influence of Combustion Noise on Acoustic Instabilities, Air Force Research Laboratory, OMB N° 0704-0188, 1997.

[14] R. Pirk, C. d´A. Souto, D. D. Silveira and C. M. Souza, 2010, Liquid rocket combustion chamber acoustic characterization, Journal of Aerospace Technology and Management.

[15] Laudien, E. and Others 1994, Experimental Procedures Aiding the Design of Acoustic Cavities, DASA- Deutsche Aerospace AG, Liquid Rocket Engine Combustion Instability, Progress in Astronautics and Aeronautics, Chapter 14, volume 169

[16] Natanzon, M. S., 1999, Edited by Culick, F. E. C., California Institute of Technology

[17] Troclet, B., Analysis of Vibro-acoustic Response of Launchers in the Low-Frequency and High-Frequency Range, Proc. of NOVEM 2000, (Lyon, France, 2000).

[18] Khlybov, V. I. and Mak´hankov, S. A., 2009, The Preparation to Experimental Studies and Acoustic Sensors Testing Program Development, Technical Report.

[19] Kinsler, Lawrence E.; Frey, Austin R. Fundamentals of Acoustics, John Wiley & Sons, 2nd edition, 1962

[20] W. Desmet, D. Vandepitte, 2001, Finite Element Method in Acoustics, Course Graduate School in Mechanics – Advanced Acoustics, Katholieke Universiteit Leuven

[21] Coyette, J-P, Lecomte, C. & Meerbergen, K., 1997, Treatment of Random Excitations using SYSNOISE Rev. 5.3.1 – Documentation Theoretical Manual, Users Manual and Validation Manual, LMS Numerical Technologies NV.

[22] Klos, J. et al ,2003, Sound Transmission Through a Curved Honeycomb Composite Panel, AIAA Conference Paper AIAA-2003-3

[23] Desmet, W, Vandepite, D., Boundary Elements in Acoustics, Katholieke Universiteit Leuven (KUL), 2001.

[24] Lyon and DeJong, Theory and Application of Statistical Energy Analysis, Second Edition, 1995

[25] Thinh, T. Do, Vibro-acoustic Modeling Study of the Delta II 10-Foot Composite Fairing, Journal of the IEST, November/December 1999.

[26] Allard, J. F., 1993, Propagation of Sound in Porous Media – Modeling Sound Absorbing Materials, Elsevier Applied Science, London and New York.

[27] Sutton, G. P., Biblarz, O., 2001, "Rocket Propulsion Elements", New York, John Wiley & Sons.

[28] Santana Jr., A. Investigation of Passive Control Devices to Suppress Acoustic Instability in Combustion Chambers. Thesis of doctor in science. Aeronautics Institute of Technology, São José dos Campos, Brazil, 2008.

[29] Guimarães, G.P.; Pirk, R.; Souto C.A.; Góes, L.C.S. Acoustic Modal Analysis of Cylindrical-Type Cavities. Proceedings of the 8th Intern. Confer. on Structural Dynamics – EURODYN, Leuven, Belgium, 2011, p. 3160-3167.

[30] Fahy, F. Fundamentals of Engineering Acoustics. Academic Press. London, UK. 2001.

Acoustic Waves in Microfluidics

Chapter 9

Acoustic Wave Based Microfluidics and Lab-on-a-Chip

J. K. Luo, Y. Q. Fu and W. I. Milne

Additional information is available at the end of the chapter

1. Introduction

Microfluidics refers to a set of technologies that control the flow of minute amounts of liquids, typically from a few picolitres (*pls*) to a few microlitres (*μls*) in a miniaturized system [1,2]. Lab-on-a-chip (LOC) systems typically consist of a set of microfluidics and sensors with dimensions from a few square millimetres (mm^2) to a few square centimetres (cm^2). Microfluidics handles liquids through droplet generation, transportation and mixing of liquid samples, chemical reactions etc. Sensors may include biochemical sensors, gas sensors and physical sensors such as humidity and temperature sensors, flow meter and viscometers etc. Therefore, LOCs are microsystems with a much broader meaning, and generally perform single or multiple laboratory processes and functions on a chip-scale.

Microfluidic and LOC systems have distinctive advantages:

- Low volume fluidic consumption (low reagents costs and less required sample volumes for analysis and diagnostics, less waste);
- Fast analysis and short response times due to short diffusion distances, high surface to volume ratios, small thermal capacity and fast heating rate;
- Better process control because of a faster response of the systems (e.g. thermal control for exothermic chemical reactions);
- Compactness of the systems owing to integration of many functionalities and the small dimensions of each component;
- Massive parallelization due to compactness, which allows high-throughput analysis;
- Low fabrication costs, allowing cost-effective disposable chips fabricated in mass production;
- Safe platform for chemical, radioactive or biological studies because of integration of functionality within the small systems and often on-board power generator.

Owing to their outstanding properties and great potential applications, microfluidics and LOCs have received tremendous interests from engineering, healthcare, medical research, drug-development sectors. They are regarded as the technologies of the future for great value-added manufacturing. So far, most of the LOCs and microfluidics are single function systems, the trend and demands are to develop LOCs and microfluidics with multi-functions.

Microfluidics and LOC systems based on acoustic waves generated through the piezoelectric effect have recently received a great attention, as acoustic waves can be utilized not only for actuation/microfluidics, but also for sensing/detection, allowing integration of various acoustic devices for LOCs to perform multi-functions. In this chapter, we will thus focus on the acoustic wave-based LOC systems.

1.1. Acoustic wave based microfluidics

Many microfluidic technologies have been explored and developed [1,3], including two major classes of devices: active devices such as micropumps, micromixers and droplet generators, and passive components such as microchannels, valves and microchambers. The mechanisms of micropumps vary widely. Based on the mechanisms, designs and applications, micropumps can be categorized into two main groups: mechanical and kinetic pumps. Mechanical micropumps typically represent miniaturized version of macro-sized pumps that typically consist of a microchamber, check valves, microchannels and an active diaphragm to induce displacements for liquid transportation. Thermal bimorph, piezoelectric, electrostatic, magnetic forces and shape memory mechanisms have been utilized to actuate the diaphragm in an oscillation mode [2,3]. These mechanical micropumps are complicated, expensive, typically made by multi-wafer processes, and are therefore difficult to be integrated with other microsystems such as integrated circuits (IC) for control and signal processing due to incompatible processes and structures [1,2,4]. They generally have a large dead volume, leading to an excessive waste of biosamples and reagents which are normally expensive and precious in biological analysis, especially for forensic investigations. These micropumps typically have moving parts (diaphragm and check-valves) which lead to low production yields in fabrication, high failure rates and poor reliability in operation.

The new trend is to develop non-mechanical or moving-part-free micropumps by utilizing electrokinetic forces and surface tension (or surface energy-related forces) such as the electroosmotic (EO) effect [5], electrophoresis (EP) [6], dielectrophoresis (DEP) [7,8], asymmetric electric field, electrowetting-on-dielectrics (EWOD) [9,10], electrostatic pumps [11,12] etc. Electrokinetic force based micropumps typically require electric/magnetic fields to mobilize ionic, or polarisable particles and species in a liquid which can drag the liquid through friction forces to form a continuous flow. The surface tension-based micropumps are typically droplet-based systems, which manipulate discrete droplets through modification of surface tension/energy by external stimuli [13]. Electric field, thermal and concentration gradients generated by localized heating or optical beams, photosensitization and capillary forces are used for these micropumps. Their key characteristic is that they can transport discrete droplets, acting as the so-called digital micropumps, on channel-less (or wall-less) planar surfaces without check valves [13]. The fabrication process is simple and requires no special substrate. Therefore, these pumps can be easily integrated with electronics for control and signal processing etc.

Acoustic waves utilized in microfluidics and lab-on-a-chip systems include ultrasound, bulk acoustic waves and surface acoustic waves (SAW). Ultrasound generated by external large piezoelectric (PE) transducers has been widely utilized for microfluidics which is very effective in mixing microfluidics, and has recently been utilized for transportation of liquid [14] and particle sorting etc through novel design of the systems [15,16]. Due to the different mechanism, ultrasonic wave based microfluidics and LOCs will not be discussed here.

SAW-based microfluidics and LOCs are one of the latest technologies. SAW can be used as an actuation force for pumping and mixing liquids, and for generating droplets and mist [17, 42,18,19]. SAW micropumps can not only manipulate discrete liquid droplets from pls to a few tens of μls, but can also pump a continuous fluid. SAW devices are also excellent sensors for monitoring physical parameters and detecting biochemical entities with high sensitivity. Furthermore the development of thin film SAW technology has opened the way for integration of SAW-based microfluidics and sensors with Si-based electronics on the same chip for LOCs with better functions and applications [20]. This chapter will mainly focus on acoustic wave technologies for microfluidics and lab-on-a-chip applications.

1.2. Acoustic wave resonant sensors

Sensors are a type of transducer that convert some physical stimulation such as temperature, pressure etc into electronic, optical, magnetic, or acoustic signals for quantitative measurement, while biosensors are devices that convert biological information into measurable electronic signals. Various technologies such as optical sensors, electrochemical sensors, field effect transistor based sensors, microcantilevers and acoustic resonators have been developed for sensing, particularly for biochemical sensing.

Compared with other biosensing technologies, acoustic wave based technologies have several advantages including simple operation, high sensitivity, small size, fast response and low cost, [21]. Another distinctive advantage of acoustic wave sensors over others is that they can be simply integrated with acoustic wave based microfluidics to form a LOC system driven by a single mechanism, which makes the system fabrication and operation much simpler.

There are four types of acoustic wave resonators: the quartz crystal microbalance (QCM), surface acoustic wave device (SAW), film bulk acoustic resonator (FBAR or BAW) and flexural plate wave (FPW) resonator. Acoustic wave sensors are able to detect not only mass/density changes, but also viscosity, elastic modulus, conductivity and dielectric properties. They have many applications in monitoring of pressure, moisture, temperature, force, acceleration, shock, viscosity, flow, pH levels, ionic contaminants, odour, radiation and electric fields [22,23]. By using specific gas absorbents and biological markers, the acoustic resonators can be made into gas sensors and biosensors. As the latter is the main focus of this chapter, we will only discuss acoustic wave biosensors here.

QCMs with a structure shown in Fig. 1 have a long history, and probably are the only one currently being commercialized and practically used. By cutting a quartz crystal with proper orientations, it is possible to make QCM sensors operate in a longitudinal mode as well as in a transverse (also called thickness shear) mode. The standing waves in the thickness shear-mode devices are parallel to the surface of the QCM plate, and the wave energy is largely preserved in the presence of liquids, therefore the shear mode QCM is suitable for sensing in liquid environments, a pre-request and necessary condition for most biosensing and physiological monitoring. QCM sensors have the advantages of simplicity in design and operation and a mature technology.

Figure 1. Typical QCM resonator.

The sensitivity of acoustic wave resonators is determined by the square of the resonant frequency, f_r, and base mass. The frequency shift Δf of an acoustic wave resonator induced by a mass loading, Δm, is described by [24]

$$\Delta f = \frac{2\Delta m f_r^2}{A\sqrt{\rho\mu}} \tag{1}$$

where A, ρ, μ and f_r are the area, density, shear modulus and intrinsic resonant frequency (sometimes defined as the operating frequency, f_0), respectively. QCM sensors have a fundamental limitation of low sensitivity due to the thickness of wafer and the large active area, hence low f_0 and large base mass.

SAW sensors consist of a pair of interdigitated transducers (IDTs) (Fig. 2). When a series of radio frequency (R.F.) signals are applied to one of the IDTs, surface acoustic waves are generated through the piezoelectric effect, and travel along the surface of the substrates, received by the IDT on opposite. The strength of the waves decays exponentially with depth into the substrate. Depending on the nature of the travelling acoustic waves, SAW can be longitudinal or transverse mode. The operating frequencies of SAW devices are typically in the range of 100-300 MHz, and the active base mass is much smaller than that of QCMs owing to the one wavelength depth of active area. Therefore, the sensitivity of SAW sensors could be much higher than that of the QCMs. The acoustic waves of shear mode SAW devices travel parallel to the surface with no longitudinal component with no acoustic energy dissipated into the liquid in contact. Therefore, the shear mode SAW devices are suitable for sensing in liquids. Furthermore rapid advances in thin film deposition technologies allow fabrication of high quality thin film SAW devices, resulting in the possibility for integration of SAW sensors with electronics on the same chip. Advantages of the SAW sensors are: simplicity in device structure and process, high sensitivity, small size compared to QCM, availability of thin film SAW, and the possibility for integration of SAW sensors with Si-electronics. However similar to QCM, the SAW devices are difficult to scale down, and other concerns include weak signal, relatively low quality factor and relatively low sensitivity.

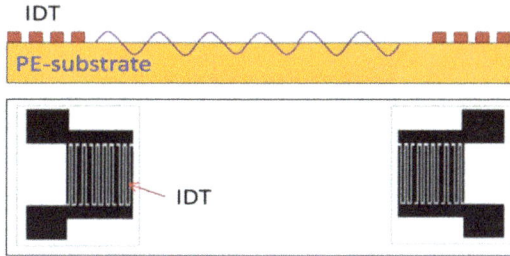

Figure 2. Typical structure of a SAW resonator.

FBARs are newly emerging acoustic nanodevices [25,26,27], with a structure similar to that of QCM as shown in Fig. 3, but thousand times smaller with a typical size from 100 μm × 100 μm to 30 μm × 30 μm and a thickness of a few micrometers. The operating frequency of FBARs is in the range from sub-gigahertz to a few GHz. The base mass of a FBAR is much smaller than those of the QCM and SAW devices owing to the much reduced area and thickness, and the sensitivity of FBAR sensors can be dramatically increased compared with other acoustic sensors as shown in Fig. 4 [28]. FBARs have the highest sensitivity, and the SAW devices are in the middle with the QCM ones the lowest.

FBARs have three basic structures: the Bragg acoustic mirror type (Fig. 3a), the back-trench type (Fig. 3b), and the air-bag type (Fig. 3c&3d). The Bragg reflector based FBARs are normally made on PE-films deposited on a solid substrate. The acoustic mirror is composed of many quarter-wavelength layers with alternating high and low acoustic impedances. Due to the high acoustic impedance ratio of the acoustic mirror, the acoustic energy is reflected and confined within the top piezoelectric layer, thus maintaining an excellent resonant bandwidth even on a solid substrate. This structure has a better mechanical robustness and a simpler fabrication process. Also cheap substrates such as glass or plastics may be used for the FBAR fabrication to reduce the cost [29]. The shortages of the Bragg reflector FBARs are long film deposition process and the difficulty in precise control of the layers thickness which may lead to poor quality factor. The back trench mode FBARs are typically made on a thin membrane with a thickness about 1-2 μm to reduce the base mass and increase the robustness. The back trenches are made either by anisotropic wet etching or by deep reactive ion etch (DRIE). The air bag types of FBARs are another type of back-trench type of FBARs, but with the large back-trench replaced by a small gaps formed through etching processes to remove the supporting material underneath. It has either a "standing-out" structure formed by a thin sacrificial layer or a "digging-in" structure.

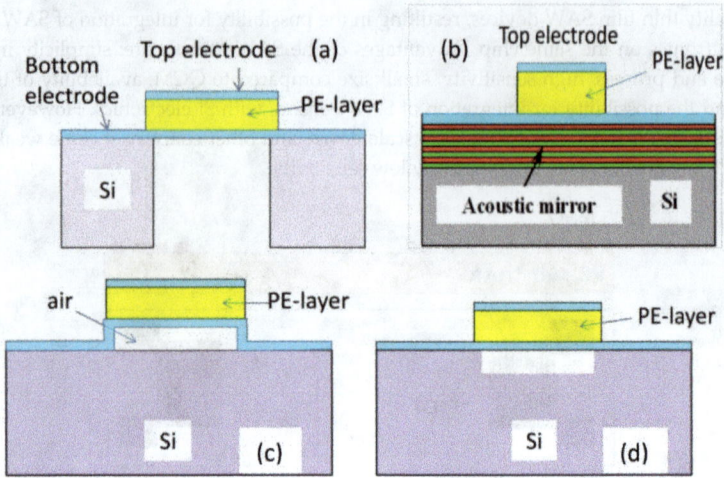

Figure 3. Various structures of FBARs. a) Back trench mode, b). Bragg reflector mode, c). Standing-out air bag mode and d). Digging-in air bag mode.

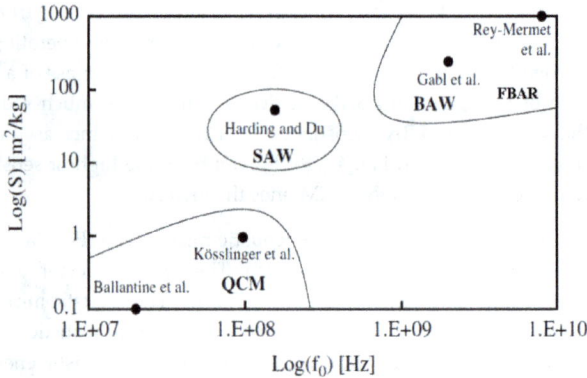

Figure 4. Comparison of sensitivity of QCM, SAW and FBAR biosensors. FBAR resonators have the highest sensitivity owing to its much reduced base mass and high operating frequency [28]. Reprinted with permission from Elsevier, Sensors and Actutars, B 114, 2006, 681.

FBARs also have extremely small size which allows the use of an array of FBARs for multiple-sensing in parallel. Also because they are based on thin film technology, the FBARs can be simply integrated with other acoustic microfluidics, microsystems and Si-electronics for LOC applications. Thickness shear mode FBARs have been tried by using off c-axis crystal materials with some success [30,31]. It is still not a simple task to obtain c-axis inclined film materials for low cost FBARs.

In a manner similar to a SAW device, flexural plate wave resonators (also called Lamb wave, and often regarded as the ultrasonic wave [32,33]) (Fig. 5) on a membrane have been developed for biochemical sensing in liquid [34]. The Lamb wave velocity in the FPW resonator is much smaller than that of the acoustic waves, and the dissipation of wave energy into the liquid is minimized, therefore it can be used for liquid sensing directly. The sensing mechanism is based on the detection of a relative change in wave magnitude induced by the perturbation on the membrane, rather than the resonant frequency shift as they would be very small. The sensitivity of the FPW devices increases as the membrane thickness becomes thinner [35,36]. The main drawback of the FPW biosensors is that there is a practical limit on the minimum film thickness due to the fragility. Compared with the other three types of acoustic wave sensors, FPW devices as sensors still need much further development.

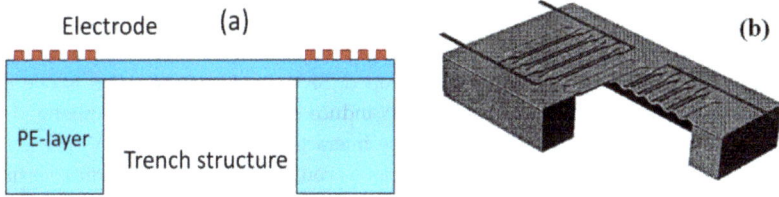

Figure 5. Typical structure of a flexural plate wave resonator.

1.3. Piezoelectric thin film technologies

QCM and SAW devices are typically fabricated using bulk materials which are expensive and cannot be integrated with electronics, microfluidics and sensors on the same substrate for applications. Various thin film-based QCMs [37,38], SAW devices [39,40,41] and FBARs [25,26,27] have been developed. PZT, ZnO and AlN piezoelectric thin films have good piezoelectric properties and high electro-mechanical coupling coefficient, k^2, thus they have been studied intensively for this purpose. They can be grown in thin film form on a variety of substrates such as Si, making these materials very promising for integration with electronic circuitry, particularly for devices aimed at low-cost and mass production for one-time use. PZT has the highest piezoelectric constant and k^2, but PZT films have very high acoustic attenuation, lower sound wave velocities, poor biocompatibility and worst of all, the requirement for extremely high temperature sintering and high electric field poling, making them unsuitable for integration with electronics. AlN and ZnO are the most common thin films used for SAW, FBARs and FPW devices. AlN is chemically inert and stable, with high acoustic velocity, but AlN thin films are relatively difficult to deposit, requiring stringent optimization for the process to obtain high quality thin films with smooth surfaces and right crystal orientations. On the other hand, ZnO PE films with a high PE quality are much easier to obtain using various deposition technologies such as sputtering, laser-ablation, chemical vapour deposition (CVD) and molecular beam epitaxy (MBE) etc, therefore, it is the most widely used PE material for thin film acoustic wave devices. The acoustic velocity of ZnO thin films is about 2700 m/s, smaller than that of AlN (11050 and 6090 m/s for longitudinal and transverse modes, respectively); hence ZnO SAW has a lower operating frequency than that of the AlN SAW devices. For wider applications, high operating frequency ZnO SAW devices have been developed by using non-PE supporting layers with high acoustic velocity such as sapphire and diamond [39, 40, 41]. Acoustic waves generated by ZnO SAW travel inside the supporting layer with high velocity, resulting in higher frequencies. Besides PZT, AlN, and ZnO, many other PE thin films have been developed, mostly for SAW device applications. Gallium arsenide, gallium nitrides, polyvinylidene fluoride (PVDF) and its copolymers are a few that have been investigated for piezoelectric applications.

2. Modelling of surface acoustic wave microfluidics

Microscale mixing and pumping are essential processes for microfluidic and LOC applications for biochemical analysis, disease diagnosis, DNA sequencing and drug development etc [42,43]. Various technologies have been developed. SAW based microfluidics is one of the most advanced technologies, which utilize SAW induced forces for pumping, mixing, droplet generating and ejecting etc. In this section, the interaction between the SAW and liquid is theoretically analysed, to show that the complicated acoustic streaming and particle sorting etc are physical phenomena which can be well addressed by theoretical model.

2.1. Navier-Stoke equation for fluid motion

It is well known that radiation of a high-intensity beam of acoustic waves into a liquid can result in acoustic streaming. Absorption of the acoustic energy results in significant attenuation of acoustic energy and moment in the fluid within a short range, meanwhile consumption of the energy leads to fluid motion. This phenomenon was observed by Lord Rayleigh in 1884 [44] and was then further studied in detail by Westervelt in 1951 [45] and Nyborg [46] in the 1960s'.

Under an external body force, F_j, induced by acoustic waves, the fundamental hydrodynamics of a steady viscous fluid is governed by Navier-stoke equation [47],

$$(V_i.\nabla)V_j = F_j - \frac{1}{\rho}\nabla p + \eta\nabla^2 V_j \qquad (2)$$

where V is the acoustic streaming velocity (or the particle velocity), p the pressure, ρ and η are the fluid density and shear viscosity coefficient, respectively. It is generally assumed that the fluid flow exhibits viscous and incompressible laminar flow. Here the subscripts i and j = 1, 2, 3 represent the x, y and z coordinates respectively for a 3-dimentional (3D) phenomenon. The nonlinear body force is correlated to the Reynolds' stress, σ, induced by the acoustic wave in the fluid with spatial variation in all three coordinates [46],

$$\sigma_{ij} = \rho\overline{v_i v_j} \qquad (3)$$

where v_i and v_j are the velocity fluctuations in x, y and z directions. The stress is a mean velocity fluctuation and density product represented by the upper bar. The relation for the force and the Reynold's stress is expressed as,

$$F_j = -\sum_{i=1}^{3} \frac{\partial \overline{v_i v_j}}{\partial x_i}$$ (4)

According to the continuity equation, the following equation applies,

$$\frac{\partial \rho}{\partial t} + \nabla.\rho V = 0$$ (5)

For a steady flow, the first term is zero and we obtain,

$$\nabla.\rho V = 0$$

This leads to zero for the right side of eq.(2),

$$F_j - \frac{1}{\rho}\nabla p + \eta \nabla^2 V_j = 0$$ (6)

The governing equation of the acoustic streaming force has been derived by Nyborg [46] for an incompressible fluid, and is given by the following equation;

$$-F_j = < V_{aj}.\nabla V_{aj} > + V_{aj} < \nabla.V_{aj} >$$ (7)

where V_a represents the acoustic wave velocity (different from the streaming velocity V), and the brackets < > indicate the time averaged value [46,48]. Therefore, the nonlinear acoustic streaming force F_j can be calculated, once the wave velocity is known.

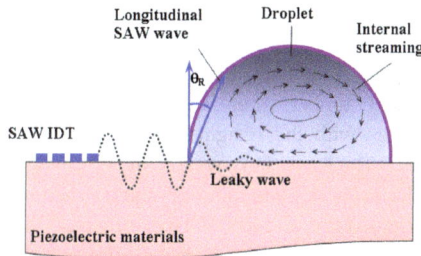

Figure 6. Schematic drawing of acoustic streaming in liquid fluid induced by SAW. The direction of SAW induced streaming is determined by the Rayleigh angle.

Acoustic waves travel in a solid medium with a relatively small attenuation. The attenuation becomes large once the waves enter the fluid medium and decays exponentially. Therefore the acoustic wave generated force is a short range force. Figure 6 schematically shows SAW induced acoustic streaming in a droplet.

The SAW travelling along the surface of the PE substrate has a small surface displacements typically less than one nanometre. The SAW changes its mode into a leaky SAW once it interacts with any liquid medium in its path. This leaky longitudinal SAW continuously travels within the liquid medium with the streaming angle determined by the Rayleigh angle Θ_R [49,50,51], as depicted in Fig. 6 according to the following equation;

$$\theta_R = \sin^{-1}\frac{V_L}{V_S} \tag{8}$$

where V_S is the Rayleigh SAW velocity on the piezoelectric substrate, and V_L is the acoustic velocity in the liquid. The leaky SAW has been well studied, and the displacement, (u_x, u_y), of the acoustic wave in the liquid is sinusoid in form with an exponential decay with distance. For a 2D case, the displacement can be expressed by [52],

$$u(x,t) = A\exp(-l(x)/l_S)\sin(\omega t - k.x + \varphi) \tag{9}$$

where A is equivalent to the SAW amplitude at the point of entry into the liquid, $\omega = 2\pi f$, the angular frequency imposed by the SAW, $l(x)$ is the path length along the wave path, and k is the wave vector of the acoustic wave. Streaming is a 3D phenomenon, especially for streaming inside a droplet; therefore displacements in all directions have to be considered. The displacements in the x and y directions are expressed by [47,53];

$$u_x = A\exp.(j\omega t).\exp(-jk_L x).\exp(-\alpha k_L y) \tag{10}$$

$$u_y = -ja A\exp.(j\omega t).\exp(-jk_L x).\exp(-\alpha k_L y) \tag{11}$$

Here, α represents the attenuation constant;

$$\alpha^2 = 1 - (\frac{V_S}{V_L})^2 \tag{12}$$

where V_S and V_L are the leaky (Rayleigh) SAW velocity and the sound velocity in liquid respectively. $K_r = 2\pi/\lambda$ is a real number, where λ is the wave length and the leaky SAW wave number $(k_L = k_r + jk_i)$ is complex with the imaginary part, jk_i, representing the SAW energy dissipation within the liquid. The leaky SAW wave number can be obtained by extending the method of Campbell and Jones [54, 47] into the solid-liquid structure assuming both stress and displacement to be continuous at y=0, and V_L=1500 m/s for water (Most of the biofluid is water-based, V_L changes if other liquid is used.). If the wave displacements (u_x, u_y) are replaced by the wave velocities using $V_j = \dfrac{\partial u_j}{\partial t}$ and substituting into eq.(7), the two components of streaming force can be obtained for an incompressible fluid as follows [47,53];

$$F_x = -(1+\alpha_1^2)A^2\omega^2 k_i \exp 2(k_i x + \alpha_1 k_i y) \tag{13}$$

$$F_y = -(1+\alpha_1^2)A^2\omega^2 k_i \alpha_1 \exp 2(k_i x + \alpha_1 k_i y) \tag{14}$$

where, $\alpha = j\alpha_1$. The total SAW streaming force F can be calculated by $F = \sqrt{F_x^2 + F_y^2}$, which is then given by;

$$F = -(1+\alpha_1^2)^{\frac{3}{2}} A^2 \omega^2 k_i \exp 2(k_i x + \alpha_1 k_i y) \tag{15}$$

The SAW force F acts on the main fluid volume as a body force, but the exponential decay of the leaky SAW limits the influence of this force within the whole fluid within the decay distance. This leads to a complete diminishing of the acoustic force within a few hundreds of micrometers from the interaction point between the SAW and the liquid.

The numerical simulation of acoustic streaming is a complicated process and should be treated case by case due to the different designs of the systems. One must solve the full set of nonlinear hydrodynamic equations consisting of the Navier-Stokes equations, the continuity equation and an equation of state for an incompressible fluid driven by the time-dependent boundary condition. The majority of researchers have implemented the modelling by using existing software such as COMSOL or ANSYS. The numerical simulation requires a few software modules to implement the multiphysics modelling for the solid components as well as the viscous liquid body, and requires the consideration of coupling between the modules. For the viscous liquid, a Finite Volume Method (FVM), OpenFOAM-1.6 CFD code (OpenCFD LTD) and Surface Evolver are often used [55, 53,].

2.2 Modelling of acoustic streaming induced by SAW

The interaction of SAW with liquid depends on the position of the SAW relative to the liquid. Figures 7a&7b show two cases of the interaction of a SAW with a droplet: (1) the droplet is in the path of surface acoustic wave with droplet size smaller than the aperture of the IDT of the SAW device; and (2) the droplet is partially on the SAW path.

The interaction of the SAW with the liquid is a dynamic process involving a transition of the flow. Figure 8 shows a comparison of the transitional and steady streaming velocities induced by a leaky SAW obtained numerically and experimentally with RF power as a function [53]. As can be seen, the streaming velocity increases with time rapidly, and becomes stable after the transitional period. The transition time is in the range up to a few hundreds of milliseconds, depending on the RF power applied.

The streaming pattern depends on the entry angle of the SAW into the liquid. Figure 9 demonstrates the different streaming patterns generated by a SAW entering from the centre

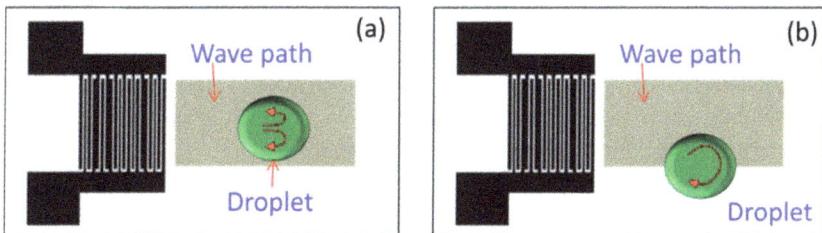

Figure 7. (a) Illustration of a droplet positioning symmetrically on surface of a SAW device; (b) asymmetric positioning of water droplet on the SAW device.

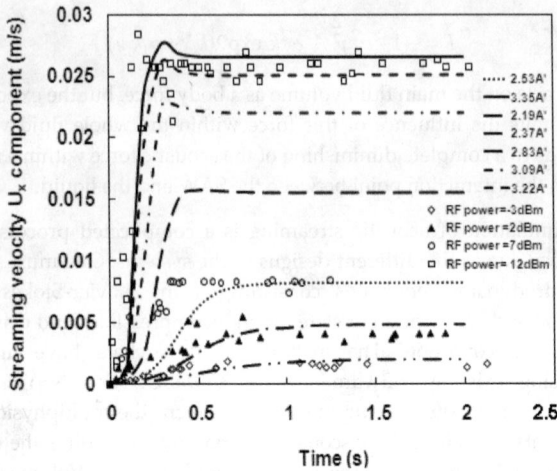

Figure 8. Streaming velocity at top centre of a 30μl droplet. SAW device has f=62MHz, an aperture of 2mm; Solid lines represent numerical results at different wave amplitudes; the symbols are experimentally measured data at different RF powers [53]. Reprinted with permission from Institute of Physics, J. Microeng. Micromech. 21, 2011, 015005.

and edge of the droplet [52]. A SAW entering in the centre of the droplet results in a symmetric shaped streaming pattern, with the highest velocity at the edge of the droplet where the SAW enters. When the SAW meets the droplet off-the central wave path, the streaming patterns is asymmetric and the degree of symmetry depends on the entry position of the SAW. A SAW entering into the droplet with a large off-centre line generates a large asymmetric flow pattern. As can be seen from Fig. 9, the numerical simulation can successfully reproduce the experimental results.

Modelling predicts that the streaming velocity is proportional to the RF power applied to the IDT electrode, in agreement with the experimental results, regardless as to whether 2D or 3D simulation was used. Figure 10 shows one of the results obtained by Alghane et al [53], demonstrating that the steady streaming velocity is approximately linearly correlated to the RF power applied.

For better understanding, 3D simulation is necessary to study the streaming phenomenon induced by leaky wave [44]. Figure 11 shows the simulated streamlines with a 3D circular flow pattern for a 30 μl droplet. The simulation results show that the highest value of a streaming velocity is located at the interaction area between the droplet and SAW as the highest momentum is delivered at this point. It attenuates rapidly as it enters the liquid droplet, in agreement with the experimental observation.

It is clear that the liquid initially moves at the Rayleigh angle before reaching the top of the droplet. It moves backward due to the constraint of the droplet boundary, forming a back flow on the two sides and the bottom of the droplet. Figure 12 is a comparison of the streaming patterns obtained experimentally and numerically, showing a good matching of

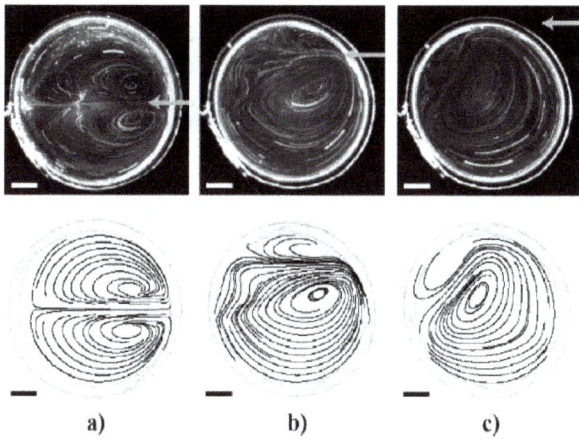

Figure 9. Comparison of streaming patterns for experiments and modelling. The narrow is SAW entering position from: (a) centre, (b) intermediate, and (c) outer part. The upper images are experimental results, while the bottom ones are the corresponding simulation results [52]. Reprinted with permission from IEEE, Transact. On Ultrasonics, Ferrelectrics and Freq. Control, 55, 2008, 2298. 1.

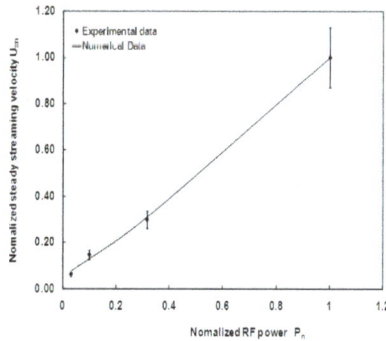

Figure 10. Flow normalized streaming velocity as a function of normalized RF power for a 30μl droplet size using a 128º YX-LiNbO₃ SAW device (IDT with 60 fingers) [53]. Reprinted with permission from Institute of Physics, J. Microeng. Micromech. 21, 2011, 015005.

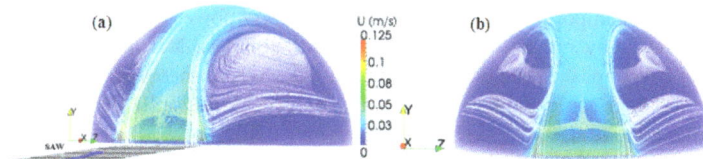

Figure 11. Acoustic streaming in 3D reproduced by numerical simulation in tilted view (a) and direct view (b). The droplet has a volume of 30ul, and the SAW enters through droplet centre [53]. Reprinted with permission from Institute of Physics, J. Microeng. Micromech. 21, 2011, 015005.

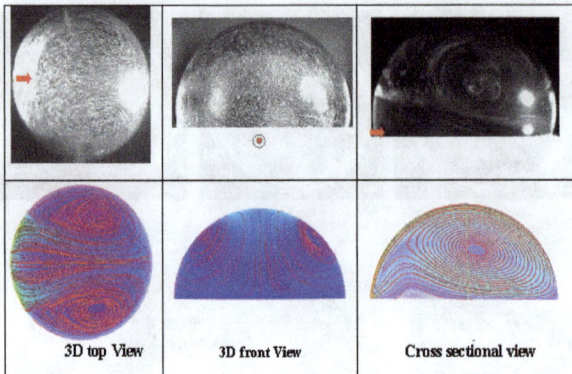

Figure 12. Comparison of experimental (upper row) and numerical modeling (bottom row) results for a 30μl droplet. SAW enters in a centre position of the droplet as shown by red arrow [53]. Reprinted with permission from Institute of Physics, J. Microeng. Micromech. 21, 2011, 015005.

the patterns. These numerical simulation results have clearly demonstrated that the chaotic acoustic streaming can be well explained by the physical laws.

3. Acoustic wave micropumps and micromixers

Although SAW devices have been commercialized over 60 years, applications in microfluidics and LOCs are only recent event. They are found to be very effective and efficient in microfluidics and LOCs [56]. SAW-based micropumps are one of them, and can transport liquid in a droplet form as well as in a continuous mode through proper design of a fluidic system. Acoustic streaming is also the most effective method of mixing liquids in small dimensions as it is quick and efficient, typically taking less than a few seconds to reach >95% mixing [57, 58]. In this section, SAW pumping and mixing are discussed.

3.1. IDT and SAW device structures

For microfluidics, one IDT electrode is enough for most pumping and mixing applications. The acoustic streaming velocity depends on the power output of the SAW devices and the amplitude of the RF signal applied to the IDT electrode. To obtain efficient acoustic streaming, it is desirable to have IDT electrodes with high and efficient power output. The SAW IDT design is important for delivering efficient SAW power output. The conventional bidirectional IDT may not be the most efficient for pumping and mixing, as the waves propagate in two opposite directions with half of the acoustic energy wasted. The simplest way is to reflect back some of the waves (Fig. 13a) by using reflector IDT. More sophisticated IDT designs include [59,60]: (a) split IDTs (Fig. 13b); (b) a SPUDT (single phase unidirectional transducer, Fig. 13c) which has the internally tuned reflectors within the IDT to form a unidirectional SAW propagation from one side of the IDT. These unidirectional acoustic wave transmission are essential for SAW microfluidics and sensors as they not only improve the performance, but maintain the SAW devices at the best operating conditions.

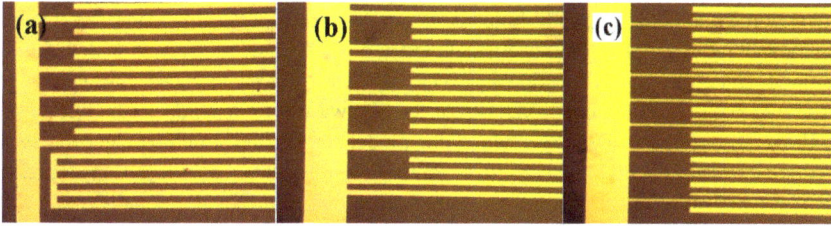

Figure 13. Common designs of IDT electrodes used for SAW devices. IDT with reflector (a), slitting electrode (b), and single phase unidirectional transducer (c).

Generally SAW IDT electrodes are so designed that the transmission spectra of the SAW devices have the highest quaility factor, Q, i.e. with a single well defined resonant peak with the highest amplitude. For microfluidic applications, the resonant frequency of a SAW device is changed slightly once liquid is loaded on the surface of the SAW device due to mass loading effect. It would be difficut to apply an RF signal with the exact frequency match to that of the SAW device, leading to the operation of SAW-microfluidics under mismatching conditions. This could be detrimental to the SAW devices as a much higher RF power is required to obtain the acoustic streaming effect, leading to overheating etc. Therefore a SAW device with a certain bandwidth of frequency would be better for SAW-microfluidic applications.

Thin film SAW devices have different transmission properties from those of SAW devices made on bulk substrates. Since the acoustic velocity of the thin film is different from that of the substrate, the acoustic waves generated by the top PE film layer may disperse into the substrate, i.e. some of waves propagate inside the substrate, resulting in two acoustic wave modes related to the PE layer and the substrate. A Sezawa mode acoustic wave is generated when the acoustic velocity in the substrate is higher than that in the PE-layer, and has been intensively studied as it can obtain SAW devices with high resonant frequencies using

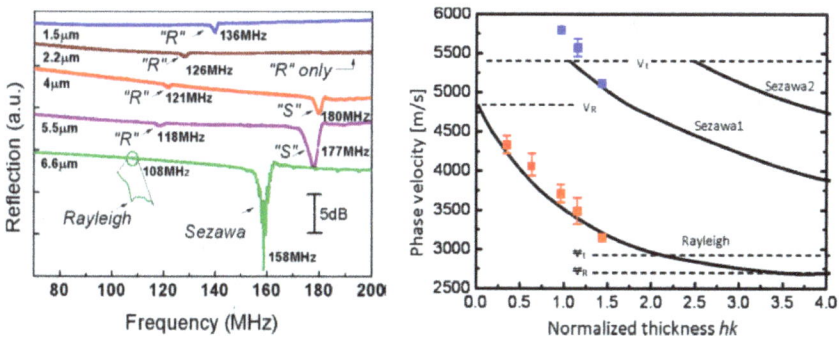

Figure 14. Reflection spectra of ZnO/Si SAWs with ZnO thickness as a parameter. The resonant frequencies of both Rayleigh and Sezawa waves decrease as the thickness increases, but the amplitudes of Sezawa waves are much larger than those of Rayleigh wave [61]. Reprinted with permission from AIP, Appl. Phys. Lett. 93, 2008, 094105.

conventional photolithography technology. Examples include ZnO films on Si, sapphire and diamond films. ZnO has a slow acoustic velocity of ~2700m/s, but ZnO SAW devices on a diamond layer can achieve a velocity over 10,000 m/s, close to those of the diamond substrate [39,40,41]. The acoustic velocity of the Sezawa waves depends on the thickness of the PE-layer, and can be well explained by the model of a layered structure. Figure 14 shows the dependence of the acoustic velocity of Rayleigh and Sezawa waves in ZnO/Si structures [61]. Sezawa waves often have much higher signal amplitude and resonant frequencies, particularly suitable for microfluidics and sensing applications.

3.2. SAW micropumps and micromixers

When the liquid is in the SAW path, acoustic moment and energy can be coupled into the liquid to induce acoustic streaming and flow. This has been utilized for fabrication of acoustic microfluidic systems, and many devices have been developed [62,63,64]. Generally speaking, both experimental and modelling results showed that acoustic waves can induce significant acoustic streaming in the liquid and result in mixing, pumping, ejection and atomization [65,66]. It was found that if the force is large enough, it can generate a significant acoustic streaming within the droplet (Fig. 15b). If the SAW device is immersed in a liquid container and an RF signal is applied to the IDT electrode, a steady flow pattern with a butterfly or quadrupolar streaming patterns can be obtained as shown in Fig. 15c. Significant acoustic streaming can facilitate internal agitation, which can speed up biochemical reactions, minimize non-specific bio-binding, and accelerate hybridization reactions in protein and DNA analysis which are commonly used in proteomics and genomics.

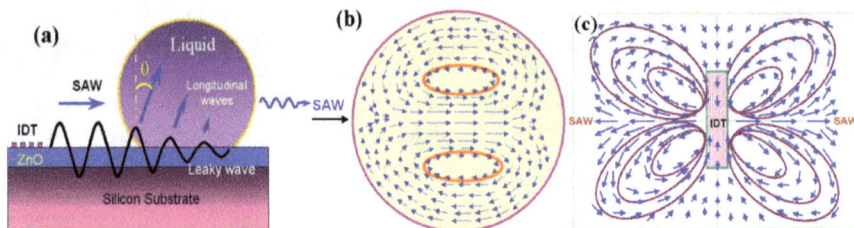

Figure 15. Patterns of acoustic streming in a droplet and in bulk liquid.

The SAW microfluidics have distinct advantages over other microfluidics, such as a simple device structure, no moving-parts, electronic control, high speed, programmability, manufacturability, remote control, compactness and high frequency response [67, 68,69]. The streaming velocity is proportional to the RF power applied, and could reach tens of centimetres per second. This is several orders of magnitude larger than other microfluidics, which are typically in the range of hundreds of micrometres to several millimetres per second [1,2,13]. Figure 16 shows the streaming velocity as a function of the amplitude of RF signal for LiNbO3 SAW devices with the IDT structure as a parameter [70]. It shows that the SAW device with a shorter wavelength has more power to induce streaming with higher velocity, and more IDT fingers are beneficial for coupling more RF power into the liquid. Du et al. also demonstrated that the third harmonic resonant waves can also be used for acoustic streaming though the streaming velocity is reduced to a third of the fundamental mode wave induced streaming [70] as higher harmonic modes have much lower RF output.

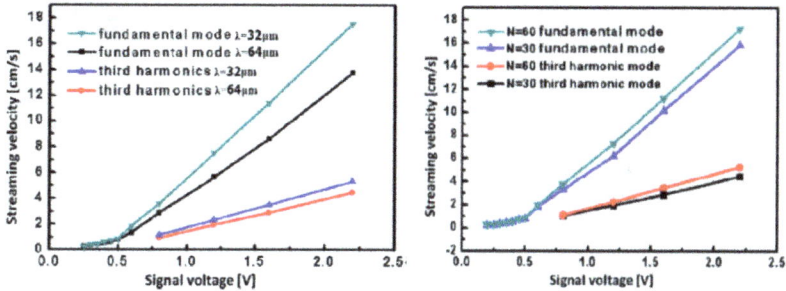

Figure 16. Acoustic streaming velocity as a function of RF signal voltage with finger pair and wave mode as parameters. [70]. Reprinted with permission from Institute of Physics, J. Microeng. Micromech. 19, 2009, 035016.

The streaming velocity depends on the RF power (or the signal voltage) applied to the IDT electrode. If the power is small enough and the droplet size is larger than the apperture of the IDT electrode, then the streaming velocity is proportional to the power as shown in Fig. 16, in agreement with the theoretical prediction as shown in Fig. 10. Once the power is over a certain range, velocity saturation is observed [71], and is believed that acoustic heating is responsible for the deviation from the linearity. Also the streaming velocity depends on the relative dimensions of the droplet and IDT aperture. At a fixed IDT aperture, a small doplet will have a low streaming velocity as the RF power can not be fully coupled into the liquid, and the streaming velocity increases with the droplet size, then reduces as the droplet size becomes larger than the aperture of the IDT. Therefore it is necessry to use an optimumal size/aperture ratio for a high performance SAW micromixer.

For streaming in a droplet, the boundary of the droplet becomes the boundary of the vortex, and the flow is confined within the droplet. This can be used for droplet-based mixing as shown in Fig. 17 by Xia et al on a LiNbO₃ SAW device [72]. Two drops with different contents are merged by a SAW, and immediate mixing occurs partially due to the kinetic energy involved and partially due to the acoustic streaming. This demonstrates the effective and efficient mixing of water and red dye droplets. The mixing process is completed in tens of ms, much shorter than those of most other micromixers.

Acoustic streaming has circulating flow patterns due to the back flow of the streaming as schematically shown in Fig.15 regardless as to whether it is a droplet or bulk liquid. This is caused by the short range of the acoustic force as discussed above. Directional flow by acoustic streaming, or liquid pumping, might be difficult due to this back flow. For pumping liquid in a specific direction, the SAW device and microchannels have to be arranged properly. Figure 18 is a schematic drawing of a SAW-based micropump. When the SAW-induced vortex size is smaller than the channel width, the SAW induces a circulating

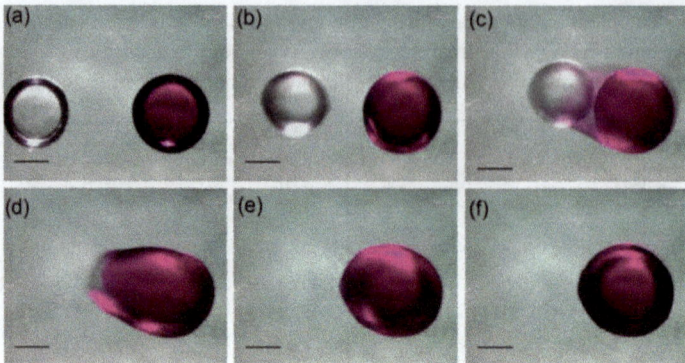

Figure 17. A water droplet and a red dye droplet mixture at (a) t=0s, (b) t = 4.066 s, (c) t = 4.4 s, (d) t = 4.466 s, (e) t = 5.000 s, and (f) t = 6.133 s [72]. Reprinted with permission from Elsevier, Talants, 84, 2011, 293.

vortex within the channel, and no net flow along the channel is produced. This can be utilized for in-channel mixing with high efficiency, but not pumping. When the size of vortex is larger than the width of the microchannel, the back flow is restricted by the channel wall, leading to a net directional flow along the channel. By using a channel with the width less than 200 µm, Yeo et al realized a SAW-based micropump [19].

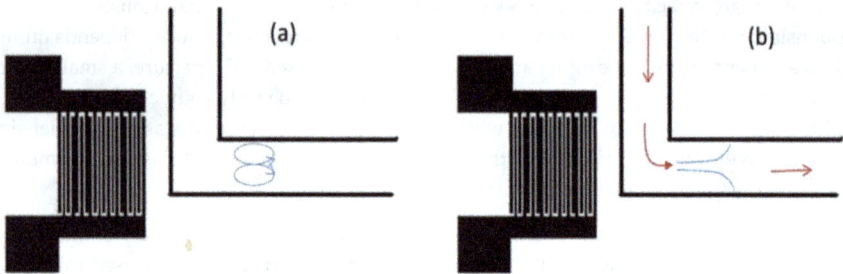

Figure 18. Principle of SAW-based micropumps with a channel. When the vortex is smaller than the channel width, the back flow occurs, suitable for in-channel mixing (a). When the vortex is larger than the channel width, a unidirectional flow is formed (b).

SAW has also been explored for in-channel mixing. Figure 19 shows two schemes used by Nguyen et al for in-channel mixing [73]. The microchannel is perpendicular to the SAW propagation direction. When a travelling acoustic wave encounters the liquid in the channel, streaming can be very effective mixing mechanism, and this mixing can be further enhanced by using curved IDT structures as shown in Fig. 19b. An alternative micromixing using SAW is to couple surface acoustic waves into liquids in a container to introduce agitation for mixing. As the streaming angle is determined by the Rayleigh angle, when a container is placed on a SAW device with a soft coupler on the surface, the SAW device can be an effective mixer [74].

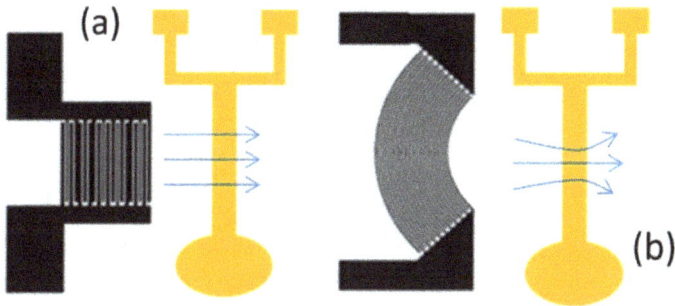

Figure 19. Schematic drawing of SAW-based in-channel micromixer.

SAW can be utilized for moving liquid droplets. If the liquid is on a hydrophobic surface with a contact angle larger than 90°, then the acoustic force may move the droplet along the wave path if the RF power is sufficiently large. Droplet-based microfluidics and LOCs are attractive as it provides the foundation for digital analysis and digital medicine with better and accurate results. The surface of most PE materials, however, is hydrophilic with water contact angle less than 90°. Surface acoustic waves are not strong enough to move the liquid on hydrophilic surfaces. Figure 20 shows the shape change of a droplet when it is subjected to an acoustic pressure on a hydrophilic surface. Water droplets cannot be moved freely on the surface under the stimulation of the SAW, but simply spreads on the wave path. The solution is to reduce the surface energy to move the droplets freely on the surface. Various surface coating technologies have been developed to increase the contact angle of the water droplet. CFx chemical vapour treatment was found to form contact angles larger than 90° and reduce the surface energy significantly [75]. Solution-based Teflon can be applied to the surface of PE substrates to form a hydrophobic surface to provide contact angle larger than 110°. The thickness of the Teflon and CFx layer has to be carefully controlled to reduce attenuation caused by the large acoustic absorption of the layer. Octadecyltrichlorosilane (OTS) was found to form a compact and hydrophobic self-assembled monolayer (SAM) with a thickness less than 10 nm. The OTS layer has a contact angle with water larger than 100° and does not damp the SAW amplitude visibly; therefore it is a good hydrophobic coating for microfluidic applications [70,71].

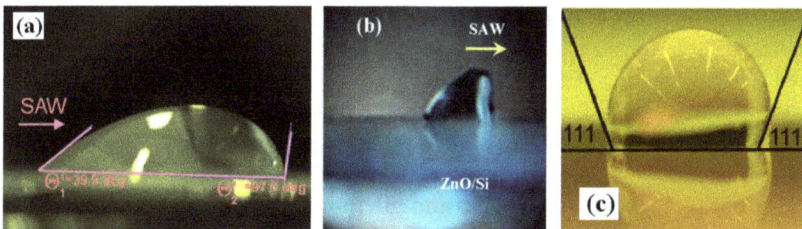

Figure 20. The shape of a droplet under acoustic pressure. The droplet deforms with the leading angle becomes larger and the trailing angle becomes smaller [70]. Reprinted with permission from Institute of Physics, J. Microeng. Micromech. 19, 2009, 035016 for (a) and (c).

Similar to streaming velocity, the droplet moving velocity driven by SAW is found to be proportional to the RF signal amplitude as shown in Fig. 21, and the velocity depends on the droplet size as well as the aperture of the IDT. However, when the droplet diameter is larger than the aperture, further increase in the power does not increase the velocity visibly as the acoustic wave does not fully coupled into the droplets as discussed in [76]. When the aperture is larger than the droplet size, it was found that the velocity decreases with increase of the droplet size as more power is needed to move a larger droplet. As can be seen, the droplet motion velocity is in the range up to 1-2 cm/sec, much larger than the velocities obtained by other methods. Although a SAW can be used to drive droplets at very high speeds, delivery of droplets to required locations with precision is more important for applications in biological analysis than driving the droplets at a high speed. This can be realized by using modulated pulsed RF signal to drive the SAW devices in a controlled manner. By adjusting the period of the on-off RF signal and the signal amplitude, it is possible to move droplets with precision distance. A moving rate of 100 μm/pulse for a droplet of 0.5 μl on a LiNbO$_3$ substrate was obtained by using the pulsed RF signal [77].

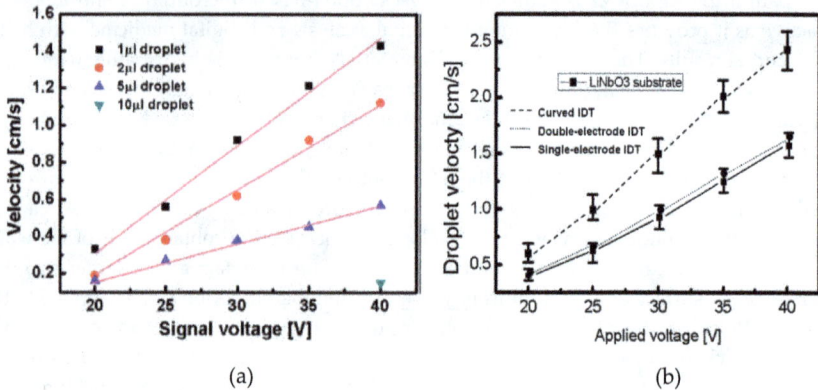

(a)   (b)

Figure 21. Droplet moving velocity vs. RF signal amplitude (a) and droplet moving capability vs. IDT structures [70]. Reprinted with permission from Institute of Physics, J. Microeng. Micromech. 19, 2009, 035016 for (a).

As discussed before, the IDT structure significantly affects the energy output, and hence the acoustic streaming and droplet motion owing to its acoustic energy distribution. Generally, a unidirectional IDT has a higher acoustic energy density than a bi-directional IDT, and the curved IDT has the highest energy density. Figure 21b shows a comparison of the droplet motion velocity for SAW devices with different IDT structures. The curved IDT SAW has the highest droplet motion velocity as expected.

3.3. Flexural plate wave micropump and micromixer

Lamb waves have also been utilized for pumping and agitating of minute volumes of liquids [32, 33] and enhancing biochemical reactions [78], based on the fluid motion induced by the travelling flexural wave in a ZnO or AlN piezoelectric membrane [79]. A novel valveless pump based on a Lamb wave was proposed by Ogawa et al [80]. The liquid in the microchannel was transported by generating a travelling wave on the channel wall, which was composed of a piezoelectric PZT thin film actuator array. A mean flow velocity of about 118 and 172 μm/s was obtained for the 200 and 500 μm wide channels, respectively.

However, due to the low frequency resulting from the thin membrane, the agitation and pumping are too small and insufficient energy is coupled into the liquid. Furthermore, the microfluidic systems contain a membrane vibrating at a high speed, and the yield for fabrication is low, and the reliability of the systems during operation is poor compared to SAW micropumps and mixers. Therefore, research and application of FPW-based microfluidics are currently very limited.

4. Acoustic droplet generator and atomizer

4.1. SAW droplet generator and manipulator

Generation of droplets with volumes from a few pls to a few μls is extremely important for modern biotechnology, life science, medical research and diagnosis. For quantitative analysis in a small volume, it is essential to measure the small volume of reagents and biosamples precisely; otherwise false results can be easily obtained. Although a few technologies have been developed to generate droplets in volumes from a few nls to a few μls, they are difficult to be integrated with microfluidic systems. By changing the acoustic force and hydrophilic and hydrophobic patterns on the surface of a SAW device, it is possible to generate droplets from a few pls to μls on a free surface. Figure 22a is a schematic drawing of a SAW-based droplet generator, consisting of SAW devices with a reservoir and hydrophilic spots surrounded by a hydrophobic surface [49]. When the liquid in the reservoir is pushed forward under the acoustic force, it makes contact with the hydrophilic

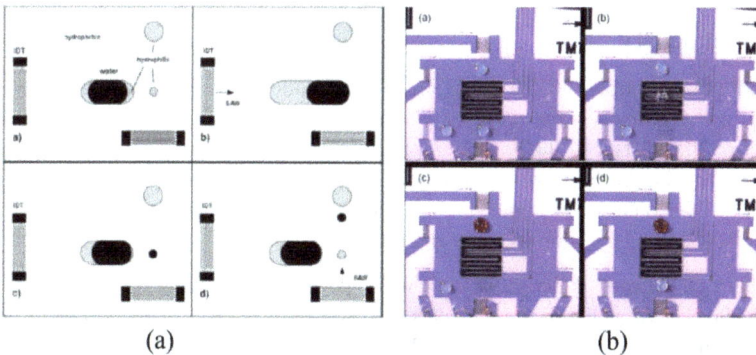

(a)   (b)

Figure 22. (a) the principle of acoustically driven nano dispenser by selective chemical modification of the wettability of parts of the chip surface and employing two SAW propagating at a right-angle to each other. (b) is a SAW driven microfluidic processor [49]. Reprinted with permission from Elsevier, Superlattice & Microstruct. 33, 2004, 389.

spot and fills it. Once the RF signal is off, the bulk liquid withdraws back to the reservoir as the acoustic force diminishes, leaving the spot filled with the liquid of a fixed volume determined by the size of the spot area and the contact angle. In this way, Woxforth et al. developed a SAW-based droplet generator [49]. Furthermore, by using the virtual containers and tracks formed by hydrophilic surfaces surrounded by hydrophobic surfaces for liquids on a chip, they have developed droplet manipulator and mixer as demonstrated in Fig. 22b.

A combination of a SAW pump with other droplet generators can realize new functions of microfluidics. Electrowetting on dielectrics (EWOD) has been combined with SAW devices to fabricate the EWOD-SAW microfluidics as shown in Fig. 23 [81]. EWOD is used to generate and separate the droplets, while the SAW device is used to move the droplets along the track. The EWOD force is employed to guide and position microdroplets precisely which can then be actuated by SAW devices for particle concentration, acoustic streaming, mixing and ejection, as well as for sensing using a shear-horizontal wave SAW device [82].

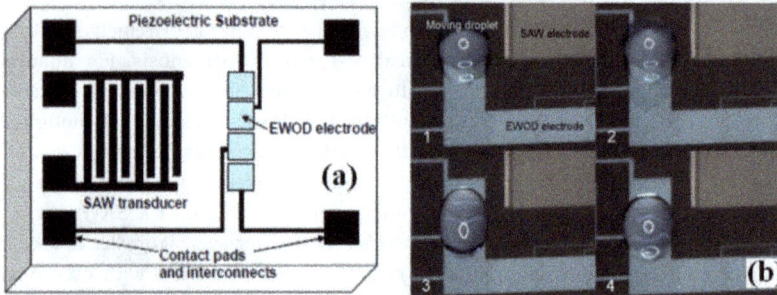

Figure 23. Schematic of integrated EWOD and SAW test structure and the droplet generation by EW and manipulation by SAW [82]. Reprinted with permission from AIP, Biomicrofluidics, 6, 2012, 012812.

4.2. SAW atomizer

If the RF power coupled to the liquid on a hydrophilic surface is sufficiently large, tiny droplets with volumes in the range of a few femtolitres (*f*ls) to *p*ls can be generated and escape from the surface of the host liquid, forming a continuous mist of droplets as shown in Fig. 24. This has been utilized for the development of SAW-based atomizers and droplet generators. Ejection of small particles and liquids has many applications ranging from inkjet printing, fuel and oil injection sprayers and propellers.

The height of the mist is dependent of the RF power applied, and could be up to 6 cm [20,71]. The ejection angle of tiny droplets escaping from the host liquid is determined by the Rayleigh angle, but affected by the RF power and height of the mist [83]. In order to generate a continuous mist on demand, there must be a continuous supply of liquid which can be realized by using a porous structure such as a filter paper linked to a large liquid reservoir as shown in Fig. 25 [65].

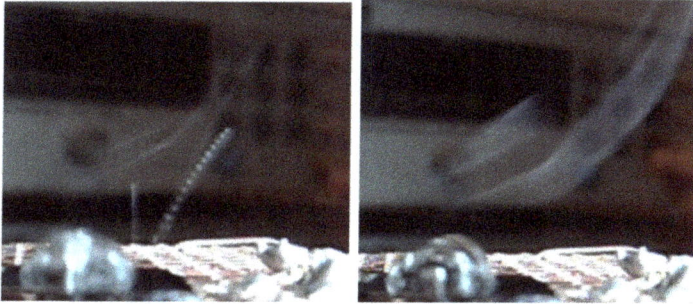

Figure 24. Acoustic streaming induced mist by a ZnO SAW device, which may reach 6mm height.

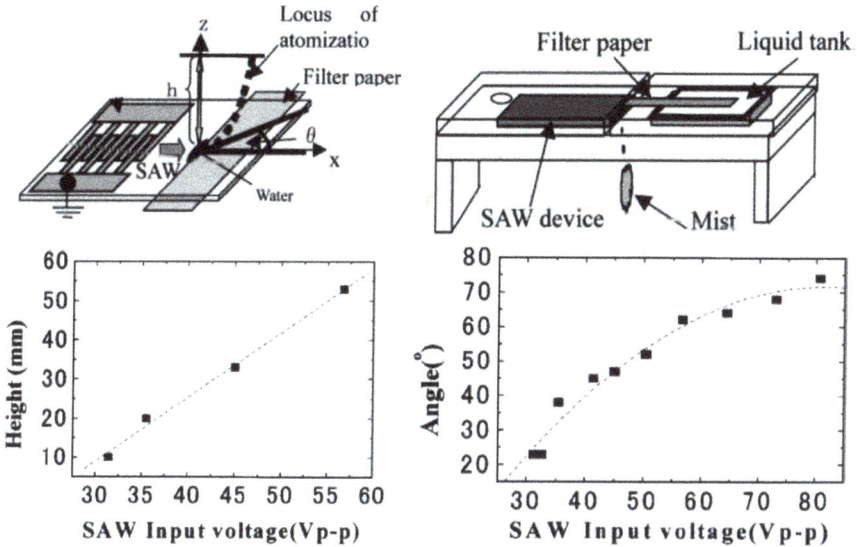

Figure 25. SAW atomization mechanism and setup for the SAW atomizer, and the mist height and angle as a function of SAW input voltage [65]. Reprinted with permission from Japan J. Appl. Phys. Lett. 43, 2004, 2987.

A proper design of SAW ejector can be utilized for nozzle-free ink applications. Tan *et al.* [84] demonstrated another method utilizing two opposite IDTs to converge the acoustic energies at point with a liquid drop ejected perpendicular to the surface, similar to the normal nozzle-based droplet ejector as schematically shown in Fig. 26. The radiation from two sides of the droplet resulted in an elongated liquid column with an angle of about 90°. These SAW ejectors do not have a nozzle head and offer a more cost effective solution when compared to the current ink ejector.

Atomization has been widely applied in pulmonary drug delivery as a promising technology to transport drug formulations directly to the respiratory tract in the form of inhaled particles. The most common methods employed for this application are jet atomization and ultrasonic atomization with difficulties to produce monodispersed particles, i.e. droplets with sizes in the range of 1~5µm in diameter. SAW atomizers are able to produce aerosol droplets with a good particle size distribution. By controlling the RF power applied to the SAW IDT, the droplet sizes can be less than 5 µm [85], suitable for the pulmonary drug delivery application.

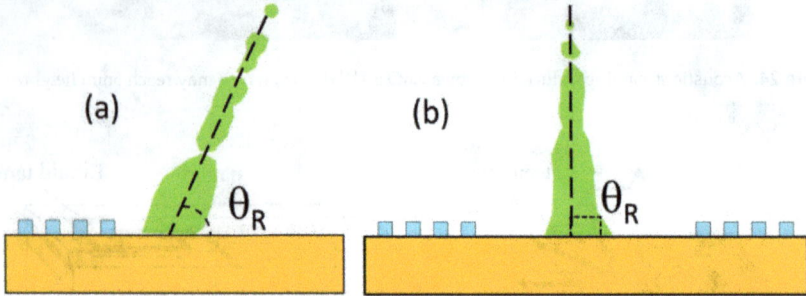

Figure 26. Droplet jetting induced by a single IDT SAW device (a), and droplet jetting induced by a pair of IDT electrodes (b).

Not just bulk SAW devices, but also thin film SAW devices can also achieve a similar atomization effect by using sufficient RF power. Figure 24 shows images of tiny liquid droplets ejected from the surface of a ZnO SAW device obtained by the authors, and the height of the mist is similar to those observed from the LiNbO₃ bulk SAW devices, even though thin film SAW devices have lower power delivered compared with the bulk devices.

5. Acoustic wave based biosensors

Most acoustic wave resonators (QCM, SAW and FBAR) can be used as sensors because all of them are sensitive to mechanical, chemical, optical or electrical perturbations on the surface of the devices [86,87]. They are versatile, sensitive and reliable, being able to detect not only mass/density changes, but also viscosity, elastic modulus, conductivity and dielectric properties etc. They have many applications such as sensing pressure, humidity, temperature, strain (stress), acceleration force, vibration, flow, pH values, radiation, electric fields etc. They are sensitive gas sensors once combined with specific gas absorption layers [88,89,90]. Development of acoustic wave based biosensors is relatively new but has a huge potential as they can detect tiny traces of biomolecules [22,23]. This can be utilized to detect viruses and genetic disorders, diagnose early stage diseases and cancers [91,92]. The principle of acoustic biosensors is similar to those of other biosensors, which are based on a specific interaction between biomarkers (also called probe molecules) deposited on the surface of the sensors with target molecules in the biosamples. Compared with other common biosensing technologies, such as surface plasmon resonance (SPR), optical fibres, and field effect transistors or cantilever-based sensors, acoustic wave based sensors have the combined advantages of simple operation, high sensitivity, small size, compact and low cost. In the following section, highlight the acoustic wave sensor technologies.

5.1. QCM sensors

Although the piezoelectric effect was discovered in the late 19th Century, quartz crystal resonators only found widespread applications in electronics, material and biological researches when it was demonstrated that there was a linear relationship between mass adsorbed on the surface and the resonant frequency of the crystal in 1959 by Sauerbrey [93]. Biosensing became possible when suitable oscillator circuits for operation in liquids were developed [94]. The QCM is one of the most developed biosensors that can be operated in a liquid environment using the thickness shear-mode. A QCM consists of a bulk piezoelectric material with typical dimensions of 1 cm in diameter and 500-1000 μm in thickness, sandwiched between two metal electrodes. When an A.C. electrical signal is applied to the two electrodes, it excites a standing wave between the two electrodes through the PE effect. The operating frequency of the QCM is determined by the thickness of the PE-layers, and is typically 5, 10 and 20MHz. As shown by eq.(1), the sensitivity of acoustic resonators is determined by the square of the frequency and the base mass. With decreasing thickness of the Quartz layer, the frequency of the QCM has been increased significantly, thus its sensitivity has been dramatically increased. There has been intensive research recently to develop thin film based QCMs with operating frequencies of several hundreds of MHz [37,38] which demonstrated their great potential for biosensing with better sensitivity.

For biosensing, QCM biosensors have been used to detect the interaction between protein-protein, DNAs, protein-DNA, viruses, bacteria etc, and demonstrated their usefulness, versatility and robustness with high sensitivity. For practical applications, more QCM sensors have been integrated with other structures and devices such as molecular imprint polymers [95], sensors [96] and microfluidics [97] for multi-task detection and monitoring with better accuracy and more functionality. There are hundreds of published papers on QCM biosensors; a review of QCM biosensors is beyond the scope of this book chapter. Readers can find more information in refs. [98,99].

5.2. SAW sensors

SAW devices are not only used for microfluidics, but are very good sensors. Since SAW devices have a much smaller active mass and much higher operating frequency than those of QCMs, the sensitivity of the SAW devices increases dramatically as shown in Fig. 3. Furthermore by using advanced photolithograph technology, especially e-beam writing techniques, it is now possible to fabricate SAW devices with operating frequencies up to the Gigahertz [100,101], and the sensitivity of SAW sensors can be further increased.

The longitudinal mode SAW device has a substantial surface-normal displacement that rapidly dissipates the acoustic wave energy into a liquid, leading to excessive damping, and hence poor sensitivity and noise for biodetection in liquid. Shear horizontal (SH-) mode SAW devices have substantially reduced coupling of acoustic energy into the liquid [102,103], hence they can maintain a high sensitivity in liquids. Consequently SH-SAW devices are suitable for biodetection, especially for "real-time" monitoring of physiological conditions of a patient. To further reduce the base mass of a SAW device to improve the sensitivity, Love wave SAW devices have been developed which consist of a normal SAW device and a thin wave guide layer (typically sub-micrometers) such as SiO_2 and polymers on top of the SAW surface. Since the acoustic velocity is slow in the wave guide layers, acoustic waves are trapped in the thin wave guide layer, resulting in a drastically reduced base mass and significantly improved sensitivity and quality of the sensors [104,105]. They are therefore frequently employed to perform biosensing in liquid conditions [106,103].

For LOC applications, integratable thin film SAW sensors are more attractive and desirable. A lot of efforts have been made to develop AlN and ZnO thin film SAW sensors. A ZnO/Si SAW device has been successfully used in the detection of aminohexanoic acid succinimidyl ester (DNP-X) and anti-DNP-KLH antibody [107]. The resonant frequency of the ZnO SAW devices was found to shift to lower frequencies as the PSAs are specifically immobilized on the surface-modified ZnO SAW device. A linear dependence has been measured between the resonance frequency change and the anti-DNP concentration over a range from 2 to 1000 ng/ml, and saturated as the concentration increases further due to the reduction of binding sites [107].

To realize biosensing in liquids with better sensitivity, Love wave SAWs have been studied intensively. ZnO has a shear wave velocity of ~2600 m/s, whereas that of ST-cut-quartz is about 4996 m/s. Therefore, it is reasonable to use ZnO as a guiding layer on substrates of ST-cut quartz to form Love mode biosensors. The other potential substrate materials for Love-mode ZnO sensors include $LiTaO_3$, $LiNbO_3$ and sapphire. A ZnO Love mode device of ZnO/ST-cut quartz has a maximum sensitivity up to ~18.77×10^{-8} m^2 s kg^{-1}, much higher than that of a SiO_2/quartz Love mode SAW device [108,109]. Mchale et al recently reported ZnO/SiO_2/Si SAW Love mode sensors with a sensitivity of 8.64 μm^2/mg [110], which is about 2 to 5 times that of ZnO/$LiTaO_3$ [111] and SiO_2/quartz Love sensors [112]. Another promising approach for making a ZnO based Love mode sensor is to use a polymer film (such as PMMA, polyimide, SU-8 or parylene C) on top of the ZnO layer as the guiding layer. However, this layered structure uses a polymer waveguide and has a relatively large attenuation compared with those of solid waveguide layers.

AlN SAW devices have higher acoustic velocities, for example, about 6000 m/s shear velocity for an AlN/Si SAW device, and thus it is desirable to use AlN SAW devices for sensors for high sensitivity. However there are not many reports on AlN based SAW biosensors. The reason could be the difficulties in the deposition of the thick AlN film (>4 μm) required for high quality SAW device fabrication. They normally have a large film stress and poor adhesion with the substrate. Although AlN films deposited to deposit on a $LiNbO_3$ substrate have been reported to form a highly sensitive Love mode sensing devices [113,114].

5.3. FBAR sensors

FBARs were initially developed as high frequency resonators for applications in electronics as filters, duplexer etc in 1980s' [25,26]. FBARs have been considered for biosensor application since 2000, and are considered as one of the most advanced sensors with extremely high sensitivity and very small dimensions. As the wavelength of the bulk resonators is determined by the thickness of the PE-layer, it is normal to fabricate FBARs with thin PE-layers. FBARs typically have frequencies of a few GHz, and ones with f_r of 8 GHz or higher have been demonstrated [115,116]. Owing to the small base mass and high operation frequency, attachment of a small amount of target mass is able to induce a large frequency shift – typically a few MHz. This improves the sensitivity and makes the signal easily to be detected using simple electronic circuitry. Although the sensitivity of FBARs is not as good as predicted by eq.(1), it is still about three and two orders of magnitude higher than those of QCMs and SAWs respectively as shown in Fig.4 [28,117].

A ZnO-based label free FBAR biosensor with an operating frequency of 2 GHz was used to detect DNA and protein molecules [115]. It showed a sensitivity of 2400 Hz·cm²/ng, which is approximately 2500 times higher than a conventional QCM device could achieve. A recent Al/ZnO/Pt/Ti FBAR design also showed a sensitivity of 3654 Hz·cm²/ng with a better thermal stability than that of ZnO-based FBARs [118,119].

Based on eq.(1), an increase in f_r would make FBARs of higher sensitivity. However it should be pointed out that the limitation to the sensitivity of acoustic resonators, especially the FBARs, is not only the frequency, but also the quality factor. It is easy to make FBARs of high f_r by using a thin PE layer, but the quality factor of the FBARs was found to decrease dramatically with decrease of the thickness of the PE-layer, mostly due to the relatively poor crystal quality, small grain size, poor thickness uniformity, rough surfaces and the existence of a thick transition layer (20-80 nm), resulting in a severely reduced quality factor when a thin PE layer is used. Also it was found that the electrode shape and material properties also significantly affect the quality factor. Electrodes with 90° regular angles reflect the surface travelling wave and deteriorate the Q-factor. Electrodes with high acoustic impedance and low mass are preferred for fabrication of high performance FBARs. Aluminium is one of the most popular electrode materials in the microelectronics industry. It has a low mass density, but the acoustic impedance is low and thus is not the best material for the fabrication of FBARs. Au, Pt and W have high acoustic impedance but their mass densities are high, leading to large mass loading effects. A carbon nanotube (CNT) layer was found to be the best electrode material for FBARs as it has a high elastic modulus and low density (hence a high acoustic impedance and low mass loading) and CNTs have thus been used for fabrication of FBARs with much improved quality factor. An improvement of Q-value by 5 times was demonstrated simply by using a CNT layer on top of existing metal electrodes [120]. Garcia-Gancedo et al recently fabricated FBARs with CNT layer as the top electrode and demonstrated FBARs with a quality factor over 2000 [121,122], one of the best values reported. Also the spurious ripples round a resonant peak found in Au electrode FBARs disappeared when CNTs were used as shown in Fig. 27. The FBARs with CNT electrodes showed a better sensitivity than the Au electrode ones with a mass detection limit down to 10^{-13}g, with the potential to go down to 10^{-15}g, suitable for detecting a single molecule.

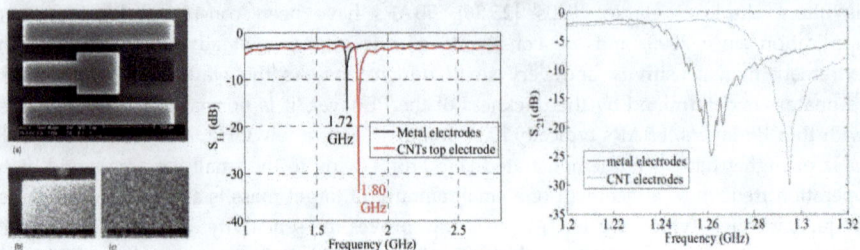

Figure 27. (a) SEM images of a fabricated SMR with CNTs layer top electrode. The middle is a typical frequency response, showing the main resonance at 1.75 GHz. The ripples for FBARs made of Au and CNTs top electrode. It disappears in CNTs-FBARs due to high acoustic impedance [122]. Reprinted with permission from Elsevier, Sensors and Actuators, B 160, 2011, 1386.

The majority of the PE thin films have crystal orientation (0002) normal to the surface of the substrates. They are suitable for fabrication of FBARs with longitudinal mode and appropriate for gas phase detection. FBARs with protein functionalized surfaces have been used as a gas phase biosensors, and demonstrated their feasibility for sensing an odorant binding protein of AaegOBP22 using N,N-diethyl-meta-toluamide (DEET) as the ligand to the odorant binding protein [123]. For sensing in liquids, novel structures, especially integrated with microfluidics, is needed. Zhang et al performed biodetection in liquids by taking advantage of the back trench structure. The trench was used as a container in which the bioreaction could take place, while a FBAR on the other side of the thin membrane was used for sensing. This demonstrated its feasibility for biodetection [124]. Similarly Wingqvist et al. used the surface of a FBAR for sensing with a built-in microchannel on the back which allows continuous flow of the biosamples or buffer solution to pass for a continuous measurement [31]. A schematic drawing is shown in Fig. 28.

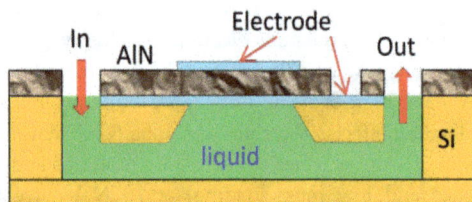

Figure 28. Typical structure of lateral field excited inclined AlN film FBAR.

For direct liquid contact sensing, it is necessary to develop ZnO or AlN films with crystal orientation inclined relative to the surface normal, allowing generation of shear waves to be used for detection in a liquid [125]. A (1120) textured ZnO film exhibits pure shear mode waves which can propagate in a liquid with little damping effect. A ZnO shear mode FBAR device has been used in a water-glycerol solution, with a high f_r of 830 MHz and a sensitivity of 1000 Hz·cm^2/ng [126]. Weber et al. fabricated ZnO FBAR devices using a ZnO film with 16° off c-axis crystal orientation, which operated in a transversal shear mode [125]. For an avidin/anti-avidin system, the fabricated devices had a high sensitivity of 585 Hz·cm^2/ng and a mass detection limit of 2.3 ng/cm^2. The shear wave FBAR devices also showed a more stable temperature coefficient of frequency [30,31].

Another method for liquid phase biodetection is to use the lateral field excitation (LFE) for FBARs as shown in Fig. 29. This structural FBAR requires both signal and ground electrodes being in-plane and parallel on the exposed surface of the PE films [127,128,129]. Since the excitation is parallel to the surface and perpendicular to the normal c-axis crystal orientation, the FBARs exhibit a thickness shear mode operation. The devices are stable in biologically equivalent environments [130,128]. However, lateral structural FBARs normally have a low quality factor and the mechanism of the exciting resonance is not fully understood yet. A lot of research is needed before they can be effectively used as sensors. Furthermore, lateral FBARs have a very narrow active area, and it is difficult to incorporate microfluidics within the narrow channel for sensing.

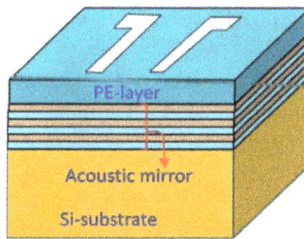

Figure 29. Schematical structure for the lateral field excited FBAR devices, which generates a thickness shear mode resonance.

5.4. Flexural plate resonant sensors

Lamb wave devices on a membrane structure have been used for biosensing in liquid [34]. Since the propagation velocity of the Lamb wave in the membrane is slower than that in the fluids on the surface, the acoustic energy is not easily dissipated, thus the Lamb wave sensors can be used for thus applications [131]. Since the resonant frequency of the FPW devices is small, it is not sensible to use the frequency as the parameter for sensing due to poor sensitivity as indicated by eq.(1), the amplitude of the resonant wave is normally used for sensing in liquid. Therefore, the sensitivity of these devices increases as the membrane thickness becomes thinner [35, 15].

A ZnO based FPW device has been used to monitor the growth of bacterium "Pseudomonas putida" in a boulus of toluene and the reaction of antibodies in an immunoassay for an antigen present in breast cancer patients [132]. Si/SiO$_2$/Si$_3$N$_4$/Cr/Au/ZnO FPW devices have been used for detecting human IgE based on the conventional cystamine SAM layer technology with a sentivity of 8.52×10^7 cm^2/g at a wave frequency of 9 MHz [133]. However, the FPW biosensor has not been widely reported because of the low sensitivity, difficulty of fabrication and high temperature sensitivity of the thin film.

6. Other SAW-based functions and lab-on-a-chip

Since the acoustic wave mechanism can be utilized for fabrication of various microfluidic devices and sensors, it would be very attractive to fabricate single acoustic wave mechanism-based lab-on-chip systems [117]. Development of such systems has been rather limited so far, as the individual acoustic technologies are yet to be fully explored, developed and optimized. Current activities have been focused mainly on the development of individual acoustic wave based devices and systems such as acoustic microheater, SAW-based polymerase chain reaction (PCR), SAW-based particle concentrator, sorting and delivery devices etc. These will be highlighted in this section.

6.1. SAW microheater

For SAW devices, an input of high RF power will induce acoustic heating through crystal vibration and absorption of acoustic energy by defects in the substrates. For sensing, the input RF signal normally has a low power and acoustic heating is not a problem. For acoustic microfluidics, especially droplet-based pumps, acoustic heat may increase the surface temperature of the SAW device over 100 °C which will damage most of the cells and bio-molecules and reduce their biological integrity. Acoustic heating can be suppressed by using a pulsed RF signal to maintain the temperature below 40 °C. Although acoustic heating has many negative effects for biological and electronic applications, controlled acoustic heating can be utilized as a remote microheater for many applications, such as in polymerase chain reaction (PCR) to amplify DNA concentration for detection or to accelerate bioreaction.

Figure 30 shows the surface temperature as a function of RF signal voltage measured for a ZnO thin film SAW device. The temperature at a position 5 mm away from the IDT on the wave path was monitored. The temperature increases with the signal amplitude and the duration of the RF signal, and decreases with the distance from the IDT [77]. The maximum temperature can reach at 140 ºC for a signal voltage of 60 V. The temperature was found to have a distribution along the wave path. It is higher near the IDT, and decreases away from the IDT due to attenuation of the wave. Acoustic heating has been utilized to construct PCR as will be discussed later [74,134]. For normal gas sensing, initialization of sensors requires a high temperature to remove all absorbed substances. Acoustic heating could be utilized for self-initialization for SAW gas sensors without need of an additional microheater, which greatly simplifies the system and reduces the cost and fabrication process. Effective utilization of acoustic heating would generate compact, useful microsystems with many functions, and many applications are yet to be explored.

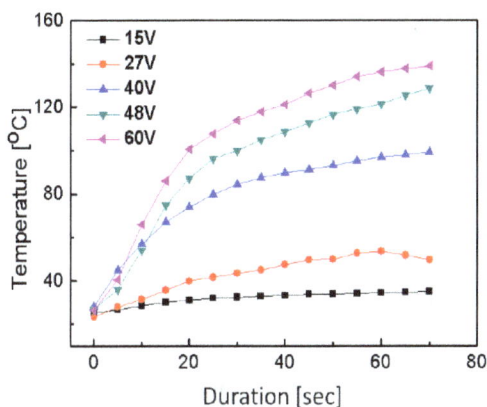

Figure 30. Acoustic heating induced temperature rise as a function of RF signal duration by a SAW microheater [77]. Reprinted with permission from AIP, J. Appl. Phys, 105, 2009, 024508.

6.2. SAW particle concentrator

Transportation and concentration of particles or bio-substances such as cells are important issues for biological applications. As above, the SAW acting on the edge of a droplet can generate a shear force within the droplet. This shear force can generate circulating streaming, moving particles towards the centre as shown in Fig. 31. From a side view, the fluid can be observed to be pushed upward just above the SAW propagation area which results in primary azimuthal rotation within the droplet periphery. Raghavan et al [135] reported that the flow phenomenon within liquid droplets due to SAW asymmetric positioning are similar to that obtained by the flow field between stationary and rotating disks. This azimuthal rotation phenomenon has been utilized for particle concentration.

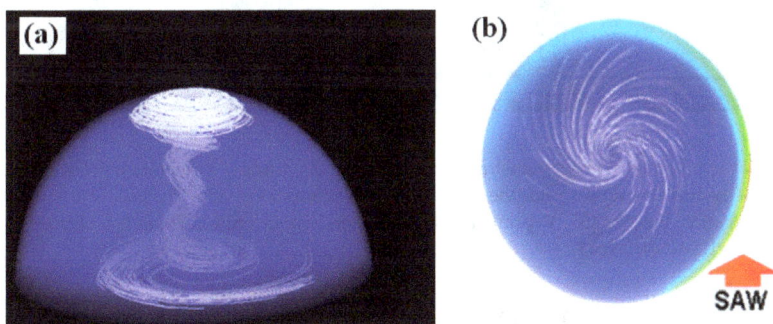

Figure 31. Side view (a) and overview (b) of simulated circulating streaming patterns induced by acoustic wave for a 10μl droplet [53].

Figure 32 shows the frame images of starch particles captured during the concentration process within a 20 μl droplet after applying an RF signal to the IDT electrode for a few seconds. Initially, the particles are uniformly dispersed in the water, and circulating streaming is induced once the RF signal is applied. The flow circulation rapidly establishes a particle cluster towards the centre of vortex, in the form of a conical shape similar to that depicted theoretically. The shear force induced particle migration is due to the gradient in the azimuthal streaming velocity in the droplet, resulting in particle motion from a higher shear force area at the droplet periphery to a lower shear force area at the bottom of the droplet centre. The shear velocity is large on the edge of the droplet, and gradually decreases on approaching the centre of the droplet. The particles circulate with the liquid in the droplet and simultaneously migrate from the high to the low shear velocity regions [53,136]. The concentration effect is dependent of the RF power as well as the properties of the particles. At a low power, it is not sufficient to generate a gradient in the azimuthal streaming velocity, whereas a high power produces a strong turbulent streaming, dispersing particles within the droplet randomly without any concentration effect [53]. The particle size is also critical for the efficient concentration. Particles with certain sizes can be easily agglomerated, whereas small particles can flow inside the liquid for very long time before being forced into the central region. Therefore it is possible to utilize this to separate particles with different sizes using a SAW device.

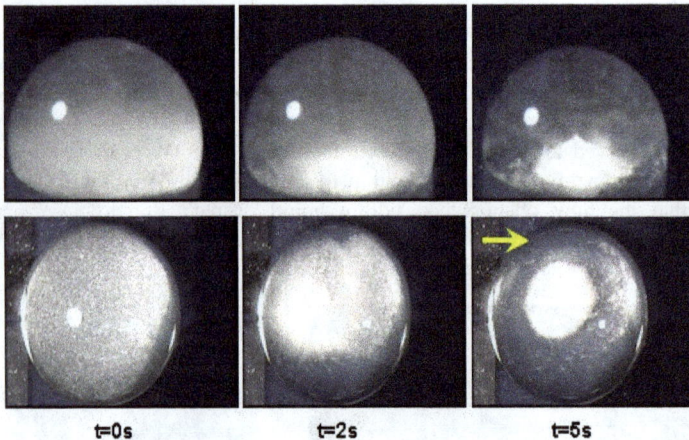

t=0s   t=2s   t=5s

Figure 32. Captured video images illustrating the rapid starch particles concentration process for a 30 μl droplet. The first row shows a side view, while the second row a top view. The yellow arrow is the SAW propagation direction [53]. Reprinted with permission from Institute of Physics, J. Microeng. Micromech. 21, 2011, 01500581.

6.3. Particle sorting and manipulator

Particle sorting, separation and counting are frequently used in biological and medical analysis. Biological samples contain various cells such as blood plasma, and red and white cells. For analysis, these cells are to be separated and counted. Cell separation and counting are powerful tools being used for quantitative analysis. Acoustic waves can be utilized for particle sorting, separation, counting etc, mostly relying on the nodes and antinodes generated by standing waves [137]. Figure 33 shows the principle of a SAW-based particle sorting device. A pair of IDTs is arranged face to face with a distance between them equal to an integer of the half wavelength, $n\lambda/2$. Upon application of an RF signal, standing waves between the two IDT electrodes can be formed. If a channel is fabricated perpendicular to the wave path of one wavelength wide, the pressure node in the channel will confine the particles within the pressure node, generating streaming with particles confined within the line.

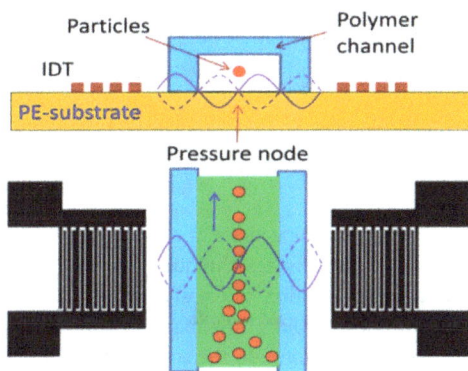

Figure 33. Formation of pressure node and anti-node by standing waves used for focusing particles inside a channel. A single wave node in a PDMS/SAW device and focus of the particles in a line.

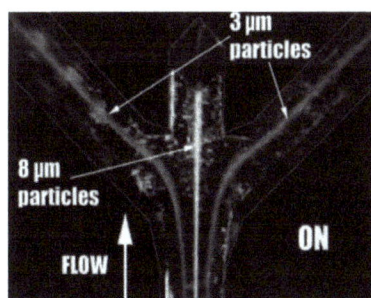

Figure 34. Microscope image of microparticle line induced by acoustic wave node [137]. Reprinted with permission from RSC Publishing, Chem. Soc. Rev. 36, 2007, 492.

Shi *et al.* employed this scheme to obtain focused particle lines using PDMS as the microchannel wall [138]. They have demonstrated fluorescent polystyrene particles sorting for a solution with a density of 1.176×10^7 beads/ml. The particle size is around 1.9 µm, and the flow containing the particles passes through a channel, and eventually forms a stable particle flow. By incorporating channels with different exits, particles with different sizes can be separated. Fig. 34 is the photo image of the particles captured by a camera [138].

It should be pointed out that standing waves have a long range force compared with the sizes of the microparticles; therefore it should be considered as a coarse manipulator or tweezers, suitable for simultaneous handling of a large group of particles or cells. It would be difficult to manipulate particles precisely down to the micrometre scale individually. For precision manipulation of particles, it would be best to integrate acoustic wave devices with other mechanisms. Wiklund et al [139] has integrated dielectrophoretic microfluidics mechanisms with ultrasonic particles concentrators. The ultrasonic standing wave delivers a long range force for high through-put particle manipulation, while the short-range dielectrophoretic forces are used for precision control to realize individual cell manipulation for bioanalysis.

Acoustic forces can be utilized to drive particles to places in order to realize scaffolding for cell growth. An in vitro cell culture is a technique used to grow cells in extra-cellular matrices and potentially for organ farming. Successful growth of cells depends on uniform distribution of seed cells into the scaffold of the matrices and the efficiency of the seeding process. The moving of a cell suspension into the scaffold material, typically a polymer, in the absence of external driving forces is exceptionally slow due to the large capillary resistance which may take from hours to days. SAW device can move a droplet with strong internal streaming and agitation as discussed, and can be used to deliver particles into the polymer matrices for cell growth. Yeo at al. have investigated the effect of SAW agitation on the efficiency of suspended fluorescence particles in a polycaprolactone (PCL) scaffold [140]. They demonstrated that efficiency up to 90% can be achieved on a SAW device within a few seconds, and the particles are uniformly distributed within the polymer matrix [140].

6.4. Other acoustic wave based functions

The acoustic wave technique has been used for biodegradable polymeric nanoparticle generation [141]. A polymeric incipient was dissolved into a solvent drop, and then atomized by a SAW. The solidified polymeric particles left behind are monodispersed. With this technique, 150–200 nm polymer spherical clusters were formed with sub-50 nm particulates. Periodically ordered polymer has also been patterned on a substrate by SAW atomization [142]. When a polymer solution was spread over the surface of a SAW device with two IDTs perpendicular to each other, the surface displacements induced by the standing waves displace the polymer film, breaking up the film in both the transverse and longitudinal directions producing evenly spaced solidified polymer microstructures. The spaces between the nanostructures in the X and Y-directions are approximately half the SAW wavelength in both directions with a pattern as shown in Fig. 35(left) schematically.

This can also be utilized to fabricate fine microwires of soft matter with nanoparticles uniformly dispersed in a solvent.

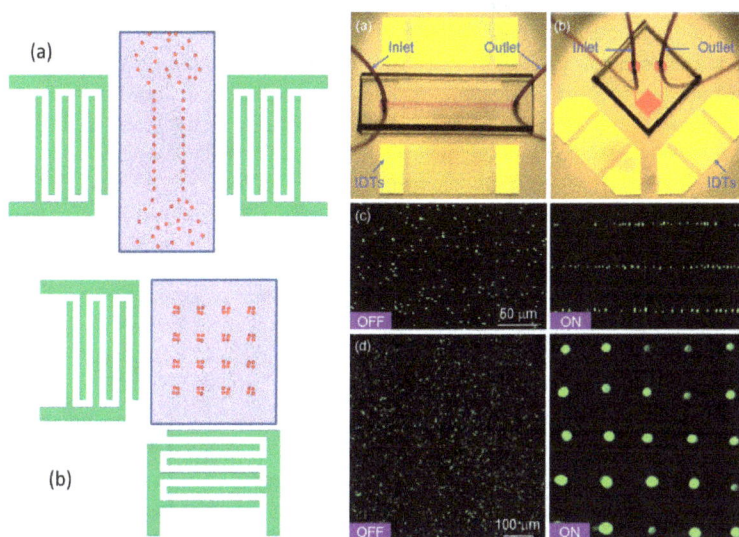

Figure 35. (Left) Particle lines formed in a fluid induced by standing waves from a pair of SAW IDTs; (Right) 2D array of particles form by a pair of IDTs in perpendicular [138]. Reprinted of (b) with permission from RSC Publishing, Lab on Chip, 9, 2009, 2890.

6.5. SAW based lab-on-a-chip

By using a combination of streaming induced mixing, enhanced biochemical reaction, droplet delivery and cell sorting etc by acoustic waves with other microfluidic and sensing functions, people have realized SAW based LOCs for various applications. A PCR system based on the combination of a SAW and resistance microheaters has been developed [74,134]. A schematic of the suggestion is shown in Fig. 36. The SAW-PCR system is a droplet-based DNA amplifier with droplets of the sample embedded in oil. SAW devices are used for moving and aligning the droplets between the zones with different temperatures through virtual tracks formed by chemical modification of the surface hydrophobicity. Once PCR amplification is completed, the droplet is moved to another heater for hybridization.

The chip is able to perform a fast and specific PCR with a small volume of 200 nl within 10 min. A single nucleotide polymorphism (SNP) responsible for the Leiden Factor V syndrome from human blood was successfully amplified by the PCR system and detected [134]. A SAW-based chip has several advantages over microfluidic channel systems. It can avoid the problems of clogging, large pressure drop and vaporization of liquid from the solid surface. Furthermore, the SAW streaming can also help to speed up the binding reaction and to get a more homogeneous fluorescence in hybridization.

Furthermore, attempts have been made to develop stand alone biodetection systems with integrated SAW microfluidics and sensors. In these cases, SAW microfluidics is mainly used for transporting liquid and agitating to minimize non-specific binding and speed-up the reaction. Due to the small dimensions, liquid in microchannels is dominated by a laminar flow, and the biochemical reaction is limited to mass transportation. The process is very slow, and the reaction is incomplete, thus additional agitation to speed up the reaction is therefore required. A SAW device is an ideal planar device to be integrated in the system for these purposes. Figure 37 is a schematic drawing of a SPR detection system with integrated SAW microfluidics [143]. By utilizing the abilities of moving droplets and acoustic streaming by SAW device, a droplet based SPR system for real-time sensing was realized.

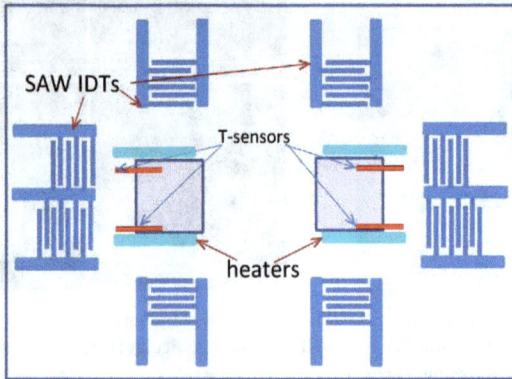

Figure 36. A schematic drawing the SAW based PRC system.

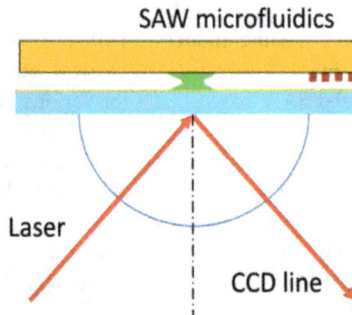

Figure 37. Schematic drawing of the set-up of the droplet-based SPR with SAW integrated for streaming to allow the real time monitoring of interactions.

7. Conclusions

Lab-on-a-chip systems for medical research, drug development and healthcare etc typically consist of a set of microfluidics and sensors. In most cases, the mechanisms for microfluidics and sensors are different and these devices are mostly assembled together to form LOCs, making the LOCs big and difficult to operate. Acoustic wave based lab-on-a-chip systems are a miniaturized microsystems with typical sizes of few square centimetres. LOCs may provide a single or multi-function such as transporting biosamples and sensing on a single chip.

Bulk and surface acoustic waves have found tremendous applications in LOCs. Surface acoustic waves have strong forces and can be utilized for fabrication of micropumps, mixers, droplets and mist generators for handling liquids and biosamples effectively and efficiently. On the other hand, bulk and surface acoustic wave based resonators (QCMs, SAWs and FBARs) are extremely sensitive to traces of absorbed mass and hence can be utilized for development of high sensitivity biosensors. Also acoustic waves can be utilized for generating many other functions such as remote heating, cell concentration and delivery. These unique functions of acoustic waves make it possible to develop lab-on-a-chip systems with a single actuation and sensing mechanism. Furthermore all these acoustic devices can be realized by using thin film technology, hence opening the way for integration of acoustic wave based LOCs with Si-based electronics on the same substrate.

Author details

J. K. Luo
Dept. Info. Sci. & Electron. Eng., Zhejiang University, China
Inst. Of Renew. Energy & Environ. Technol., Bolton University, UK

Y. Q. Fu
Thin Film Centre, University of West of Scotlant, Paisley, Scotland

W. I. Milne
Dept. of Eng. University of Cambridge, UK

Acknowledgement

The authors would like to acknowledge the partial financial support from the Leverhulme Trust under the grant No. of F/01431, The Knowledge Centre for Materials Chemistry under Grant No. of X00680PR, Canergie Trust Funding and Royal Society of Edinburgh, the Natural Science Foundation of China (No. 61274037) and Zhejiang Provincial Natural Science Foundation, China (No. Z1110168).

8. References

[1]   Nguyen N T, Huang X and Chuan T K; J. Fluids Eng. 124; 2002; 384.
[2]   Laser D J and Santiago J G; J. Micromech. Microeng. 14; 2004; R35.
[3]   Woias P; Sens. Actuat. B105, 2005, 28.
[4]   Truong T Q and Nguyen N T; J. Micromech.Microeng. 14, 2003, 632.
[5]   Cho S K, Moon H and Kim C-J; J. Microelectromech. Syst. 12, 2003, 70.

[6] Lacharme F. and Gijs M A; SensorsActuators A 117, 2006, 384.

[7] Darabi J, Rada M, Ohadi M and Lawler J; J. Microelectomech. Syst. 11, 2002, 684.

[8] Jones T B; J. Electrostat. 51/52, 2001, 290.

[9] Pollack M G, Fair R B and Shenderov A D; Appl. Phys. Lett. 77, 2000, 1725.

[10] Pollack M G, Shenderov A D and Fair R B; Lab Chip 2, 2002, 96.

[11] Torkkeli A, Saarilahti J, Haara A, Harma H, Soukka T and Tolonen P; Proc. IEEE MEMS. pp.475–78, 2001.

[12] Mugele F and Herminghaus S; App. Phys. Lett. 81, 2002, 2303.

[13] Luo J K, Fu Y Q, Li Y, Du X-Y, Flewitt A J, Walton A J and Milne W I, J. Micromech. Microeng. 19, 2009, 054001.

[14] Chao C, Cheng C H, Liu Z B, Yang M, Leung W F; Proc. IEEE Int. Ultrason. Symp. 2008, pp.521.

[15] Luginbuhl P, Collins S D, Racine G A, Gretillat M A, De Rooij N F, Brooks K G and Setter N; J. Microelectromech. Syst. 6, 1997, 337.

[16] Kwon J W, Yu H, Zou Q, Kim E S; J. Micromech. Microeng. 16, 2006, 2697.

[17] Wixforth A, Strobl C, Gauer C, Toegl A, Scriba J and Guttenberg Z V; Anal. Bioanal. Chem. 379, 2004, 982.

[18] Friend J R and Yeo L Y, Rev. Modern Phys. 83, 2011, 647.

[19] Yeo L Y and Friend J R, Biomicrofludics, 33, 2009, 012002.

[20] Fu Y Q, Luo J K, Du X Y, Flewitt A J, Li Y, Markx G H, Walton A J and Milne W I, Sens. Actuat. B143, 2010, 606.

[21] Lange K, Rapp B E and Rapp M; Anal. Bioanal. Chem. 391, 2008, 1509.

[22] Shiokawa S, Kondoh J; Jap. J. Appl. Phys., 43, 2004, 2799.

[23] Ballantine D S Jr, White R M, Martin S J, Ricco A J, Frye G C, Zellars E T, Wohltjen H; 1997. Acoustic Wave Sensor—Theory, Design, and Physico-Chemical Applications, Academic Press, San Diego.

[24] Buttry D A and Ward M D; Chem. Rev. 92, 1992, 1355.

[25] Lakin K M and Wang J S; Appl. Phys. Lett. 38, 1981, 125.

[26] Lakin K M; IEEE Trans. Ultrason. Ferroelectr. & Freq. Control, 52, 2005, 707.

[27] Ruby R; Proc. IEEE Ultrasonics Symp., 1-6, pp.1029-1040, 2007

[28] Rey-Mermet S, Bjurstrom J, Rosen D and Petrov I; IEEE Trans. Ultrason. Ferroelectric and Freq. Control; 51, 2004, 1347.

[29] Kang Y R, Kang S R, Paek K K, Kim Y K, Kim S W and Ju B K, Sens. & Actuat. A117, 2005, 62.

[30] Link M, Schreiter M, Weber J, Gabl R, Pitzer D, Primig R, Wersing W, Assouar M B and Elmazria O; J. Vac. Sci. Technol. A24, 2006, 218.

[31] Wingqvist G, Bjurstrom J, Liljeholm L, Yantchev V and Katardjiev I; Sens. & Actuat. B123, 2007, 466.

[32] Nguyen N T and White R MSens. & Actuat. 77, 1999, 229.

[33] Meng A H, Nguyen N T and White R M, Biomed. Microdevice, 2:3, 2000, 169.

[34] Muralt P, Ledermann N, Baborowski J; IEEE Trans. Ultrason. Ferroelectr. Freq. Control. 52, 2005, 2276.

[35] Nguyen N T and White R M; IEEE Trans. Ultrason. Ferroelect. Freq. Control. 47, 2000, 1463.

[36] Lucklum R and Hauptmann P; Meas. Sci. Technol. 14, 2003, 1854.

[37] Kao P, Doerner S, Schneider T, Allara D, Hauptmann P and Tadigadapa S, J. Microelectromech. Syst. 18, 2009, 522.

[38] Kao P and Tadigadapa S, Sens. & Actuat. A149, 2009, 189.

[39] Wu T T and Wang W S; J. Appl. Phys. 96, 2004, 5249

[40] Benetti M, Cannatà D, Pietrantonio F D and Verona E, Proc. IEEE Ultrason. Symp. pp.1738, 2003.

[41] Hachigo A, Nakahata H, Itakura K, Fujii S and Shikata S, Proc. IEEE Ultrason. Symp. pp.325, 1999.

[42] Wixforth A; J. Associat. Lab. Automat., 11, 2006, 399.

[43] Josse F, Bender F and Cernosek R W; Anal. Chem. 73, 2001, 5937.

[44] Rayleigh L; On the circulation of air observed in Kundt's tubes, and on some allied acoustical problems, Trans. R. Soc. 25, 1884, 224.

[45] Westervelt P J; J. Acoust. Soc. Am. 25, 1953, 60.

[46] Nyborg W L; Acoustic streaming Physical Acoustics, Edited by W P Mason (New York: Academic) 1965, pp.265–331

[47] Shiokawa S, Matsui Y and Ueda T; Proc. Ultrasonic Symp. pp. 645, 1989.

[48] Nyborg W L Acoustic streaming Nonlinear Acoustics, Edited by M F Hamilton and D T Blackstock (New York: Academic), pp.207–331, 1998

[49] Wixforth A; Superlattices & Microstruct. 33, 2004, 389.

[50] Shiokawa S, Matsui Y and Morizum T, Jpn. J. Appl. Phys. 28, 1989, 126.

[51] Shiokawa S, Kondoh J; 2004. Jap. J. Appl. Phys. 43 2004, 2799.

[52] Frommelt T, Gogel D, Kostur M, Talkner P, Hanggi P; IEEE Transact. Ultrasonics, Ferroelectrics & Freq. Control. 55, 2008, 2298.

[53] Alghane M, Fu Y Q, Li Y, Luo J K, Bobbili B, Feng Y, Liu Y F, Hao Z C, Chen B X, Markx G, Wang C H and Walton A J; J. Microeng. Micromech. 21, 2011, 015005.

[54] Campbell J C and Jones W R; IEEE Trans. Sonics Ultrason. 15, 1968, 209.

[55] Trujillo F J and Knoerzer K, Proc. 7th Int. Conf. CFD in minerals & Process Ind. CSIRO, Melbourne, Australia, 9, Dec. 2009.

[56] Luo J K, Fu Y Q, Li Y F, Du X Y, Flewitt A J, Walton A J, Milne W I; J. Micromech. Microeng.19, 2009, 54001.

[57] Sritharan K, Strobl C J, Schneider M F, Wixforth A and Guttenburg Z; Appl. Phys. Lett. 88, 2006, 054102.

[58] Shilton R, Tan M K, Yeo L Y and Friend J R; J. Appl. Phys.104, 2008, 014910.

[59] Nakamura H, Yamada T, Ishizaki T and Nishimura K; IEEE Trans. On Ultrasonics. Ferroelectric. Freq. Control. 49, 2004, 761.

[60] Lehtonen S, Plessky V P, Hartmann C S and Salomaa M; IEEE Trans. Ultras. Ferroelectri. Freq. Control. 51, 2004, 1697.

[61] Du X Y, Fu Y Q, Tan S C, Luo J K, Flewitt A J, Milne W I, Lee D S, Maeng S, Kim S H, Park N M, Park J and Choi Y J; Appl. Phys. Lett. 93, 2008, 094105.

[62] Toegl A, Kirchner R, Gauer C, Wixforth A; J. Biomed. Technol. 14, 2003, 197.

[63] Wixforth A, Strobl C, Gauer C, Toegl A, Sciba J, Guttenberg Z V; Anal. Biomed. Chem. 379, 2004, 982.

[64] Newton M I, Banerjee M K, Starke T K, Bowan S M, McHale G; Sensor & Actuat. 76, 1999, 89.

[65] Chono K, Shimizu N, Matsu Y, Kondoh J, Shiokawa S; Jap. J. Appl. Phys. 43, 2004, 2987.

[66] Murochim N, Sugimoto M, Matui Y, Kondoh J; Jap. J. Appl. Phys. 46, 2007, 4754.

[67] Renaudin A, Tabourier P, Zhang V, Camart J C, Druon C; Sensor & Actuat. B113, 2006, 387.

[68] Toegl A, Scribe J, Wixforth A, Strobl C, Gauer C, Guttenburg Z V; Anal. Bioanal. Chem. 379, 2004. 69.

[69] Franke T and Wixforth A; Chem Phys Chem. 9, 2008, 2140.

[70] Du X Y, Swanwick M, Fu Y Q, Luo J K, Flewitt A J, Lee D S, Maeng S, Milne W I; J. Micromech. Microeng. 19, 2009, 035016.

[71] Fu Y Q, Garcia-Gancedo L, Pang H F, Porro S, Gu Y W, Luo J K, Zu X T, Placido F, Wilson J I B, Flewitt A J and Milne W I; Biomicrofluidics, 6, 2012, 024105.

[72] Zhang A L, Wu Z Q, Xia X H; Talanta, 84, 2011, 293.

[73] Luong T D, Phan V N and Nguyen N T; Microfluidic Nanofluid.10, 2011, 619.

[74] Wixforth A; JALA, 18, 399, 2006,

[75] D.Beyssen, L.Le.Brizoual, O.Elmazria and P.Alnot; Sens. & Actuat. B118, 2006, 380.

[76] M. Alghane, Y. Q. Fu1, B. X. Chen1, Y. Li, M. P. Y. Desmulliez and A. J. Walton, Microfluid. & Nanoflud. Submitted.

[77] Du X Y, Fu Y Q, Luo J K, Flewitt A J and Milne W I; J. Appl. Phys. 105, 2009, 024508.

[78] Nguyen N T and White R T; Sens. & Actuat. 77, 1999, 229.

[79] Moroney R M, White R M, Howe R T; Appl. Phys. Lett. 59, 1991, 774.

[80] Ogawa J, Kanno I, Kotera H, Wasa K, Suzuki T; Sens. & Actuat. A52, 2009, 211.

[81] Li Y, Flynn B W, Parkes W, Liu Y, Feng Y, Ruthven A D, Terry J G, Haworth L I, Bunting A, Stevenson J T M, Smith S, Bobbili P, Fu Y Q and Walton A J; Proc. European Solid State Device Res. Conf. pp.14-18, 2009.

[82] Li Y, Fu Y Q, Brodie S, Mansuor R and Walton A, Biomicrofluid. 6, 2012, 012812.

[83] Bennès J, Alzuaga S, Ballandras S, Chérioux F, Bastien F and Manceau J F; Proc. IEEE Ultrason. Symp. Rotterdam, pp.823-826, 2005.

[84] Tan M K, Friend J R and Yeo L Y; Phys. Rev. Lett. 103, 2009, 024501.

[85] Qi A, Friend J R and Yeo L Y; Proc. 2nd Micro/nanoscale heat & mass Transf. Int Conf. Shanghai, pp.1-8, 2009.

[86] Lucklum R and Hauptmann P; Meas. Sci. Technol. 14, 2003, 1854.

[87] Grate W J, Martin S J, White R M; Anal Chem. 65, 1993, 940.

[88] Cote G L, Lec R M, Pishko M V; IEEE Sens. J. 3, 2003, 251.

[89] Kuznestsova L A and Coakley W T; Biosens. & Bioelectron. 22, 2007, 1567.

[90] Teles F R R and Fonseca L P; Talanta, 77, 2008, 606.

[91] Vellekoop M J; Ultasonics. 36, 1998, 7.

[92] Gizeli E; Smart. Mater. Struct. 6, 1997, 700.

[93] Sauerbrey G. Verwendung von Schwingquarzen zur Wagung dunner Schichten und zur Microwagang. Z. Phys. 155, 1959, 206.

[94] Nomura T, Okuhara M; Anal. Chim. Acta. 142, 1982, 281.

[95] Lin T Y, Hu C H and Chou T C; Biosens. & Bioelectron. 20, 2004, 75.

[96] Maturos T, Wong-ek K, Sangworasil M, Pintavirooj C, Wisitsora-at A and Tuantranont A; Proc. Int. Conf. Electric/Electron. Eng. Comput, Telecom. & Info Tech. Vol.1, 478, 2009.

[97] Han J H, Zhang J P, Xia Y T, Li S H and Jiang L; Colloid. & Surf. A: Physicochem. Eng. Aspects 379, 2011, 2.

[98] Dickert F L, Hayden o, Blindeus R, Mann K J, Blaas D and Waigmann E, Anal Bioanal Chem, 378, 2004, 1929.

[99] Wolfbeis O S, Editor, Springer Series on Chemical Sensors and Biosensors, Methods and applications, 2007, Volume 5, Part C, 371-424,

[100] Wang Q J, Pflugl C, Andress W F, Ham D, Capasso F and Yamanishi M; J. Vac. Sci. Technol. B26, 2008, 1848.

[101] Kirsch P, Assouar M B, Elmazria O, Mortet V and Alnot P; Appl. Phys. Lett. 88, 2006, 223504.

[102] Barie N and Rapp M; Biosensors & Bioelectron. 16, 2001, 978.

[103] Kovacs G and Venema M; Appl. Phys. Lett. 61, 1992 639.

[104] Josse F, Bender F, Cernosek R W; Anal. Chem. 73, 2001, 5937.

[105] Mchale F G; Meas. Sci. Technol, 14, 2003, 1847.

[106] Lindner G; J. Phys. D. 41, 2008, 123002.

[107] Lee D S, Lee J H, Luo J K, Fu Y Q, Milne W I, Maeng S and Jung M Y; J. Nanosci. Nanotechnol. 9, 2009, 7181.

[108] Jian S J, Chu S Y, Huang T Y and Water W; J. Vac. Sci. Technnol. A22, 2004, 2424.

[109] Krishnamoorthy S and Iliadis A A; Solid-State Electronics, 50, 2006, 1113.

[110] Mchale G, Newton M I, Martin F; J. Appl. Phys. 91, 2002 9701.

[111] Powell D A, Zadeh K K, Wlodaiski W; Sens. & Actuat. A115, 2004, 456.

[112] Du J and Harding G L ; Sens. & Actuat. A65, 1998, 152.

[113] Kao K S, Cheng C C, Chen Y C and Lee Y H; Appl. Phys. A76, 2003,1125.

[114] Kao K S, Cheng C C, Chen Y C, et al., Appl. Surf. Sci. 230, 2004, 334.

[115] Gabl R, Feucht H D, Zeininger H, Eckstein G and Wersing W; Biosens. Bioelectron. 19, 615 (2004).

[116] Lanz R and Muralt P; IEEE Trans. Utralson, Ferroelectric & Freq. Control, 52, 2005, 936.

[117] Luo J K, Ashley G M, Garcia-Gancedo L, Kirby P B, Flewitt A J and Milne W I; Int. J. Nanomanufact. 7, 2011, 448.

[118] Lin R C, Chen Y C, Chang W T, Cheng C C, Koo K S; Sens.& Actuat. A147, 2008, 425.

[119] Yan Z, Song Z, Liu W, Appl. Surf. Sci. 253, 2007, 9372.

[120] Muller A, Neculoiu D, Vasilache D, Konstantinidis G, Grenier K, Dubuc D, Bary L, Plana R and Flahaut E; Appl. Phys. Lett. 89, 2006, 143122.

[121] García-Gancedo L, Al-Naimi F, Flewitt A J, Milne W I, Ashley G M, Luo J K, Zhao X B and Lu J R; IEEE Trans Ultrason. Ferroelectric & Freq. Control. 58, 2011, 2438.

[122] García-Gancedo L, Zhu Z, Iborra E, Clement M, Olivares J, Flewitt A J, Milne W I, Ashley G M, Luo J K, Zhao X B and Lu J R; Sens. & Actuat. B160, 2011, 1386.

[123] Zhao X B, Ashley G M, Garcia-Gancedo L, Jin H, Luo J K, A J Flewitt and Lu J R; Sens. & Actuat. B163, 2012, 242.

[124] Zhang H, Marma M S, Kim E S, McKenna C E and Thompson M E; J. Micromech. Microeng. 15, 2005, 1911.

[125] Weber J, Albers W M, Tuppurainen J, Link M, Gabl R, Wersing W, Schreiter M; Sensors & Actuat. A128, 2006, 84.

[126] Link M, Webber J, Schreiter M, Wersing W, Elmazria O, Alnot P; Sensors & Actuat. B121, 2007, 372.

[127] Ho G K, Abdolv R and Ayazi F; Digest of MEMS 2007, Kobe, Japan, pp.791-795, 2007.

[128] Corso C D, Dickherber A and Hunt W D; J. Appl. Phys. 101, 2007, 054514.

[129] Ho G K, Abdolv R, Sivapurapu A, Humad S and Ayazi F; J. Microelectromech. Syst. 17, 2008, 512.

[130] Dickherber A, Corso C D and Hunt W D; Sens. & Actuat. A144, 2008, 7.

[131] Wenzel S and White R; Sens. & Actuat. A21–23, 1990, 700.

[132] White R M; Faraday Discuss. 107, 1997, 1.

[133] Huang I Y and Lee M C; Sens.& Actuat. B132, 2008, 340.

[134] Guttenberg Z, Muller H, Habermuller H, Geisbauer A, Pipper J, Felbel J, Kielpinski M, Scriba J and Wixforth A, Lab Chip. 5, 2005, 308.

[135] Raghavan R V, Friend J and Yeo L Y; Microfluid. Nanofluid. 8, 2009, 0452.

[136] Wood C D, Evens S D, Cumingham J E, O'Rorke R, Walti C and Davies A G, Appl. Phys. Lett. 92, 2008, 044104.

[137] Laurell T, Petersson F and Nilsson A; Chem. Soc. Rev. 36, 2007, 492.

[138] Shi J J, Ahmed D, Mao X L, Lin S C S, Lawit A and Huang T J; Lab Chip. 9, 2009, 2890.

[139] Wiklund M, Gunther C, Lemor R, Jager M, Fuhr G and Hertz H M; Lab Chip. 6, 2006,1537.

[140] Li H Y, Friend J R and Yeo L Y; Biomaterials, 28, 2007, 4098.

[141] Friend J R, Yeo Y L, Arifin D R and Mechler A; Nanotechnol. 19, 2008, 145301.

[142] Alvarez M, Friend J R and Yeo L Y; Langmuir, 24, 2008, 10629.

[143] Galopin E, Beaugeois M, Pinchemel B, Camart J, Bouazaoui M and Thomy V; Biosens. & Bioelectron. 23, 2007, 746.

Acoustics in the Oceans

Chapter 10

Ray Trace Modeling of Underwater Sound Propagation

Jens M. Hovem

Additional information is available at the end of the chapter

1. Introduction

Modeling acoustic propagation conditions is an important issue in underwater acoustics and there exist several mathematical/numerical models based on different approaches. Some of the most used approaches are based on ray theory, modal expansion and wave number integration techniques. Ray acoustics and ray tracing techniques are the most intuitive and often the simplest means for modeling sound propagation in the sea. Ray acoustics is based on the assumption that sound propagates along rays that are normal to wave fronts, the surfaces of constant phase of the acoustic waves. When generated from a point source in a medium with constant sound speed, the wave fronts form surfaces that are concentric circles, and the sound follows straight line paths that radiate out from the sound source. If the speed of sound is not constant, the rays follow curved paths rather than straight ones. The computational technique known as ray tracing is a method used to calculate the trajectories of the ray paths of sound from the source.

Ray theory is derived from the wave equation when some simplifying assumptions are introduced and the method is essentially a high-frequency approximation. The method is sufficiently accurate for applications involving echo sounders, sonar, and communications systems for short and medium short distances. These devices normally use frequencies that satisfy the high frequency conditions. This article demonstrates that ray theory also can be successfully applied for much lower frequencies approaching the regime of seismic frequencies.

This article presents classical ray theory and demonstrates that ray theory gives a valuable insight and physical picture of how sound propagates in inhomogeneous media. However, ray theory has limitations and may not be valid for precise predictions of sound levels, especially in situations where refraction effects and focusing of sound are important. There exist corrective measures that can be used to improve classical ray theory, but these are not discussed in detail here. Recommended alternative readings include the books. [1-4] and the articles [5-6].

A number of realistic examples and cases are presented with the objective to describe some of the most important aspects of sound propagation in the oceans. This includes the effects of geographical and oceanographic seasonal changes and how the geoacoustic properties of the sea bottom may limit the propagation ranges, especially at low frequencies. The examples are based on experience from modeling sonar systems, underwater acoustic communication links and propagation of low frequency noise in the oceans. There exist a number of ray trace models, some are tuned to specific applications, and others are more general. In this chapter the applications and use of ray theory are illustrated by using Plane Ray, a ray tracing program developed by the author, for modeling underwater acoustic propagation with moderately range-varying bathymetry over layered bottom with a thin fluid sedimentary layer over a solid half with arbitrary geo-acoustic properties. However, the discussion is quite general and does not depend on the actual implementation of the theory.

2. Theory of ray acoustics

The theory of ray acoustics can be found in most books and [1-4] will not be repeated here, but instead we follow a heuristic approach based on Snell's law, which is expressed by.

$$\xi = \frac{\cos\theta(z)}{c(z)} = \frac{\cos\theta_0}{c_0}. \tag{1}$$

Figure 1 shows a small segment of a ray path and the coordinate system. The segment has horizontal and vertical components dz and dr, respectively, and has the angle θ with the horizontal plane. When the speed of sound varies with depth the ray paths will bend and the rays propagate along curved paths. The radius of curvature R is defined as the ratio between an increment in the arc length and an increment in the angle

$$R = \frac{ds}{d\theta}. \tag{2}$$

Figure 1 shows that the radius of curvature is

$$R = \frac{1}{\sin\theta}\frac{dz}{d\theta}. \tag{3}$$

When the sound speed varies with depth the ray angle θ is a function of depth according to Snell's law. Taking the derivative of Eq. (3) with respect to gives the ray's radius of curvature at depth z expressed as

$$R(z) = -\frac{c(z)}{\cos\theta(z)}\frac{1}{g(z)} = -\frac{1}{\xi g(z)}. \tag{3}$$

The ray parameter ξ is defined in Eq. (1) and $g(z)$ is the sound speed gradient.

$$g(z) = \frac{dc(z)}{dz}.$$ (4)

At any point in space, the ray curvature is therefore given by the ray parameter ξ and the local value of the sound speed gradient $g(z)$. The positive or negative sign of the gradient determines whether the sign of R is negative or positive, and thereby determines if the ray path curves downward or upward.

A ray with horizontal angle θ_{in} strikes a plane with inclination α, the reflected ray is changed to θ_{out}.

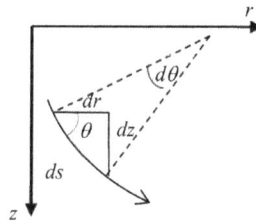

Figure 1. A small segment of a ray path in a isotropic medium with arc length ds.

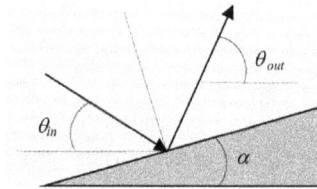

Figure 2. A ray with horizontal angle θ_n strikes a plane with inclination α, the reflected ray is changed to θ_{out}.

The ray parameter is not constant when the bathymetry varies with range. The change in ray direction is illustrated in Figure 2 showing that after reflection the angle θ_n of an incoming ray is increased by twice the bottom inclination angle α.

$$\theta_{out} = \theta_{in} + 2\alpha.$$ (5)

Consequently, after the ray is reflected, its ray parameter must change from ξ_{in} to ξ_{out}, which is expressed as

$$\xi_{out} = \frac{\cos(\theta_{out})}{c} = \frac{\cos(\theta_{in} + 2\alpha)}{c}$$

$$= \xi_{in}\cos(2\alpha) \pm \frac{\sqrt{1 - \xi_{in}^2 c^2}}{c}\sin(2\alpha).$$ (6)

The coordinates of a ray, starting with the angle θ_1 at the point (r_1, z_1), where the sound speed is c_1, as shown in Figure 3. For the coordinates of the running point at (r_2, z_2) along the ray path, the horizontal distance is

$$r_2 - r_1 = \int_{z_1}^{z_2} \frac{dz}{\tan\theta(z)} = \int_{z_1}^{z_2} \frac{\cos\theta(z)\,dz}{\sqrt{1-\cos^2\theta(z)}} = \int_{z_1}^{z_2} \frac{\xi c(z)\,dz}{\sqrt{1-\xi^2 c^2(z)}}. \tag{7}$$

The travel time between the two points is obtained by integrating the quantity $1/c$, the slowness, along the ray path:

$$\tau_2 - \tau_1 = \int_{z_1}^{z_2} \frac{ds}{c(s)} = \int_{z_1}^{z_2} \frac{dz}{c(z)\sin\theta(z)} = \int_{z_1}^{z_2} \frac{dz}{c(z)\sqrt{1-\xi^2 c^2(z)}}. \tag{8}$$

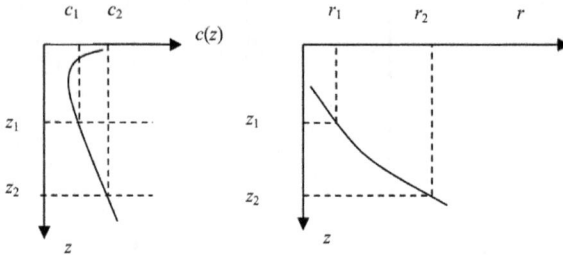

Figure 3. Left: The sound speed profile. Right: A portion of a ray traveling from point (r_1, z_1) to (r_2, z_2).

The acoustic intensity of a ray can, according to ray theory, be calculated using the principle that the power within a ray tube remains constant within that ray tube. This is illustrated in Figure 4 showing two rays with a vertical angle separation of $d\theta_0$ that define a ray tube centered on the initial angle θ_0. At a reference distance r_0 from the source, the intensity is I_0. Taking into consideration the cylindrical symmetry about the z axis, the power ΔP_0 within the narrow angle $d\theta_0$ is

$$\Delta P_0 = I_0 2\pi r_0^2 \cos\theta_0 d\theta_0. \tag{9}$$

At horizontal distance r, the intensity is I. In terms of the perpendicular cross section dL of the ray tube, the power is

$$\Delta P = I 2\pi r dL. \tag{10}$$

Since the power in the ray tube does not change, we may equate Eq.(9) and eq. (10) , and solve for the ratio of the intensities:

$$\frac{I}{I_0} = \frac{r_0^2}{r}\cos\theta_0 \left|\frac{d\theta_0}{dL}\right|. \tag{11}$$

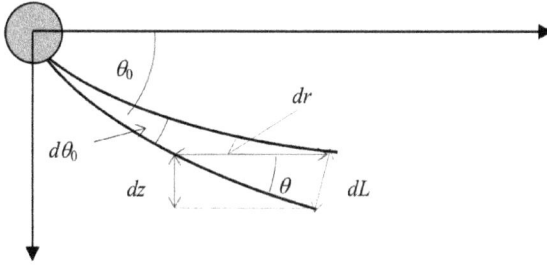

Figure 4. The principle of intensity calculations: energy radiated in a narrow tube remains inside the tube; r_0 represents a reference distance and θ_0 is the initial ray angle at the source; $d\theta_0$ is the initial angular separation between two rays; dr is the incremental range increase; θ is the angle at the field point; dz is the depth differential; and dL is the width of the ray tube.

Instead of using Eq.(11), it may be more convenient to use the vertical horizontal ray dr, which is

$$dr = \left| \frac{dL}{\sin \theta} \right|, \tag{12}$$

resulting in

$$\frac{I}{I_0} = \frac{r_0^2}{r} \left| \frac{\cos \theta_0}{\sin \theta} \frac{d\theta_0}{dr} \right| = \left(\frac{r_0^2}{r} \right) \left(\frac{c_0}{c} \right) \left| \frac{\cos \theta}{\sin \theta} \frac{d\theta_0}{dr} \right|. \tag{13}$$

The last expression in Eq.(13) is obtained by assuming that the ray parameter is constant and by using Snell's law. The absolute values are introduced to avoid problems with regard to the signs of the derivatives and of $\sin \theta$.

With respect to the reference distance r_0, the transmission loss TL is defined as

$$TL = -10\log\left(I / I_0 \right). \tag{14}$$

By inserting Eq.(13) into Eq.(14) the transmission loss becomes

$$TL = 10\log\left(r / r_0^2 \right) + 10\log \left| \frac{dz}{d\theta_0} \right| + 10\log \left(\frac{c}{c_0} \right). \tag{15}$$

The term c_0/c is close to unity in water and can be ignored in most cases.

In this treatment the transmission loss includes only the geometric spreading loss. Therefore bottom and surface reflection losses and sea water absorption loss must be included separately.

The geometric transmission loss in Eq.(15) consists of two parts. The first term represents the horizontal spreading of the ray tube and results in a cylindrical spreading loss. The second and third terms represent the vertical spreading of the ray tube and are influenced by the depth gradient of the sound speed.

Eq.(13) predicts infinite intensity under either of two conditions: when $\theta = 0$ or when $dr / d\theta_0 = 0$. The first condition signifies a turning point where the ray path becomes horizontal; the second condition occurs at points where an infinitesimal increase in the initial angle of the ray produces no change in the horizontal range traversed by the ray. The locus of all such points in space is called a caustic. In both cases there is focusing of energy by refraction and where classical ray theory incorrectly predicts infinite intensity. Caustics and turning points will be discussed further in section 8.2.

3. A recipe for tracing of rays

A simple receipt for a ray tracing algorithm is to divide the whole water column into a large number of layers, each with the same thickness Δz. Within each layer, the sound speed profile is approximated as linear so that, in the layer $z_i < z < z_{i+1}$, the sound speed is taken to be

$$c(z) = c_i + g_i(z - z_i). \tag{16}$$

where c_i is the speed at depth z_i, and g_i is the sound speed gradient in the layer. From Eq. (7) and Eq. (8) the range and travel time increments in the layer are given by

$$r_{i+1} - r_i = \frac{1}{\xi g_i}\left[\sqrt{1 - \xi^2 c^2(z_i)} - \sqrt{1 - \xi^2 c^2(z_{i+1})}\right], \tag{17}$$

and

$$\tau_{i+1} - \tau_i = \frac{1}{|g_i|}\ln\left(\frac{c(z_{i+1})}{c(z_i)}\frac{1 + \sqrt{1 - \xi^2 c^2(z_i)}}{1 + \sqrt{1 - \xi^2 c^2(z_{i+1})}}\right). \tag{18}$$

When $\xi^2 c^2(z_{i+1}) \geq 1$, the ray path turns at a depth between z_i and z_{i+1}, and Eq.s (17) and (18) must be replaced by the following expressions:

$$r_{i+1} - r_i = \frac{2}{\xi g_i}\sqrt{1 - \xi^2 c^2(z_i)}, \tag{19}$$

and

$$\tau_{i+1} - \tau_i = \frac{2}{|g_i|}\ln\left(\frac{1 + \sqrt{1 - \xi^2 c^2(z_i)}}{\xi c(z_i)}\right). \tag{20}$$

These equations give the trajectories and the travel times for any ray's path to the desired range. By applying Eqs. (13) and (14), the geometrical transmission loss is also determined.

The simplicity of this method lies in the approximation of the sound speed profiles with straight-line segments and the ray path's subsequent decomposition into circular segments. The method's accuracy is determined by how well the linear fit matches the actual profile. In practice, the sound speed profile is often given as measured sound speeds at relatively few depth points. It is therefore advisable to use an interpolation scheme that is consistent with the usual behavior of the sound speed profile to increase the number of depth points to an acceptable high density.

The examples in this article are generated using the ray trace program PlaneRay that has been developed by the author [7-8]. However, any other ray programs with similar capabilities could have been use and the discussion is therefore valid for ray modeling in general. Other models frequently used and are the Bellhop model [9], and the models [10-11].

Figure 5 shows an example of ray modeling. The sound speed profile is shown at the left panel and the rays from a source at 50 m depth is shown in the right panel, which also shows the bathymetry and the thickness of the sediment layer over the solid half space.

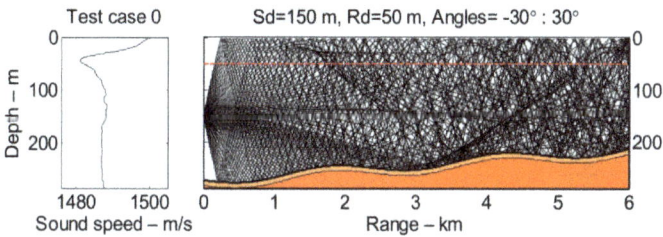

Figure 5. Sound speed profile and ray traces for a typical case. The source depth is 150 m and the red dotted line indicates a receiver line at a depth of 50 m. The initial angles of the rays at the source are from –30º to 30º.

4. Eigenray determination

To calculate the acoustic field it is necessary to have an efficient and accurate algorithm for determination of eigenrays. An eigenray is defined as a ray that connects a source position with a receiver position. In most case with multipath propagation there are many eigenrays for a given source/receiver configuration, which means that finding all eigenrays is not a trivial task.

The PlaneRay model uses a unique sorting and interpolation routine for efficient determination of a large number of eigenrays in range dependent environments. This approach is described by the two plots in Figure 6, which displays the ray history as function of initial angle at the source. All facts and features of the acoustic fields such as the transmission loss, transfer function and time responses are derived from the ray traces and their history The two plots show the ranges and travel times to where the rays cross the receiver depth line (marked by the red dashed line in Figure 5). A particular ray may intersect the receiver depth line, at several ranges. For instance at the range of 2 km, there are 11 eigenrays and from Figure 6 the initial angles of these rays are approximately found to be 5.9°, 9.6°, 22°, 24° for the positive (down going) rays and–2.0°,–3.6, °– 7-4°– 15.0°– 17.0°– 25.0°,–27.0°, for the negative (up-going waves). However, the values found in this way are often not sufficiently accurate for the determination of the sound field. Further processing may therefore be required to obtain accurate results.

The graphs of Figure 6 are composed of independent points, but it is evident that the points are clustered in independent clusters or groups. This property is used for sorting the points into branches of curves that represents different ray history. These branches are in most case relatively continuous and therefore amenable to interpolation. An additional advantage of this method is that the contribution of the various multipath arrivals can be evaluated separately, thereby enabling the user to study the structure of the field in detail.

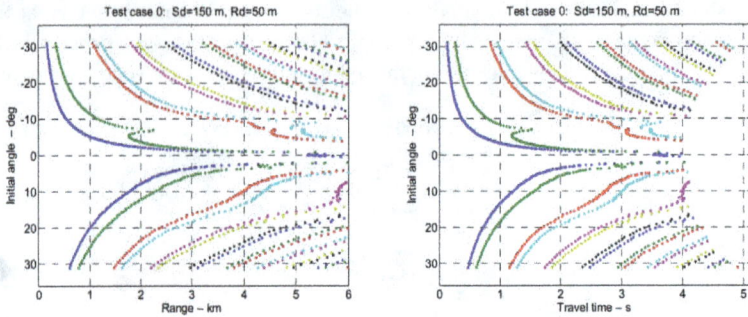

Figure 6. Ray history of the initial ray tracing in Figure 5 showing range (left) and travel time (right) to the receiver depth as function of initial angel at the source.

In most cases the eigenrays are determined by one simple interpolation yields values that are sufficiently accurate for most application, but the accuracy increases with increasing density of the initial angles at the cost of longer computation times.

Figure 7 shows examples of eigenrays traces with rays a receiver located at 2.5 km from the source for the scenario shown in Figure 5. To this receiver there are a total of 12 eigenrays, spanning the range of initial angles from -30° to 29°.

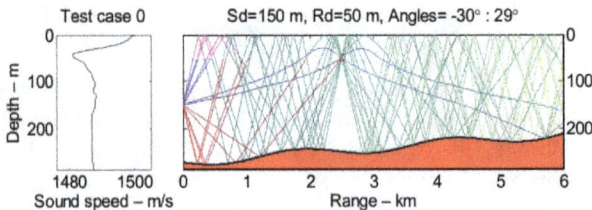

Figure 7. Eigenrays from a source at 150 m depth to a receiver at 50 m depth and distance of 2.5 km from the source.

5. Acoustic absorption in sea water

Sound absorption is important for long range propagation especially at higher frequencies. The absorption increases with frequencies and is dependent on temperature, salinity, depth and the pH value of the water. There exists several expressions for acoustic absorption in sea water; one of the preferred options is the semi-empirical formulae by Francoise and Garrison [12]. Figure 8 shows sound absorption as function of frequency in sea water using this expression for the values given in the figure caption.

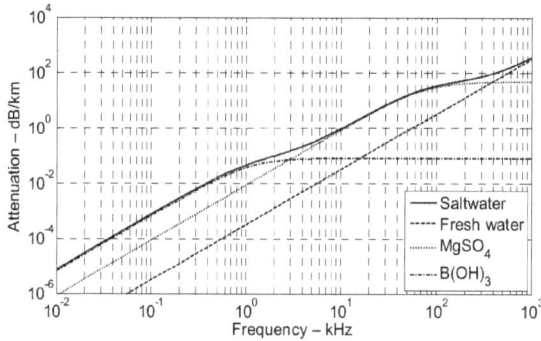

Figure 8. Acoustic absorption (dB/km) for fresh water and saltwater, plotted as a function of frequency (kHz) for water temperature of 10°C, atmospheric pressure of one atmosphere (surface), salinity of 35 pro mille, and pH value of 7.8. The various contributions to the absorption are also indicated.

6. Boundary conditions at the surface and bottom interfaces

Ray tracing is greatly simplified when no rays are traced into the bottom, but stops at the water-bottom interface. This avoids tracing of multiple reflections in layered bottoms. Instead the boundary conditions at the sea surface and the bottom can be approximately satisfied by the use of plane wave reflection coefficient.

A simple and useful bottom model is assuming a fluid sedimentary layer over a homogeneous solid half space. The reflection coefficient of a bottom with this structure is

$$R_b = \frac{r_{01} + r_{12} \exp\left(-2i\gamma_{p1}D\right)}{1 + r_{01}r_{12} \exp\left(-2i\gamma_{p1}D\right)}, \tag{21}$$

where γ_{p1} is the vertical wave number for sediment layer and D is the thickness of the sediment layer. The reflection coefficient between the water and the sediment layer, r_{01}, is given as

$$r_{01} = \frac{Z_{p1} - Z_{p0}}{Z_{p1} + Z_{p0}}, \tag{22}$$

and r_{12} is the reflection coefficient between the sediment layer and the solid half space,

$$r_{12} = \frac{Z_{p2}\cos^2 2\theta_{s2} + Z_{s2}\sin^2 2\theta_{s2} - Z_{p1}}{Z_{p2}\cos^2 2\theta_{s2} + Z_{s2}\sin^2 2\theta_{s2} + Z_{p1}}. \tag{23}$$

In Eq (15) and (16) Z_{ki} is the acoustic impedance for the compressional ($k = p$) and shear ($k = s$) waves in water column ($i = 0$), sediment layer ($i = 1$) and solid half-space ($i = 2$), respectively. The grazing angle of the transmitted shear wave in the solid half-space is denoted θ_{s2}.

Figure 9 shows an example of the bottom reflection loss as function of angle and frequency for a bottom with a sediment layer with the thickness $D = 2$ m with sound speed of 1700 m/s and density 1800 kg/m³ over a homogenous solid half space with compressional speed 3000 m/s, shear speed 500 m/s and density 2500 kg/m³. The wave attenuations are 0.5 dB/wavelength. The critical angle changes from 60° at very low frequencies to about 28° at high frequencies, the two angles are given by the sound speed in the water and the two bottom sound speed of 3000 m/s and 1700 m/s. The small, but significant, reflection loss at lower angles is caused by shear wave conversion and bottom absorption In this case the attenuation is about 1 dB in the frequency band around 50 Hz to 100 Hz.

The reflection coefficient of a flat even sea surface is –1 for. For a sea surface with ocean waves there will be diffuse scattering to all other direction than the specular direction, which result in a reflection loss that in the first approximation can be modeled by the coherent rough surface reflection coefficient

$$R_{coh} = \exp\left[-2\left(\frac{2\pi}{\lambda}\sigma_h \sin\theta\right)^2\right]. \tag{24}$$

In this expression θ is the grazing angle and σ_h is the rms. wave height and λ, is the acoustic wavelength, both in meters.

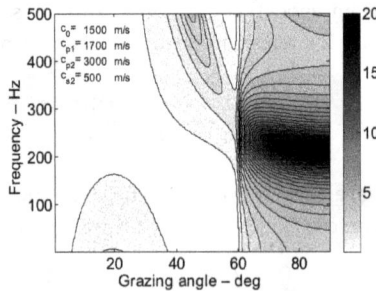

Figure 9. Bottom reflection loss (dB) as function of frequency and incident angle for a 2 m sediment layer over solid rock. The parameters are given in the text.

The reflection loss associated with reflection from a rough sea surface is

$$RL = -20\log 10 \left(R_{coh} \right).$$ (25)

The same rough surface reflection coefficient may also be applied to a rough bottom.

Figure 10 shows the rough surface reflection loss as function of grazing angle, calculated for a wave height of 0.5 m and the frequencies of 50 Hz, 100 Hz, 200 Hz and 400 Hz.

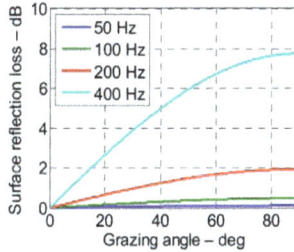

Figure 10. Reflection loss (dB) of rough surface with rms. wave height of 0.5 m as function of grazing angle, for the frequencies in the legend

7. Synthesizing the frequency domain transfer function and the time responses

The total wave field at any receiving point is calculated in the frequency domain by coherent summation of all the eigenray contributions. The first step in the calculation is to determine the geometrical transmission loss of each of the multipath contributions by applying Eq. (13) and Eq.(14) to the sorted and interpolated range-angle values. The frequency domain transfer function and the transmission loss are obtained by adding the multipath contributions coherently in frequency domain taken into account the phase shifts associated the travel times from the interpolated history of the travel times. The frequency dependent acoustic absorption of sound in water is included at this point in the process. The transfer function $H(\omega, r)$ can be expressed as

$$H\left(\omega,r\right) = \sum_{n} A_n B_n\left(\omega\right) S_n\left(\omega\right) T_n \exp\left(i\omega\tau_n\right).$$ (26)

Eq. (26) expresses the transfer function $H(\omega, r)$ to a distance r from the source at the at angular frequency ω as a sum over the n eigenrays that are included in the synthesis. A_n is the geometrical spreading loss factor, defined as the square root of the expression in Eq. (13). B_n, and S_n, are the combined effects of all bottom reflections and surface reflections, respectively, T_n, is $-90°$ phase shift associated with caustics and turning points, and τ_n is the travel time.

The synthesis of the received signals is performed in the frequency domain by multiplying the frequency spectrum of the source signal with the transfer function of each of the eigenrays and summing the contributions. The time domain response is obtained after multiplication with the frequency function of a source signal followed an inverse Fourier transform of the product. This requires the choice of a source signal, a sampling frequency (f_s) and a block length (N_{fft}) of the Fourier transform.

The total duration of the time window (T_{max}) after Fourier transform is

$$T_{max} = \frac{N_{fft}}{f_s}. \tag{27}$$

It is important to select the values of N_{fft} and f_s such that Fourier time window, T_{max}, is larger than the actual length or duration of the signal. In reality the real time duration of the received signal is often not known in advanced and therefore the user may have to experiment with different values to find appropriate values for of N_{fft} and f_s.

Figure 11 shows an example where the transmission loss (in dB) as function of range has been calculated for the frequencies of 100 Hz and 200 Hz. The dashed black line indicates the geometrical spreading loss, which is added for comparison and given by,

$$TL_{geo}(r) = 10\log_{10}\left[r^2\left(1+\frac{r^2}{r_t^2}\right)^{-1/2}\right]. \tag{28}$$

This expression yields a transmission loss proportional to $20\log(r)$ when $r < r_t$ and proportional to $10\log(r)$ for $r > r_t$. This approximation to the geometrical transmission loss may be used for approximate calculations of transmission loss for flat bottom and simple sound speed profiles. In the case shown in Figure 11 r_t is set equal to the water depth at source location, which in this case is 200 m.

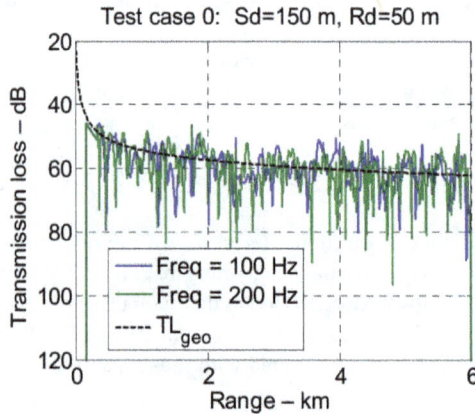

Figure 11. Transmission loss as function of range calculated for 100 Hz and 200 Hz The dashed black line is values of Eq.(28)

Figure 12 shows the synthesized time response at receivers spaced at 200 m separations in range up to 6 km. The sound speed and bathymetry is the same as in Figure 5 with the source at 150 m and all receivers at 50 m depth. The time scale is in reduced time to remove the gross transmission delay between the source and receiver. The reduced time is defined as

$$t_{red} = t_{real} - \frac{r}{c_{red}}. \tag{29}$$

In Eq. (29), t_{real} and t_{red} are the real and reduced times, respectively, r is range and c_{red} is the reduction speed. The actual value of c_{red} is not important as long as the chosen value results in a good display of the time responses.

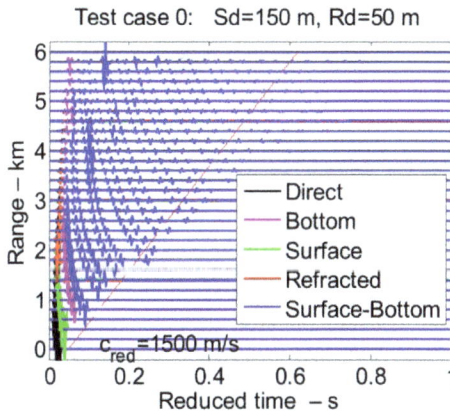

Figure 12. Received time signals as function of range and reduced time.

In the example shown above, the time signal and calculated assuming a narrow band-limited source signal in the form of a Ricker pulse. An example of a Ricker pulse and its frequency spectrum are shown in Figure 13.

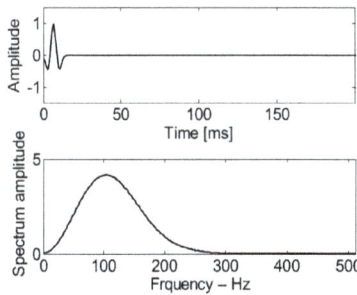

Figure 13. Ricker time pulse and frequency function the center frequency of 100 Hz

The time responses in Figure 12 are sorted according to the history of their eigenrays and color coded to allow for studying the various multipath contributions. This is particularly useful when dealing with transient signal and broad band signal, especially when knowledge of the multipath structure is important. In many such situations only the direct arrival or the refracted arrivals in the water column may carry the useful signals and all the other arrivals represent interference. In this case there are direct arrivals, followed by surface reflected and refracted arrivals at the turning points. Notice the high sound pressure values caused by the caustics at 3 km, 6, km and 7 km, which are apparent in both plots, this issue is discussed in section 8.2.

The red dotted line in Figure 12 represents an estimate of the duration of the cannel impulse response. This time duration is mainly given by the bottom reflection coefficient and the critical angle. Rays that propagate at angles closer to the horizontal plane than the critical angle experience almost no bottom reflection loss and may therefore propagate to long distances. Rays with steeper angles will experience higher reflection losses and die out more rapidly with range. Thus the time duration of the impulse is directly determined by the ratio of sound speeds in the water and the bottom as

$$t_{red} = \frac{r}{c_0}\left(\frac{1}{\cos\theta_{crit}} - 1\right) = r\left(\frac{c_b - c_0}{c_0^2}\right) \tag{30}$$

This estimate of the time duration of the channel impulse response assumes that the bottom is fluid, homogenous and flat, but the estimate may also be useful in other cases with moderately range dependent depth and with solid or layered bottom.

8. Special considerations

8.1. Frequency of applications

Ray tracing is a high frequency approximation to the solution of the wave equation and in principle more valid for high than for low frequency applications. However, high resolution prediction of higher frequency acoustic fields is difficult both for numerical and physical reasons. Principally most important is the physical limitation caused by the fact that the sound speed and the environment are generally not known in sufficient detail. This can be illustrated by a simple example. Consider coherent communication using a frequency of 10 kHz with wavelength of 10 mm. The required accuracy in order to be correct at a distance of 1 km is that the sound speed is known and stable with a relative error less 10^{-5}, an impossible requirement to satisfy in practice regardless of the numerical accuracy of the computer model.

8.2. Caustics and turning points

As mentioned before, the locations where $dr/d\theta_0=0$ are called caustics where the ray phase is decreased by 90° and the where the intensity, according to ray theory, goes to infinity. In reality the intensity is high, but finite, and the basic ray theory breaks down at these points. There exists theories to amend and repair the defects of ray theory at these points [1, 2, 13], but that is not discussed here.

Figure 14 shows details of the field at a showing the rays with initial angles in the range of −6° to −1°. The scenario is the same as in of Figure 5, but for clarity the tracing of rays have been stopped after the first bottom reflection and the figure concentrates on the details the field at the caustic at 1760 m range for a ray with initial angle of −5.6°. Figure 15 shows the time responses for ranges in the interval from 1.6 km to 1.9 km. In this case, the source signal is a Ricker pulse with a peak frequency of 200 Hz. There is a first direct arrival (black color) at all ranges. From the range 1760 there is also a refracted arrival a little later than the direct, but with higher amplitude, in particular near the range of 1760 m. Notice the effect of the 90° phase shift for ranges beyond the caustic at 1760 m and that the amplitude at this range is considerable higher than at the other ranges.

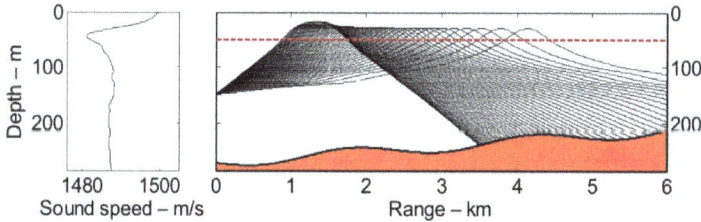

Figure 14. Rays through a caustic

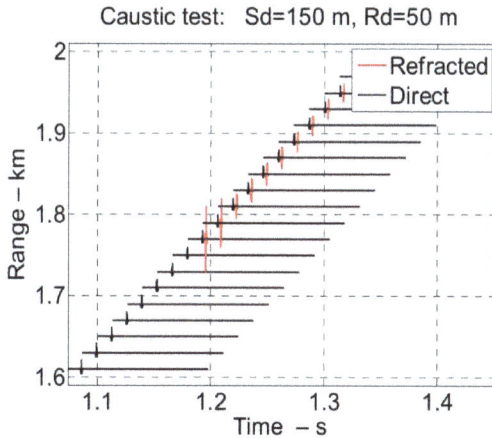

Figure 15. Time responses around the caustic at 1.76 km. The transmitted signal is a Ricker pulse with peak frequency of 200 Hz.

8.3. The principle of reciprocity and its validity in ray modeling

The principle of reciprocity is an important and useful property of linear acoustics and systems theory. The principle is very general and valid also in cases where the wave undergoes reflections at boundaries on its path from source to receiver [14]. The reciprocity principle is correctly represented in ray modeling, as can easily be understood from the eigenray plots of Figure 7. The eigenrays from a source position to the receiver position are the same as when source and receiver changes positions. The reflections at the bottom and at the sea surface are also symmetric in angles and consequently the acoustic fields are the same. However, it should be noted that the reciprocity principle applies to a point-to-point situation. This means that, for instance, that the development of the transmission loss as function source-receiver separation is generally not the same for the two directions.

8.4. The validity of using plane wave reflection coefficients

The accuracy of any ray model depends on the validity and limitation of ray theory and the implementation. A fundamental assumption of model is that the interactions with the boundaries are adequately described by plane wave reflection coefficient. In this section the validity of this assumption is investigated.

The general expression for the reflected field is given in text books, for instance in [13], over horizontal wave numbers k, as

$$\Phi_{ref}\left(r,z_r,\omega\right) = \frac{S(\omega)}{8\pi i}\int_0^\infty \Re(k)\frac{\exp\left(i\gamma\left|z_r+z_s\right|\right)}{\gamma}kH_0^{(1)}\left(kr\right)dk \ . \tag{31}$$

$\Phi_{ref}(r, z_r \omega)$ is the reflected field due to point source with frequency ω and source strength $S(\omega)$. $\Re(k)$ is the reflection coefficient, $H_0^1(kr)$ is the Hankel function of first kind, which represents a wave progressing in the positive r-direction. The horizontal wave number k and the vertical wave number γ are related to the sound speed, frequency and the angle θ by

$$k = \frac{\omega}{c}\cos\theta,$$
$$\gamma = \frac{\omega}{c}\sin\theta. \tag{32}$$

Eq.(31) states that the field is given as an integral over all horizontal wave numbers, or as consequence of Eq.(32), integration over all angles both real and the imaginary.

Consider now the situation where $\Re(k)=\Re$ is constant and independent of k or the angle. The integral in Eq.(31) becomes a standard integral and

$$\Phi_{ref}\left(r,z_r,\omega\right) = \frac{S(\omega)}{4\pi R}\Re\exp(ikR). \tag{33}$$

$$R = \left[r^2 + \left(z_s + z_r \right)^2 \right].$$ (34)

According to Eq. (33) the reflected wave is the same as the outgoing spherical wave from the image of the source in the mirror position of the real source and modified by the constant reflection coefficient $\Re$. The situation with a constant reflection coefficient is valid for perfectly flat sea surface where the reflection coefficient is equal to -1 for all angles of incidence. Thus the reflection from a smooth sea surface is accurately described plane wave reflection coefficients.

In the general case, and for reflections from the bottom, the reflection coefficient $\Re(k)$ is not constant and the integral can only be solved approximately or numerically. In order to obtain an approximation of the integral in Eq.(31) , the Hankel function is expanded in a power series with the first terms being

$$H_0^1 (kr) \approx \sqrt{\frac{2}{\pi kr}} \exp \left[i \left(kr - \frac{\pi}{4} \right) \right] \left[1 + \frac{1}{8ikr} + \cdots \right].$$ (35)

Restricting the integral of Eq.(31) to the first term yields

$$\Phi_{ref} \left(r, z_r, \omega \right) =$$
$$= \frac{S(\omega)}{4\pi} \frac{1}{\sqrt{2\pi r}} \int_{-\infty}^{\infty} \Re(k) \frac{\sqrt{k}}{\gamma} \exp \left[ikr + i\gamma \left(z_r + z_s \right) \right] dk .$$ (36)

The exponential in the integrand will normally be a rapid varying function and therefore the value of the integral will be small except when the phase term of Eq.(36) is nearly constant. The phase term of Eq.(36) is

$$\alpha = i\gamma \left(z_r + z_s \right) + ikr .$$ (37)

The stationary points are defined to the values of the horizontal wave number k where the derivative of the phase with respect to k is equal to zero, that is where $d\alpha/dk=0$, giving the stationary point as

$$r = \frac{\left(z + z_s \right)}{\tan \left(\theta_0 \right)} .$$ (38)

The interpretation of this result is quite simple; the reflected wave field is equal to that of the image source multiplied with the reflection coefficient at the specular angle θ_0.

There are however situations where this approximation is not sufficient in practice. This is discussed in [13] and in the following their results are cited without proof. The accuracy of the approximation depends on the source or receiver distance from the bottom interface. The result of the analysis is that the distance z from the bottom must satisfy

$$z \gg \frac{\lambda}{2\pi} \frac{\frac{\rho_b}{\rho_w}}{\sqrt{\left(\frac{c_b}{c_w}\right)^2 - 1}}. \tag{39}$$

With the water parameters of ρ_w = 1000 kg/m³ and c_w=1500 m/s, and the bottom parameters of ρ_b = 1500 kg/m³ and c_b=1700 m/s. Equation (39) requires than the distance from the bottom satisfy $z \gg$ 0.5 λ for the validity of using plane wave reflection coefficient at the bottom interface. A harder bottom with ρ_b = 1800 kg/m³ and c_b=3000 m/s, gives the requirement that $z \gg$ 1.0 λ. Hence the condition for validity is somewhat easier to satisfy for a soft bottom than for a hard bottom.

8.5. Bench marking ray modeling

The wave number integration model OASES [15] has been used to validate the accuracy and the limitation of the ray trace model using the simple case with constant water depth of 100 m and constant sound speed of 1500 m/s.

Figure 16 show the calculated transmission loss for the frequencies of 25 Hz, 50 Hz, 100 Hz and 200 Hz. The agreements between the results are very god for the higher frequency, but with some discrepancies for the lower frequencies, in particular for 25 Hz. The discrepancy is mainly a phase shift in the interference patterns of the two results, most pronounced for low frequencies and long ranges. This observation agrees with the theory outlined earlier. The seriousness of this discrepancy or errors may not very important in practice since the mean level is nearly the same as shown by the comparison with the OASES model.

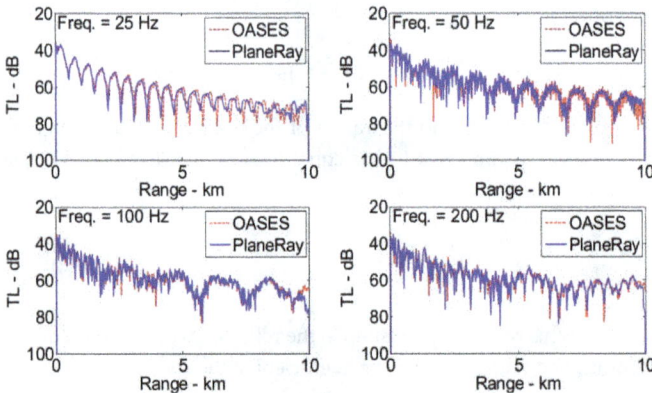

Figure 16. Comparison of the transmission loss as function of range for selected frequencies by PlaneRay (solid blue line) and OASES (dotted red line) for Pekeris' waveguide with a homogenous solid bottom with compressional wave speed of 3000 m/s and shear wave speed 500 m/s. Both wave attenuations have the values of 0.5 dB/wavelength.

9. Case studies

In the following we present two case studies that are relevant application of the modeling techniques descried in this article. The first if these is in connection with acoustic underwater communication and the transmission of digital information. In this case the multipath communication may be a significant problem causing intersymbol interference and significant degradation of reliability and performance. The second case is related to studies on the propagation of low frequency sound and the effect such noise may affect marine life, sea mammals and fish.

9.1. Seasonal variations of communication links

In connection with a study of underwater acoustic communication the propagation over a 6 km track has been modeled for the various seasonal sound speed profiles.

 The sound speed some months are shown in Figure 17. The sound speed profiles depend on the sea water temperature, the salinity and the depth. In the present case the sea water temperature variation with depth and the seasons is the main reason for changes in sound speed profile. During winters the surface water is cold and the sound speed is low, in the summer the surface water temperature and the sound speed is higher. The seasonal heating and cooling of the surface water propagates also to deeper depths, but with diminishing temperatures changes. At very large depths the water temperature is nearly the same at all seasons and the sound speed increases linearly and slowly with depth.

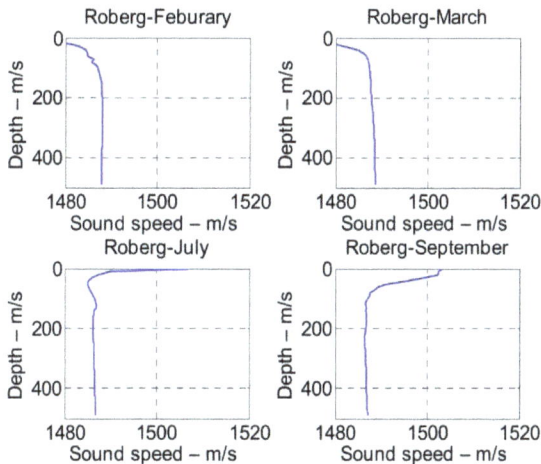

Figure 17. Sound speed profiles measured at specific dates for the months given in the figures

Figure 18 shows ray tracing results are for the same profiles as displayed in Figure 17. The purpose of the study was to investigate the possibility of communication to positions beyond a sea mount and to study the multipath arrival structure as function of range and depth.

There is a seamount with a peak at about 3 km from the transmitting station. In order to simplify the interpretation ray tracings in these plots have been terminated after 6 bottom reflections, but all rays are included in the calculation of the acoustic field, but rays with so many bottom reflections, or more, will in most case not be useful for data communication because of the reflection loss and reduced coherence.

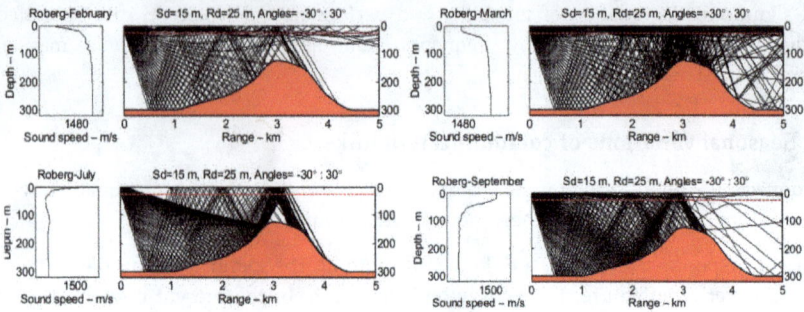

Figure 18. Ray tracing plots assuming a source depth of 15 meter for four monthly conditions at the Roberg test site. The sound speed profiles are the same as shown in Figure 17.

Figure 19 shows examples of received time responses at 25 m depth using a Ricker pulse as source signal. The different multipath contributions are color coded for clarity. At distances from the source over 1.5 km the first arrivals is follow paths surface reflected and upward refracted paths

Figure 20 shows the channel responses at a fixed range as function of depth down to 50 m. This figure shows the total response after adding all the individual multi path contributions. The plots demonstrate that the surface channel consists of deep refracted path and a number of paths reflected from the surface and deeper upwards refractions. The stability of these paths may be uncertain and subject to rapid changes in the environmental conditions near the surface due to temperature wind and current.

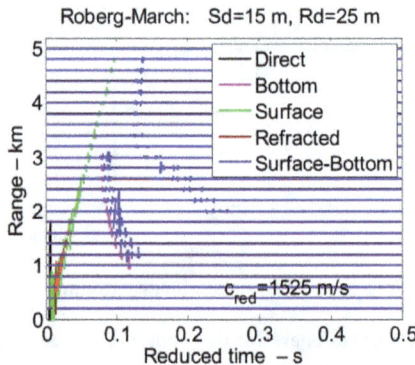

Figure 19. Time responses as function of range for receivers at depths of 25 m with a source at 15 m.

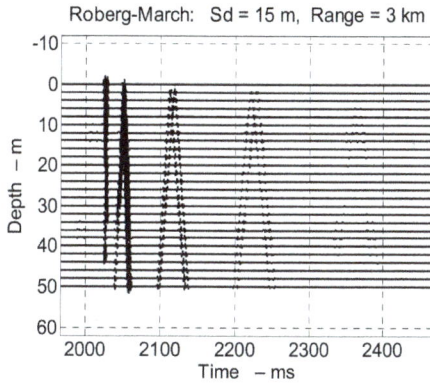

Figure 20. *Time responses as function of receiver depth at a fixed horizontal distance of 3 km from a source at 15 m depth.*

9.2. Seismic noise propagation

In many areas of the world anthropogenic noise often dominates over the natural ambient noise, especially in the low frequency band from approximately from 10 Hz and upwards to 1000 Hz, or more. This frequency band coincides approximately with the frequencies of perception of sea mammals and fish and may therefore be harmful to their natural activities, or even cause physical damages. An example is the case of the seismic exploration for oil and gas in certain areas where there is important commercial fishing interest. The propagation and distribution of acoustic noise depends the environmental conditions, in particular the oceanographic parameters, the topography of the seafloor and the acoustic properties of the bottom. In this section some of examples are presented to illustrate how the environment may affect the distribution of sound and noise. This study and discussion is also relevant for passive sonar applications to detect and track submerged vehicles and objects base emitted acoustic noise

The effects of bathymetric are illustrated in Figure 21 showing ray traces of upslope and downslope conditions for typical summer conditions at the Halten Bank in the Norwegian Sea. With downslope propagation there is a thinning the ray density with distance and upslope propagation gives a concentration of rays as the water depth diminishes.

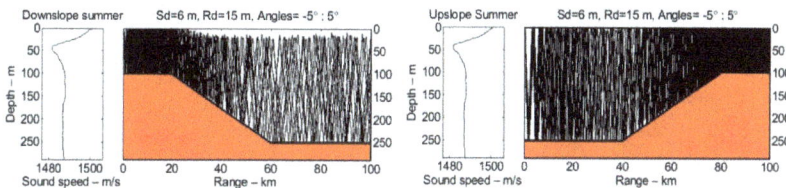

Figure 21. The effect of up and down sloping bottoms on the acoustic field distribution calculated for the typical summer condition in the month of July.

Figure 22 and Figure 23 show the calculated sound pressure level as function of range for the downslope and upslope propagation. The sound pulse from an airgun array is modeled as s a Ricker pulse with a peak pressure of 260 dB rel. 1μPa, centered on the frequency of 50 Hz, The horizontal dashed line is the assumed threshold value for fish reaction to sound. The bottom is modeled with a 2 m thick sedimentary layer over solid rock. The sound speed in the sediment layer is 1700 m/s and the density is 1800 kg/m³. The compressional sound speed in the rock is 3000 m/s, and density is 2500 kg/m³. The results in Figure 22 and Figure 23 are obtained under two conditions: (a) with a shear speed of 500 m/s, and (b) with no shear wave in the rock, i.e. the shear speed is zero. The absorptions are assumed to be 0.5 dB per wavelength for all the waves in the sediment layer and the rock. In the first case (a) the bottom reflection loss is as shown in Figure 9 with a significant low frequency reflection loss at angles lower than the critical angle caused by absorptions and conversion to shear wave in the bottom, which draws energy for the reflected wave. In the case of Figure 22 this results in a low-frequency and low-angle reflection loss of about 1 dB. For long ranges and many reflections this adds up to a significant total propagation loss. With no shear conversion the reflection loss is considerably reduced and the sound propagates easier to long ranges. The difference between the sound level at 50 Hz and 100 Hz is partly a result of increase attenuation at the higher frequency and partly that the source level in this case is higher for 50 Hz than for 100 Hz.

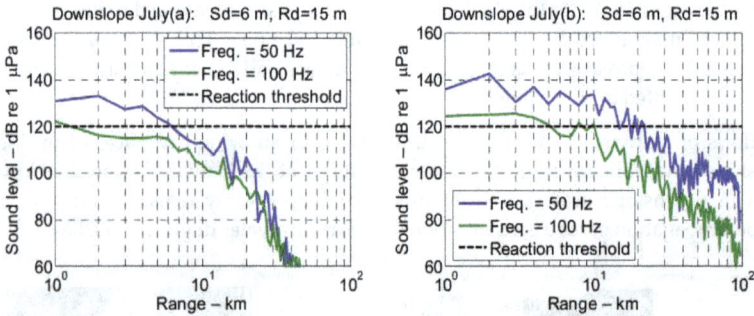

Figure 22. Sound pressure level as function of range for downslope propagation and July conditions. Left: With shear wave conversion (500 m/s). Right: No shear wave conversion.

Figure 24 and Figure 23 show similar results for downslope and upslope propagation for typical winter conditions represented by a sound speed profile measured in the month of February. For downslope conditions the sound level decrease rapidly with increasing depth and much more rapidly with shear wave conversion (Figure 24a) than without shear (Figure 24b). With upslope propagation (Figure 23) the sound levels are near independent of shear conversion except at the very long rages where the water depth becomes constant. The examples demonstrate that sound propagation in the ocean is strongly influenced by both by the oceanographic conditions and the geophysical properties of the bottom. Reliable prediction of acoustic propagation condition requires modeling tool that can that can handle both bottom and water properties.

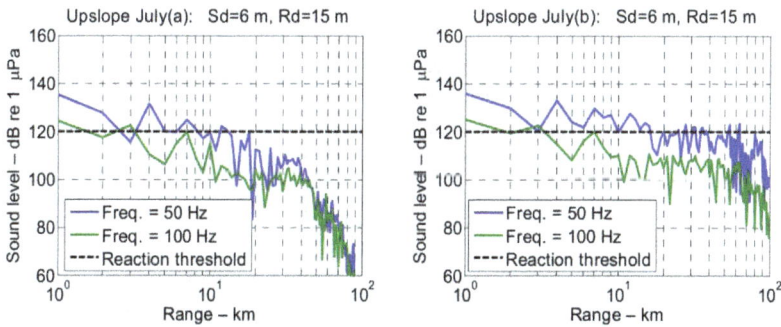

Figure 23. Sound pressure level as function of range for upslope propagation and July conditions. Left: With shear wave conversion (500 m/s). Right: No shear wave conversion.

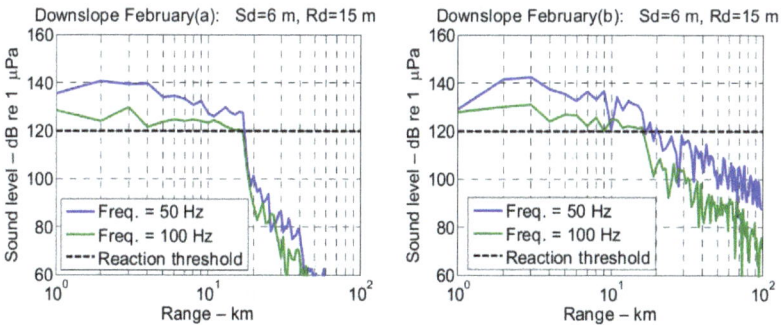

Figure 24. Sound pressure level as function of range for downslope propagation and February conditions. Left: With shear wave conversion (500 m/s). Right: No shear wave conversion.

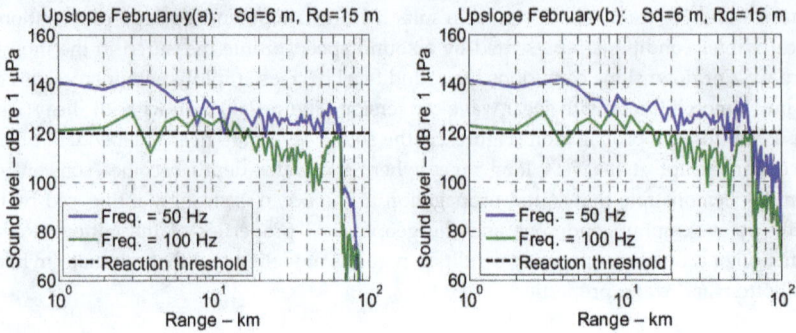

Figure 25. Sound pressure level as function of range for upslope propagation and February conditions. Left: With shear wave conversion (500 m/s). Right: No shear wave conversion

10. Summary

The article has outlined the theory of ray modeling and described how the theory can be applied to study acoustic wave propagation in the ocean. The complete acoustic fields are calculated by coherent addition of the contributions of a large number of eigenrays. In this method no rays are traced into the bottom, but the bottom interaction is modeled by plane wave reflection coefficients. Ray tracing is, by definition, frequency independent and therefore the ray trajectories through the water column are valid for all frequencies. Frequency dependency is introduced by reflections from the sea surface and the bottom, including loss associated with absorption and diffuse scattering of a rough ocean and bottom interfaces. Ray tracing is therefore a computational effective method for modeling broad of frequency band wave fields and for calculation of time responses.

Ray tracing is high-frequency approximation to the solution of the wave equation and the accuracy and validity at lower frequencies may be questioned, in particular the use of plane ray reflection coefficient to represent the bottom effects. This problem has been considered both theoretically and by simulations and comparison with more accurate model. The results of this study shows that source and receiver should be at a height above the bottom of at least half a wavelength, but there is no similar requirement to the distance from the sea surface. Less fundamental is the limitation of the numerical accuracy of the determination of the eigenrays, which is most serious in the calculation of the ray amplitude and the transmission loss. These inaccuracies are of more practical nature and can be reduced by refinements in the calculations.

Examples relevant for application in acoustic underwater communication and active sonar have been presented. The propagation of low frequency sound to large distances has been presented showing the effect of the bathymetry and the acoustic properties of the bottom. An important conclusion is the effect of bathymetry and the sound speed structure interacts and that accurate modeling of sound propagation requires information about the oceanography, the bathymetry and the geology of the bottom.

Author details

Jens M. Hovem

Norwegian University of Science and Technology, Norway

11. References

[1] Officer, C. B. (1958). *Introduction to the theory of sound transmission.* McGraw-Hill, New York City.

[2] Jensen F. B., W. A. Kuperman, M. B. Porter, and H. Schmidt, (2011). *Computational Acoustics,* Second edition, Springer, New York.

[3] Kinsler, L. E., A. R. Frey, A. B. Coppens, and J. V. Sanders. 2000. *Fundamentals of acoustics,* 4th ed. New York: John Wiley & Sons.

[4] Hovem, J. M. (2010). *Marine Acoustics: The Physics of Sound in Underwater Environments.* Peninsula Publishing, Los Altos, California, USA. ISBN 9780932146656

[5] Westwood, E. K and P. J. Vidmar. (1987). *Eigenray finding and time series simulation in a layered-bottom ocean,* J. Acoust. Soc. Am. 81, pp. 912-924.

[6] Westwood, E. K and C. T. Tindle. (1987). *Shallow water time simulation using ray theory.* J. Acoust. Soc. Am. 81, pp. 1752 -1761.

[7] Hovem Jens M. (2008). *PlaneRay: An acoustic underwater propagation model based on ray tracing and plane wave reflection coefficients,* in Theoretical and Computational Acoustics 2007, Edited by Michael Taroudakis and Panagiotis Papadakis, Published by the University of Crete, Greece, pp. 273-289, (ISBN: 978-960-89785-4-2).

[8] Hovem, Jens M. *Ray trace modeling of underwater sound propagation.* SINTEF Report A21539, 2011.11.23, ISBN978 82 14 04997-8.

[9] Porter, M. B. (1991). *The Kraken normal mode program.* Rep.SM-245. (Nato Undersea Research Centre), La Spezia, Italy. (1991).

[10] Abrahamsson, L. (2003). *RAYLAB---a ray tracing program in underwater acoustics,* Sci. Rep. FOIR--1047--SE, Division of Systems Technology, Swedish Defence Research Agency, Stockholm, Sweden.

[11] Ivansson, S. (2006). *Stochastic ray-trace computations of transmission loss and reverberation in 3-D range-dependent environments,* ECUA 2006, Carvoeiro, Portugal, pp. 131-136.

[12] Francois, R. E., and G. R. Garrison. 1982. Sound absorption based on ocean measurements: Part II. Boric acid contribution and equation for total absorption. *J. Acoust. Soc. Am.* 72(6), 1879–1890.

[13] Brekhovskikh, L. M., and Yu. Lysanov. (2003). *Fundamentals of ocean acoustics,* 3rd ed. Springer-Verlag, New York City.

[14] Landau, L. D., and E. M. Lifshitz. (1959). *Fluid Mechanics.* Pergamon Press, Oxford UK.

[15] Schmidt H. (1987). *SAFARI:* Seismo-acoustic fast field algorithm for range independent environments. User's guide, SR-113, SACLANT Undersea Research Centre, La Spezia, Italy.

Chapter 11

Underwater Acoustics Modeling in Finite Depth Shallow Waters

Emerson de Sousa Costa, Eduardo Bauzer Medeiros
and João Batista Carvalho Filardi

Additional information is available at the end of the chapter

1. Introduction

Ocean Acoustics is the science which studies the sound in the sea and covers not only the study of sound propagation, but also its masking by the phenomena of acoustic interference [1].

Recent developments in underwater acoustic waves modeling have been influenced by changes in global geopolitics. These changes are evidenced by strategic shifts in military priorities as well as by efforts to transfer defense technologies to non-defense applications.

Despite the restrictiveness of military security, an extensive body of relevant research accumulated in the open literature, and much of this literature addressed the development and refinement of numerical codes that modeled the ocean as an acoustic medium [2].

One of the most important properties of the oceans as far practical applications are concerned lies in their high sensitivity to the propagation of acoustic signals with frequencies in the range of 1Hz to 20kHz that, different types of electromagnetic radiation, bring together a significant amount of information on the marine environment [3]. Another reason for the practical interest in acoustic propagation in the ocean is the distance the sound can spread, reaching several hundred kilometers.

Some properties of the seabed, such as the propagation velocities and compressional attenuation, density, among others, contribute to the spread in shallow waters significantly, making it interesting to perform a quantitative estimation of their values.

Underwater acoustic models are designed to simulate in detail the acoustic wave characteristics, thus enabling the prediction of the of the relevant phenomena behaviour. However a number of limitations are inherent to these models and often have to do with the medium characteristics, e.g., depth variation, number of degrees of freedom, just to mention a few. Other effects such as dispersion, are influenced by a different set of conditions such as surface irregularities, presence of substances derived from natural or artificial, and others.

2. Shallow waters

Acoustic wave propagation in conditions differing from the ideal infinite conditions for wave propagation, normally described as shallow water environments will be discussed in the following text.

2.1. What is shallow water

The term "shallow waters" is used when the ocean environment model is restricted by its surface at the top and by the seabed at the bottom. An important feature of this configuration is to allow the trapping of sound energy between these two interfaces which also favours the propagation of sound over long distances.

The existing criteria for defining the regions of what is "shallow" is based not only on the properties of sound propagation in the medium, but mainly by the frequency of the sound source and the interactions of sound with the background, resulting in a ratio linking the wavelength with the dimensions of the waveguide. Moreover, according to the hypsometric criterion [2], related to the depths, we define "shallow" as the waters of the continental shelf[1]. Since the average depth of the platform along the slope is usually found to be around 200m, the regions of "shallow" are defined as having depths less than 200m.

Additionally, ocean areas beyond the continental shelf can be considered to be "shallow" when the propagation of a signal with very low frequencies is accompanied by numerous interactions with the surface and the bottom Also, in practical terms, for a given frequency, " water regions are considered to be "shallow when the "shallow boundaries and reflective effects have a majour effect on the propagation and the energy is distributed in the form of a cylindrical divergence, getting trapped between the surface and the bottom.

2.2. Model of a shallow-water sound channel

The shallow-water acoustic communication channel can be classified as a multipath fading channel. It generally exhibits a long multipath delay spread, which can lead to intersymbol interference (ISI) if the spread exceeds the symbol time of communication system.

The main characteristic of sound propagation in the "shallow" is the profile setting the speed of sound, which usually has a negative or approximately constant gradient along the depth. This means that the spread over long distances due almost exclusively to the interactions of sound with the bottom and surface. Because each reflection at the bottom there is a large attenuation, spread over long distances is associated with large losses of acoustic energy [2].

---

[1] The ocean shelf is the zone around a continent, stretching from the low-water line to depths at which there is a sharp increase in the slope of the bottom in the direction of great depths [5].

The emission frequency of the source is also a crucial parameter. As in most regions of the ocean the bottom is composed of acoustic energy absorbing material, this will become more transparent to the energy in waves of low frequencies, which reduces the energy trapped in the waveguide Thus, for the lower frequencies, greater penetration of sound in the background is observed and therefore, exhibiting a greater dependence of propagation in relation to the parameters geoacoustics. At high frequencies (> 1kHz), sensitivity to the roughness of the interfaces and the marine life is greater, resulting in a greater spread, that is, a lower penetration of the bottom and a larger volume attenuation [4]. Spread over long distances therefore occurs in the range of intermediate frequencies (100 Hz to about 1 kHz) and is strongly dependent on the depth and the mechanisms of attenuation. Figure 1 shows the attenuation of sound absorption in seawater as a function of frequency. According to [2], the dependence with frequency can be categorized into four major regions, in increasing order of frequency: absorption in the background, the boric acid relaxation, relaxation of magnesium sulfate and viscosity.

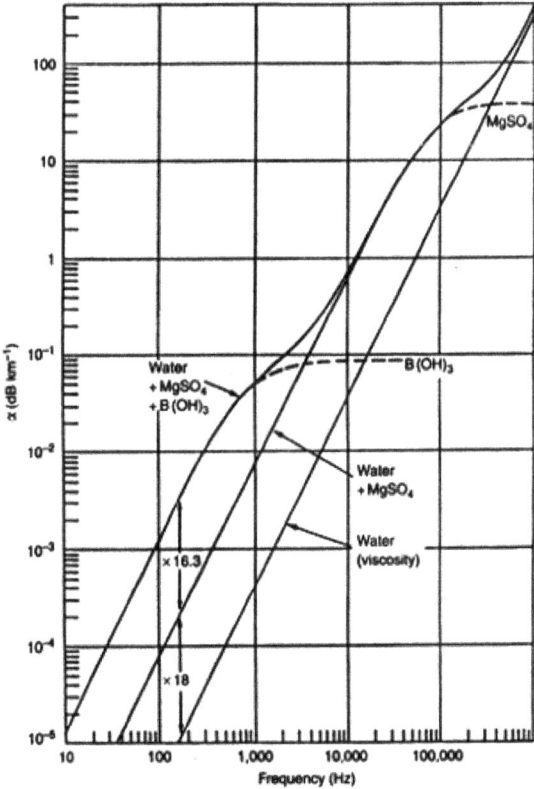

Figure 1. Absorption Coefficients for sea water [2].

2.2.1. Water layer

In seawater, the sound speed is measured using special devices or computed from special empirical formulae, using measured values of the temperature, salinity and hydrostatic pressure.

In summer, the sound channel is mainly near the bottom, so that the propagation of sound takes place by sequential reflections of refracting rays from the bottom, causing high propagation losses.

The propagation of sound in winter, takes place either in a channel with a constant sound speed, when it is described by bottom-surface rays, or in a near-surface channel.

The propagation of low-frequency sound is most affected by random inhomogeneities with characteristic sizes $\geq$ 1m. Random onhomogeneities with vertical scales of 1-10 m and horizontal scales of 100-1000 m are mainly due to the fine thermohaline stricture an to internal waves [5].

In shallow water, the field of internal waves has a number of very specifics characteristics:

- A considerable inhomogeneity and non-stationarity, due to the characteristic trains of intense soliton-like internal waves against a relatively weak background;
- Clearly expressed anisotropy, determined by the bottom relief, when waves propagate mainly in the direction towards the coast, perpendicularly to the outside edge of the shelf;
- Synchronicity of the vertical fluctuations of all fluid layers, attesting to a predominance of the first gravitational mode.

As far the fine thermohaline structure is concerned, according to [5], the ocean is a finely stratified medium, in which there exist layers with thickness from tens of centimeters to tens of meters, with comparatively homogeneous properties, separated from each other by thin boundary layers with sharp changes in the thermodynamic characteristics (the vertical gradients of the physical properties in these layers may be 10-100 times greater than their average values). Using special sound-speed meters with a large resolution, the existence of sharp fine-scale changes in the vertical dependence of the sound speed was established. Roughnesses of the water layer boundaries (roughnesses of the bottom and a disturbed surface) may also have a marked effect on the propagation of sound in shallow water. Losses on the propagation of low-frequency (up to a few kHz) sound are due to various mechanisms (absorption, scattering, geometric divergence). The absorption of sound in clear water, propagation at low frequencies, is physically mainly due to the conversion of the sound energy to heat and is the result of the chemical composition of seawater, which is a complex electrolyte. A change in sound pressure leads to a periodic change in its ionic composition, which affects the volume viscosity. The absorption relaxation mechanism in this case is well described by a formula [5], which yields the absorption coefficient β

$$\beta = \frac{0{,}11f^2}{1+f^2} + \frac{44f^2}{4100+f^2} \tag{1}$$

where the frequency is measured in kHz, and the absorption coefficient is given in dB/km. In particular, in the region of interest to us, namely approximately from 100 to 1000 Hz, the coefficient increases monotonically from 10^{-3} to 0,06 dB/km.

2.2.2. Layer of sediments

This layer consists mainly of bottom deposits of the mud type, denser sedimentary rocks or basic rocks (granite, basalt, etc.). Its parameters, like the parameters of the other layers, depend on the geographical region.

In the layer of unconsolidated sediments, one characteristic property is the existence of abrupt random inhomogeneities. These include layered (intermittent and tapered) structures of length up to tens of kilometers, and vertical channels, associated with the venting of gases and diapers (dome-shaped folds, in which rocks with a high plasticity are extruded from below).

In the layer of semi-consolidated sediments, the speed of longitudinal waves is $(2 - 3) \times 10^3$ m/s, where a small positive gradient with respect to depth is possible. This layer is also absorptive and the absorption coefficients for longitudinal and transverse waves differ and are distinguished by a large spread.

In the layer of consolidated sediments basement is characterized by a high speed of both longitudinal ($c \approx (4 - 6) \times 10^3$ m/s) and transverse ($c_s \approx (1 - 3) \times 10^3$ m/s) waves where the attenuation coefficients are estimated as $\alpha \approx 0,1$ dB/(km.Hz), $\alpha_s \approx 0,01 - 0,1$ dB/(km.Hz).

In the theory of the propagation of sound sediments are seen as a two-component medium, consisting of a solid skeleton and a fluid component. These theories use a large number of parameters (porosity, average grain size, mean-square deviation from the average size, etc.), which determine the acoustic properties of a porous medium. This model allows one to consider the speed and attenuation coefficients of different types of waves. One of the most important characteristics of sediments, which can be calculated, is the frequency dependence of the attenuation coefficient of a longitudinal or transverse wave.

2.3. The sound field

Sound propagation in the ocean is conveniently described by the wave equation, having parameters and boundary conditions which are able to describe the ocean environment. There are essentially four types of computational models (computer solutions to the wave equation) normally used to describe sound propagation in the sea: Ray Theory, Fast Field Program (FFP), Normal Mode (NM) and Parabolic Equation (PE). All of these modes enable the ocean environment to vary with the depth. A model that also allows horizontal variations in the environment, i.e., sloping bottom or spatially variable oceanography, is termed "range dependent". For high frequencies (few kilohertz or above), ray theory, the infinite frequency approximation, is still the most practical whereas the other three model types become more and more applicable and useable below, say, a kilohertz [7].

The wave equation for an acoustic field of angular frequency ω is

$$\nabla^2\phi(r,z) + K^2(r,z)\phi(r,z) = -\delta^2(r-r_s)\delta(z-z_s); \; K^2(r,z) = \frac{\omega^2}{c^2(r,z)}, \tag{2}$$

where the subscript "s" denotes the source coordinates. The range dependent environment manifests itself as a coefficient, $K^2(r,z)$, of the partial differential equation for the sound speed profile and the range dependent bottom type and topography appears as both coefficients (elasticity effects are an added complication) and complicated boundary conditions.

Throughout the theoretical development of these five techniques, the potential function G normally represents the acoustical field pressure. When this is the case, the Transmission Loss (TL) can easily be calculated as:

$$TL = 10\log_{10}\left[\phi^2\right]^{-1} = 20\log_{10}\left[\phi\right]. \tag{3}$$

3. Approximate methods in shallow-water acoustics

The various physical and mathematical models all have inherent limitations in their applicability. These limitations are usually manifested as restrictions in the frequency range or in your specification of the problem geometry. Such limitations are collectively referred to as "domains of applicability," and vary from model to model. The model selection criteria are provided to guide potential users to those models most appropriate to their needs [2].

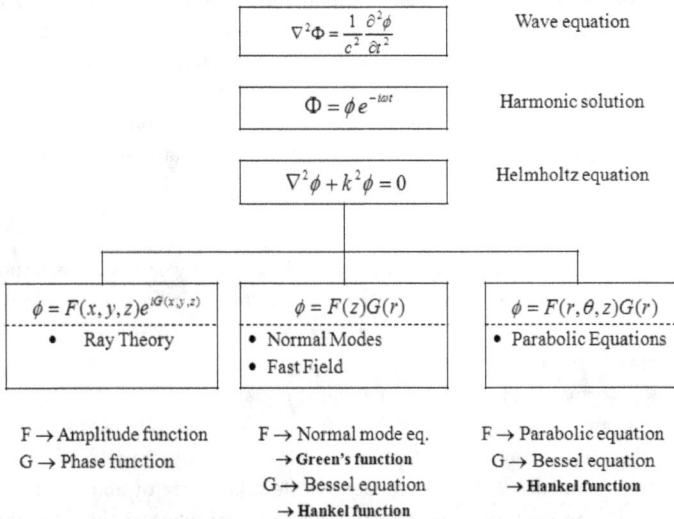

Figure 2. Summary relationship among theoretical approaches for propagation modeling.

A further subdivision can be made according to range-independent and range-dependent models. Range independence means that the model assumes a horizontally stratified ocean in which properties vary only as a function of depth. Range dependence indicates that some properties of the ocean medium are allowed to vary as a function of range (r) and azimuth (θ) from the receiver, in addition on a depth (z) dependence. Such range-varying properties commonly include sound speed and bathymetry, although other parameters such as sea state, absorption and bottom composition may also vary. Range dependence can further be regarded as two dimensional (2D) for range and depth variations or three dimensional (3D) for range, depth and azimuthal variations [2].

3.1. Ray-theory models

Ray-theoretical models, a geometrical approximation, calculate TL on the basis of ray tracing. Ray theory starts with the Helmholtz equation. The solution for ϕ is assumed to be the product of a pressure amplitude function $A = A(x,y,z)$ and a phase function $P = S(x,y,z)$: $\phi = Ae^{iS}$, where the exponential term allows for rapid variations as a function of range and $A(x,y,z)$ is a more slowly varying "envelope" which incorporates both geometrical spreading and loss mechanisms. Substituting this solution into the Equation (2) and separating real and imaginary terms yields:

$$\frac{1}{A}\nabla^2 A - \left[\nabla S\right]^2 + K^2 = 0 \tag{4}$$

and

$$2\left[\nabla A \cdot \nabla S\right] + A\nabla^2 S = 0. \tag{5}$$

Equation 4 contains the real terms and defines the geometry of the rays. Equation 5, also known as the transport equation, contains the imaginary terms and determines the wave amplitudes. The separation of functions is performed under the assumption that the amplitude varies more slowly with position than does the phase (geometrical acoustics approximation). The geometrical acoustics approximation is a condition in which the fractional change in the sound-speed gradient over a wavelength is small compared to the gradient c/λ, where c is the speed of sound and λ is the acoustic wavelength [2]. Specifically

$$\frac{1}{A}\nabla^2 A \ll K^2. \tag{6}$$

In other words, the sound speed must not change much over one wavelength. Under this approximation, Equation 4 reduces to

$$\left[\nabla S\right]^2 = K^2. \tag{7}$$

Equation 7 is referred to as the eikonal equation. Surfaces of constant phase (S = constant) are the wavefronts, and the normal to these wavefronts are the rays. Eikonal refers to the acoustic path length as a function of the path endpoints. Such rays are referred to as eigenrays when the endpoints are the source and receiver positions. Differential ray equations can then be derived from the eikonal equation.

The ray trajectories are perpendicular to surfaces of constant phase, S, and may be expressed mathematically as follow:

$$\frac{d}{dl}\left[K\frac{dR}{dl}\right] = \nabla K, \tag{8}$$

where l is the arc length along the direction of the ray and R is the displacement vector. One can determine that the direction of average flux (energy) follows that the trajectories and the amplitude of the field at any point can be obtained from the density of rays. Once S is obtained, the Equation 5 yields the amplitude. We mention here, also, that "corrected" ray theory assumes that A can be expanded in powers of inverse frequency-the leading term is the infinite-frequency result with the additional terms being frequency corrections [7].

The ray theory method is computationally rapid, extends to range dependent problems and the ray traces give a very physical picture of the acoustic paths. It is helpful in describing how noise redistributes itself when propagating long distances over paths that include shallow and deep environments and/or mid latitude to Polar Regions. The disadvantage of ray theory is that it does not include diffraction and such effects that describe the low frequency dependence ("degree of trapping") of ducted propagation.

3.2. Fast Field Program (FFP)

In the underwater acoustics, fast-field theory is also referred to as "wavenumber integration." Range independent wave theory solves the wave equation exactly when the ocean environment does not change in range. One of the possible derivations of the solution technique is to Fourier decompose the acoustic field an infinite set of horizontal waves,

$$\phi(r,z) = \frac{1}{2\pi}\int_{-\infty}^{\infty} d^2 k g(k,z,z_s) e^{ik(r-r_s)}. \tag{9}$$

and from Equation 2, the depth dependent Green's function, $g(k, z, z_s)$, satisfies

$$\frac{d^2 g}{dz^2} + \left(K^2(z) - k^2\right)g = -\frac{1}{2\pi}\delta\left(z - z_s\right). \tag{10}$$

Assuming azimuthal symmetry, we can integrate Equation 9 over the angular variable to Hankel functions and their asymptotic form reduces Equation 9 to (for simplicity, we take $r_s = 0$)

$$\phi(r,z) = \frac{e^{-i\pi/4}}{(2\pi r)^{1/2}} \int_{-\infty}^{\infty} dk (k)^{1/2} g(k,z,z_s) e^{ikr}. \tag{11}$$

Note that the factor $r^{1/2}$ arises from cylindrical spreading. We now discretize the above integral and transform to a form amenable to the FFT technique by setting $k_m = k_0 + m\Delta k$; $r_n = r_0 + n\Delta r$ where n, m = 0, 1, ..., $N-1$. The additional condition $\Delta r\Delta k = 2\pi/N$ and N is an integral power of two. The discretization scheme limits the solution to outgoing waves and Equation 10 becomes

$$\phi(r_n,z) = \frac{\Delta k e^{i(k_0 r_n - \pi/4)}}{(2\pi r)^{1/2}} \sum_{m=0}^{N-1} X_m e^{2\pi i m n/N},$$

$$X_m = (k_m)^{1/2} g(k_m, z, z_s) e^{imr_0\Delta k}. \tag{12}$$

The above equation is now easily evaluated using the FFT algorithm with the bulk of the effort going into evaluating g by solving Equation 10. Although the method is labeled "fast field" it is fairly slow because of the time required to calculate the g's. However, it has advantages when one wishes to calculate the "near field" region or to include shear wave effects in elastic media. Because of this latter capability, it can be used as the propagation component of a description of (micro) seismic noise. The FFP method is often used as a benchmark for others less exact techniques. One such technique, not applicable to the near field but exact for a large class of range independent far-field problems is the computationally faster normal mode method [7].

3.3. Normal Mode Model (NM)

Normal-mode solutions are derived from an integral representation of the wave equation. In order to obtain practical solutions, however, cylindrical symmetry is assumed in a stratified medium (i.e. the environment changes as a function of depth only). The solution for the potential function ϕ can be written in cylindrical coordinates as the product of a depth function $F(z)$ and a range function $S(r)$:

$$\phi(z, r) = F(z) \cdot S(r). \tag{13}$$

Next, a separation of variables is performed using ξ^2 as the separation constant. The two resulting equations are:

$$\frac{d^2 F}{dz^2} + (k^2 - \xi^2)F = 0 \tag{14}$$

$$\frac{d^2 S}{dr^2} + \frac{1}{r}\frac{dS}{dr} + \xi^2 S = 0 \tag{15}$$

Equation 14 is the depth equation, better known as the normal mode equation, which describes the standing wave portion of the solution. Equation 15 is the range equation, which describes the traveling wave portion of the solution. Thus, each normal mode can be viewed as traveling wave in the horizontal *(r)* direction and as a standing wave in the depth *(z)* direction [2].

The normal-mode Equation 14 poses an eigenvalue problem. Its solution is known as the Green's function. The range Equation 15 is the zero-order Bessel equation. Its solution can be written in terms of a zero-order Hankel function $\left(H_0^{(1)}\right)$. The full solution for ϕ can be expressed by an infinite integral, assuming a monochromatic (single-frequency) point source:

$$\phi = \int_{-\infty}^{\infty} G(z, z_0; \xi) \cdot H_0^{(1)}(\xi r) \cdot \xi \, d\xi \tag{16}$$

where G is the Green's function, $H_0^{(1)}$ a zero-order Hankel function of the first kind and z_0 the source depth. Note that ϕ is a function of the source depth (z_0) and the receiver (z). To obtain what is known as the normal-mode solution to the wave equation, the Green's function is expanded in terms of normalized mode functions.

The advantages of the Normal Modes procedure are: that once value problem is solved one has the solution for all source and receiver configurations, and, that is easily extended to moderate range dependent conditions using the adiabatic approximation.

3.4. Parabolic Equation Model (PE)

The parabolic approximation method was successfully applied to microwave waveguides, laser beam propagating, plasma physics, seismic wave propagation and underwater acoustic propagation.

The PE is derived by assuming that energy propagates at speeds close to a reference speed – either the shear speed or the compressional speed, as appropriate [2].

The PE method factors an operator to obtain an outgoing wave equation that can be solved efficiently as an initial-value problem in range. This factorization is exact when the environment is range independent. Range-dependent media can be approximated as a sequence of range-independent regions from which backscattered energy is neglected. Transmitted fields can then be generated using energy-conservation an single-scattering corrections [2].

The basic acoustic equation for acoustic propagation can be rewritten as:

$$\nabla^2\phi + k_0^2 n^2 \phi = 0 \tag{17}$$

where k_0 is the reference wavenumber (ω/c_0), $\omega(=2\pi f)$ the source frequency, c_0 the reference sound speed, $c(r,\ \theta,\ z)$ the sound speed in range (r), azimuthal angle (θ) and depth (z), n the refraction index (c_0/c), ϕ the velocity potential and ∇^2 the Laplacian operator.

Equation 17 can be rewritten in cylindrical coordinates as:

$$\frac{\partial^2\phi}{\partial r^2}+\frac{1}{r}\frac{\partial\phi}{\partial r}+\frac{\partial^2\phi}{\partial z^2}+k_0^2n^2\phi = 0 \tag{18}$$

where azimuthal coupling has been neglected, but the index of refraction retains a dependence on azimuth. Further, assume a solution of the form:

$$\phi = \Psi(r,z)\cdot S(r) \tag{19}$$

and obtain:

$$\Psi\left[\frac{\partial^2S}{\partial r^2}+\frac{1}{r}\frac{\partial S}{\partial r}\right]+S\left[\frac{\partial^2\Psi}{\partial r^2}+\frac{\partial^2\Psi}{\partial z^2}+\left(\frac{1}{r}+\frac{2}{S}\frac{\partial S}{\partial r}\right)\frac{\partial\Psi}{\partial r}+k_0^2\Psi\right]=0 \tag{20}$$

Using k_0^2 as a separation constant, separate Equation 20 into two differential equations as follows:

$$\left[\frac{\partial^2S}{\partial r^2}+\frac{1}{r}\frac{\partial S}{\partial r}\right]=-Sk_0^2 \tag{21}$$

and

$$\left[\frac{\partial^2\Psi}{\partial r^2}+\frac{\partial^2\Psi}{\partial z^2}+\left(\frac{1}{r}+\frac{2}{S}\frac{\partial S}{\partial r}\right)\frac{\partial\Psi}{\partial r}+k_0^2n^2\Psi\right]=\Psi k_0^2 \tag{22}$$

Rearrange terms and obtain:

$$\frac{\partial^2S}{\partial r^2}+\frac{1}{r}\frac{\partial S}{\partial r}+k_0^2S=0 \tag{23}$$

which is the zero-order Bessel equation, and:

$$\frac{\partial^2\Psi}{\partial r^2}+\frac{\partial^2\Psi}{\partial z^2}+\left(\frac{1}{r}+\frac{2}{S}\frac{\partial S}{\partial r}\right)\frac{\partial\Psi}{\partial r}+k_0^2n^2\Psi-k_0^2\Psi=0. \tag{24}$$

The solution of the Bessel equation 24 for outgoing waves is given by the zero-order Hankel function of the first kind:

$$S = H_0^{(1)}\left(k_0r\right) \tag{25}$$

For $k_0r \gg 1$ (far-field approximation):

$$S \approx \sqrt{\frac{2}{\pi k_0 r}} e^{i(k_0 r - \pi/4)} \tag{26}$$

which the asymptotic expansion for large arguments. The equation for $\Psi(r,z)$ (Equation 24) can be simplified to:

$$\frac{\partial^2 \Psi}{\partial r^2} + \frac{\partial^2 \Psi}{\partial z^2} + 2ik_0 \frac{\partial \Psi}{\partial r} + k_0^2(n^2 - 1)\Psi = 0. \tag{27}$$

Further assume that:

$$\frac{\partial^2 \Psi}{\partial r^2} \ll 2k_0 \frac{\partial \Psi}{\partial r} \tag{28}$$

which is the paraxial approximation. Then, Equation 27 reduces to:

$$\frac{\partial^2 \Psi}{\partial z^2} + 2ik_0 \frac{\partial \Psi}{\partial r} + k_0^2(n^2 - 1)\Psi = 0. \tag{29}$$

which is the parabolic wave equation. In this equation, n depends on depth (z), range (r) and azimuth (θ). This equation can be numerically solved by "marching solutions" when the initial field is known [2]. The computational advantage of the parabolic approximation lies in the fact that a parabolic differential equation can be marched in the range dimension whereas the elliptic reduced wave equation must be numerically solved in the entire range-depth region simultaneously Typically, a Gaussian field or a normal-mode solution is used to generate the initial solution.

4. Final considerations

A brief description of underwater acoustic propagation models for the "shallow environment has been considered. The choice of appropriate model depends on the simplifications needed for the environment in question. All the discussed models are well known and have been successfully developed by several authors for a variety of conditions.

The application of underwater acoustics is mostly sensor-based, including ocean sampling networks, environmental monitoring, undersea explorations, disaster prevention, assisted navigation, speech transmission between divers, distributed tactical surveillance, and mine reconnaissance.

Acoustical transmission is more flexible than others approaches, as it can be deployed in a wide variety of configurations, including networks consisting of both mobile and stationary nodes. It is not, however, free of complexity, In fact, certain aspects of underwater acoustic communications are more difficult than those RF terrestrial networks, especially high propagation delay. In general, underwater acoustic communications are influenced by transmission loss, bubbles, stratification, multipath propagation, Doppler spread, noise, and high propagation delay.

Transmission loss describes the weakening intensity of sound over a distance and is comprised of losses from both spreading and attenuation. Spreading loss is a geometrical effect that represents the weakening of sound as the wave moves outward from the source. It can be further classified as spherical spreading, cylindrical spreading, or a variant with properties somewhere between the two. Attenuation loss encompasses the effects of absorption, scattering, and leakage out of a sound channel. Absorption, a true loss of acoustic energy that results from the conversion of that energy into heat, accounts for the majority of attenuation. Bubbles produced by breaking waves at the surface can influence the propagation of high frequency signals. No bubble-induced losses were discovered for waves produced with wind speeds of 6 m/s or less [6].

The propagation of sound waves in the ocean is a somewhat complex process, particularly when there are multiple interactions with the seabed, which is often difficult to model. The theory of wave propagation is physical basis for the study of underwater acoustics and descriptive, and this we considered that the proper domain of the theory in simple environments is essential for proper understanding in solving problems realistic sound propagation in shallow water.

Author details

Emerson de Sousa Costa
CEFET-MG – Federal Center of Technological Education of Minas Gerais – Brazil
Postgraduate Program in Department of Mechanical Engineering –UFMG-Brazil

Eduardo Bauzer Medeiros and João Batista Carvalho Filardi
Postgraduate Program in Department of Mechanical Engineering –UFMG-Brazil

5. References

[1] Maia, L. P., Acoustic inversion and passive source location in shallow waters Master Thesis, Ocean Engineering. COPPE/UFRJ, Brazil, 2010. *(in Portuguese)*.

[2] Etter, P. C., Underwater Acoustic Modeling and Simulation, Spon Press, 2003.

[3] Rodríguez, O. C., Submarine Acoustic Propagation Models: Comparison with the results of the with the analytical solution of the 3 layer problem. Signal processimg laboratory, Universidade do Algarve, Portugal, 1995 *(in Portuguese)*.

[4] Xavier, B. C., Shallow Water Acoustic Propagation Models, MSc Dissertation, Ocean Engineering, COPPE/UFRJ, Brazil, (2005). *(in Portuguese)*.

[5] Katsnelson, B. G. and Petnikov, V.G., Shallow water acoustics, Springer-Praxis books in geophysical sciences, 2002.

[6] Borowski, B. S., Doctoral Dissertation of Faculty of the Stevens Institute of Technology, 2010.

[7] Kerman, B. R., Sea Surface Sound, 253-272, Kluwer Academic Publishers, 1988.

Permissions

The contributors of this book come from diverse backgrounds, making this book a truly international effort. This book will bring forth new frontiers with its revolutionizing research information and detailed analysis of the nascent developments around the world.

We would like to thank Marco G. Beghi, for lending his expertise to make the book truly unique. He has played a crucial role in the development of this book. Without his invaluable contribution this book wouldn't have been possible. He has made vital efforts to compile up to date information on the varied aspects of this subject to make this book a valuable addition to the collection of many professionals and students.

This book was conceptualized with the vision of imparting up-to-date information and advanced data in this field. To ensure the same, a matchless editorial board was set up. Every individual on the board went through rigorous rounds of assessment to prove their worth. After which they invested a large part of their time researching and compiling the most relevant data for our readers. Conferences and sessions were held from time to time between the editorial board and the contributing authors to present the data in the most comprehensible form. The editorial team has worked tirelessly to provide valuable and valid information to help people across the globe.

Every chapter published in this book has been scrutinized by our experts. Their significance has been extensively debated. The topics covered herein carry significant findings which will fuel the growth of the discipline. They may even be implemented as practical applications or may be referred to as a beginning point for another development. Chapters in this book were first published by InTech; hereby published with permission under the Creative Commons Attribution License or equivalent.

The editorial board has been involved in producing this book since its inception. They have spent rigorous hours researching and exploring the diverse topics which have resulted in the successful publishing of this book. They have passed on their knowledge of decades through this book. To expedite this challenging task, the publisher supported the team at every step. A small team of assistant editors was also appointed to further simplify the editing procedure and attain best results for the readers.

Our editorial team has been hand-picked from every corner of the world. Their multi-ethnicity adds dynamic inputs to the discussions which result in innovative

outcomes. These outcomes are then further discussed with the researchers and
contributors who give their valuable feedback and opinion regarding the same.
The feedback is then collaborated with the researches and they are edited in a
comprehensive manner to aid the understanding of the subject.

Apart from the editorial board, the designing team has also invested a significant
amount of their time in understanding the subject and creating the most relevant
covers. They scrutinized every image to scout for the most suitable representation
of the subject and create an appropriate cover for the book.

The publishing team has been involved in this book since its early stages. They
were actively engaged in every process, be it collecting the data, connecting with
the contributors or procuring relevant information. The team has been an ardent
support to the editorial, designing and production team. Their endless efforts to
recruit the best for this project, has resulted in the accomplishment of this book.
They are a veteran in the field of academics and their pool of knowledge is as vast
as their experience in printing. Their expertise and guidance has proved useful
at every step. Their uncompromising quality standards have made this book an
exceptional effort. Their encouragement from time to time has been an inspiration
for everyone.

The publisher and the editorial board hope that this book will prove to be a valuable
piece of knowledge for researchers, students, practitioners and scholars across the
globe.

List of Contributors

Matthieu Chatras, Stéphane Bila, Sylvain Giraud, Lise Catherinot, Ji Fan, Dominique Cros and Michel Aubourg
XLIM, UMR CNRS 7262, University of Limoges, Limoges, France

Axel Flament , Antoine Frappé , Bruno Stefanelli and Andreas Kaiser
IEMN, UMR CNRS 8520, Villeneuve d'Ascq, France

Andreia Cathelin
STMicroelectronics, TR&D, Crolles, France

Jean Baptiste David and Alexandre Reinhardt
CEA-LETI, Grenoble, France

Laurent Leyssenne and Eric Kerhervé
IMS, UMR CNRS 5818, Université de Bordeaux, Talence, France

T. Baron, E. Lebrasseur, F. Bassignot, G. Martin, V. Pétrini and S. Ballandras
FEMTO-ST, Université de Franche-Comté, CNRS, ENSMM, UTBM, Département Temps-Fréquence, France

Sergey E. Babkin
Physical-Technical Institute, Ural Branch of Russian Academy of Sciences, Izhevsk, Russia

Bodong Li and Jürgen Kosel
Electrical Engineering Department, King Abdullah University of Science and Technology, Thuwal, Saudi Arabia

Hommood Al Rowais
Electrical Engineering Department, Georgia Institute of Technology, Atlanta, Georgia, USA

Abhilash Mandloi and Vivekanand Mishra
Department of Electronics Engineering, S.V. National Institute of Technology, Surat 395007, Gujarat, India

Jerzy Filipiak and Grzegorz Steczko
Institute of Electronic and Control Systems, Technical University of Czestochowa, Częstochowa, Poland

M. El Hassan
University of Balamand –Al Kura, Lebanon

E. Kerherve, Y. Deval and K. Baraka
IMS Laboratory – UMR 5218 CNRS – University of Bordeaux, France

J.B. David
CEA-Leti – Minatec – Grenoble, France

D. Belot
ST Microectronics – Crolles, France

Rogério Pirk, Carlos d'Andrade Souto and Gustavo Paulinelli Guimarães
Institute of Aeronautics and Space (IAE) and Technological Institute of Aeronautics
(ITA), São José dos Campos, Brazil

Luiz Carlos Sandoval Góes
Technological Institute of Aeronautics (ITA), São José dos Campos, Brazil

J. K. Luo
Dept. Info. Sci. & Electron. Eng., Zhejiang University, China
Inst. Of Renew. Energy & Environ. Technol., Bolton University, UK

Y. Q. Fu
Thin Film Centre, University of West of Scotlant, Paisley, Scotland

W. I. Milne
Dept. of Eng. University of Cambridge, UK

Jens M. Hovem
Norwegian University of Science and Technology, Norway

Emerson de Sousa Costa
CEFET-MG – Federal Center of Technological Education of Minas Gerais – Brazil
Postgraduate Program in Department of Mechanical Engineering –UFMG-Brazil

Eduardo Bauzer Medeiros and João Batista Carvalho Filardi
Postgraduate Program in Department of Mechanical Engineering –UFMG-Brazil

www.ingramcontent.com/pod-product-compliance
Lightning Source LLC
Chambersburg PA
CBHW050124240326
41458CB00122B/1150